Plant genetic engineering

Plant genetic engineering

Edited by John H. Dodds
International Potato Center, Lima Peru

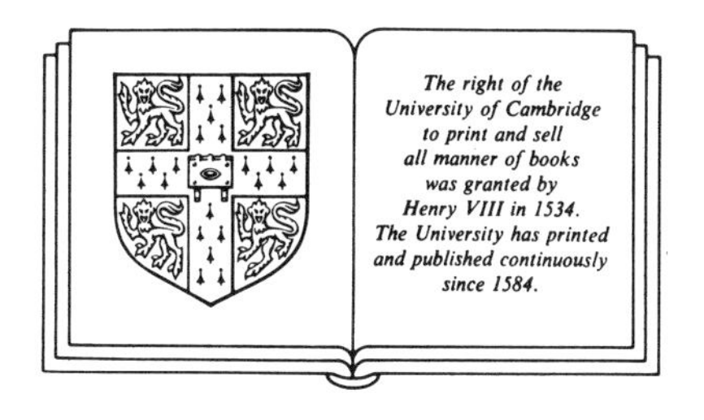

CAMBRIDGE UNIVERSITY PRESS
Cambridge
London New York New Rochelle
Melbourne Sydney

Published by the Press Syndicate of the University of Cambridge
The Pitt Building, Trumpington Street, Cambridge CB2 1RP
32 East 57th Street, New York, NY 10022, USA
10 Stamford Road, Oakleigh, Melbourne 3166, Australia

First published 1985
Reprinted 1986

Printed in the United States of America

Library of Congress Cataloging in Publication Data

Main entry under title:

Plant genetic engineering.

Includes indexes.

Contents: Introduction / J.H. Dodds – Isolation
and culture of plant protoplasts / J.H. Dodds – Fusion
of plant protoplasts / J.H. Dodds – [etc.]
1. Plant genetic engineering – Addresses, essays,
lectures. I. Dodds, John H.
QK981.5.P57 1985 581.1'5 84–23296
ISBN 0 521 25966 5

Contents

Contributors

R. R. D. Croy
Department of Botany
Science Laboratories
South Road
Durham DH1 3LE, England

M. De Block
Laboratorium voor Genetica
Rijksuniversiteit Gent
B-9000 Gent, Belgium

J. H. Dodds (editor)
International Potato Center
P.O. Box 5969
Lima, Peru

J. A. Gatehouse
Department of Botany
Science Laboratories
South Road
Durham DH1 3LE, England

L. Herrera-Estrella
Laboratorium voor Genetica
Rijksuniversiteit Gent
B-9000 Gent, Belgium

R. Hull
John Innes Institute
Colney Lane
Norwich NR4 7VH, England

M. G. K. Jones
Rothamsted Experimental Station
Harpenden
Herts AL5 2JQ, England

H. Lörz
Max-Planck Institut für
Züchtungsforschung
D-5000 Köln 30, Germany

H. J. Newbury
Department of Plant Biology
University of Birmingham
P.O. Box 363
Birmingham B15 2TT, England

J. Schell
Max-Planck Institut für
Züchtungsforschung
D-5000 Köln 30, Germany

M. Van Montagu
Laboratorium voor Genetica
Rijksuniversiteit Gent
B-9000 Gent, Belgium

P. Zambryski
Laboratorium voor Genetica
Rijksuniversiteit Gent
B-9000 Gent, Belgium

Acknowledgments

I would like to thank all the contributors to this volume for their assistance and especially for the promptness with which they submitted their contributions. I should add that I am responsible for any errors that may have crept into their chapters.

I would also like to thank all the production staff and particularly Richard Ziemacki at Cambridge University Press for the skillful way they have handled the production of this book, as well as my colleagues, too numerous to mention, for their comments.

1 Introduction

J. H. DODDS

For centuries humans have been trying to improve the quality and yield of their cultivated plants by conventional breeding and selection programs. The value of those efforts should not be underestimated, for they pointed the way toward what has come to be known as the "green revolution."

Although the history of plant tissue culture goes back to the turn of this century (Haberlandt 1902), it is only since the 1950s that practical applications have been considered. The field of plant tissue culture uses many different techniques, some of which have long been functionally applied in agricultural and horticultural science. A good example is the eradication of virus infections by meristem tip culture and subsequent micropropagation of the material. Micropropagation of potatoes and fruit trees, for example, is now a routine procedure. The meristems or shoot tips are incubated on a suitable medium to induce the outgrowth of axillary meristems, which can then be excised and subcultured again. The rates of propagation in these systems are so rapid that it is easily possible to produce a million plants from a single source in one year.

There are other areas of plant tissue culture that are still very much in experimental stages. The process of isolation, culture, and fusion of protoplasts falls into this category. It is possible to regenerate whole plants from isolated protoplasts in a number of genera; however, the rate of success is still low, and the results in cereals, for example, have been disappointing.

Nevertheless, it is these experimental areas that seem to hold enormous promise for improvements in crop plants. Over the last decade we have seen tissue culture methods used in conjunction with the molecular sciences to investigate the fundamental structures and actions of plant genomes and to analyze the possibilities of modifying plant genomes.

The new possibilities in this area of plant molecular biology have proven interesting not only to scientists in universities and government research centers but also to a wide range of industrial and agrochemical companies. It is obvious that any new development, any "novel" plant or transformed plant carrying a new characteristic, such as resistance to a herbicide or higher photosynthetic efficiency, could have significant economic value. Such new plants would be patentable, with royalty payments to the company producing them.

An important unknown factor in the area of plant genetic manipulation and engineering is the time scale of the operation. It may be only a few years before commercial companies achieve a breakthrough that will lead to large-scale production of these novel plants; on the other hand, such developments could take decades.

Recent publications have indicated that in model systems such as the tobacco plant, it is now possible to insert and translate new genetic information (Chilton et al. 1978, 1980; Herrera-Estrella et al. 1983*a*, 1983*b*, in press). We now face two immediate problems. First, the technology used in these model systems must now be transferred to economically important crop plants. This may prove to be relatively straightforward for the potato, a solanaceous plant that exhibits a high degree of plasticity in culture. However, for other plants, such as cereal grains, the situation is more difficult, and much more basic research will be required. Second, a consensus must be reached among agriculturalists, physiologists, and molecular technologists to determine what genes to include in order to produce improved plants. A number of ideas are currently being discussed. For obvious economic reasons, the agrochemical companies are greatly interested in the possibility of including genes conferring resistance to specific herbicides, in order to produce crop lines with specific resistance factors.

There is also the possibility of improving the nutritional value of a crop, either by improving the photosynthetic efficiency of the plant (see Chapter 7) or by modifying the quality of the food component, that is, their storage proteins (see Chapter 8). However, the complexities of plant improvement still pose difficult questions, and if we consider the feasibility of improving a plant simply by addition or modification of a few genes, the number of possibilities would appear to be few.

In this volume we try to put into perspective the two major aspects of these technologies. The first part deals with tissue culture or, to be more specific, protoplast culture work. It describes the problems associated with this work and presents ideas regarding the potential for these techniques in the future. The second part deals with the molecular biology involved and concentrates on two model systems: one looking at overall photosynthetic efficiency and a second looking at specific modification of an important storage protein. Chapter 9 gives an overview of the possible implications of genetic engineering for agricultural science.

It is hoped that this volume will stimulate further interest in the area of plant genetic engineering. The problems we face are considerable, but if the expected breakthroughs are achieved, their applications will be of great importance. We hope that this volume will encourage more people to join in these studies.

References

Chilton, M. D., Drummond, M. H., Merlo, D. J., and Sciaky, D. (1978). Highly conserved DNA of Ti plasmids overlaps T-DNA, maintained in plant tumors. *Nature* **275,** 147–9.

Chilton, M. D., Saiki, R. K., Yadau, N., Gordon, M. P., and Quetier, F. (1980). T-DNA from *Agrobacterium* Ti plasmid is the nuclear DNA fraction of crown gall tumor cells. *Proc. Natl. Acad. Sci. U.S.A.* **77,** 4060–4.

Haberlandt, G. (1902). Kulturversuche mit isolerten Pflanzenzellen. *Sber. Akad. Wiss. Wein* **111,** 69–92.

Herrera-Estrella, L., Depicker, A., Van Montagu, M., and Schell, J. (1983*a*). Expression of chimaeric genes transferred into plant cells using a Ti plasmid-derived vector. *Nature* **303,** 209–13.

Herrera-Estrella, L., De Block, M., Messens, E., Hernalsteens, J. P., Van Montagu, M., and Schell, J. (1983*b*). Chimeric genes as dominant selectable markers in plant cells. *EMBO* **2,** 987–95.

Herrera-Estrella, L., Van Der Broeck, G., Moenhaut, R., Van Montagu, M., Schell, J., Tunko, M., and Cashmore, A. (in press). Light inducible expression of bacterial *cat.* gene controlled by the 5′ flanking sequence of the small subunit gene of Rubisco. *Nature.*

2 Isolation and culture of plant protoplasts

J. H. DODDS

An isolated protoplast is a plant cell in which the outer wall has been mechanically or enzymatically removed. The result of this wall removal is that the plasma membrane is the only barrier between the cell cytoplasm and the external environment. Isolation of plant protoplasts is not a new technique. Mechanical isolation of protoplasts was carried out as early as 1893 (Rechinger 1893). At that time, isolation and observation of protoplasts was primarily of academic interest, and the number of protoplasts that could be liberated was extremely small, on the order of a few hundred per hour. In the early 1960s, many workers became interested the synthesis and structure of the plant cell; as a means of studying its composition, work was carried out on the activities of fungal enzymes known to cause digestion of the cell wall. The purification of these enzymes opened up the possibility of isolating plant protoplasts by an enzymatic method. By appropriate selection of the enzyme mixture it is possible to liberate billions of protoplasts in a few hours (Cocking 1960, 1972, 1973), as will be described later.

In this chapter we shall look at the culture conditions and techniques required for isolation and culture of protoplasts and eventual regeneration of a whole intact plant. The demonstrated ability of a single isolated protoplast to regenerate a whole plant (Raveh and Galun 1975; Takebe et al. 1971) is a perfect demonstration of the hypothesis of totipotency of plant cells (Haberlandt 1902). In this chapter, each stage of the culture process will be analyzed, and indications will be given regarding the problems that are normally encountered. One of the fundamental problems with this aspect of plant tissue culture, a problem that also arises in other areas, is that what will work for one plant genus often will not work for another. In general, the Solanacea family provides an excellent model system for protoplast work; this family of important crop plants includes both potatoes and tomatoes (Melchers 1978, 1982; Melchers et al. 1978; Shepard 1982; Thomas 1981). Unfortunately, success with cereal protoplasts has thus far been relatively limited (Vasil 1982).

Physiological state of the plant

The physiological state of the starting plant material for protoplast isolation apparently has a significant effect on the yield and survival of the resulting protoplasts. In general, it is advisable to grow the material in a room in which the environmental conditions can be closely controlled, and the use of systemic herbicides or pesticides should be avoided (D. Roscoe, personal communication). If possible, it is better to start with material in the form of axenic shoot culture. Axenic shoot cultures have a very thin cuticle, thus permitting easy access of the enzyme mixture, and their inherent sterility removes the need to subject the material to the possibly damaging effects of surface sterilization.

There is some evidence that plant material that is slightly drought-stressed to cause leaf wilting a couple of hours before isolation helps improve the isolation procedure. An alternative to this is to plasmolyze the detached leaves in a 13% (w/v) mannitol solution for a few minutes prior to the enzyme treatment.

Enzymology and osmoticum

Once the protoplast has been isolated, either by mechanical methods or by enzymic methods, the plasma membrane is the only boundary between the cell contents and the external bathing environment. Thus, the osmotic protection of the cell wall has been removed. To compensate for this, an osmotic agent must be included in the surrounding medium to keep the cells in an isotonic state. Various osmotic agents are available. Mannitol, an osmotic sugar alcohol, is the most common and is usually employed at a concentration of approximately 12–14% (w/v). Sorbitol can be used as an alternative osmotic sugar alcohol. It is possible to use sucrose as an osmotic agent, but the problem with using a conventional sugar is that it is metabolized by the cytoplasm of the cells, thus constantly dropping in concentration and osmotic value.

In the last few years a number of enzymes have become available

Table 2.1. *Types of enzymes available for isolation of single cells and protoplasts*

Enzyme	Source
Cellulase R-10	Kinki Yakult Biochemical, Nishinomiya, Japan
Cellulase (Cellulysin)	Calbiochemicals, San Diego, California
Cellulase (Drieselase)	Kyowa Hakko Co., Tokyo, Japan
Pectinase macerozyme R-10	Kinki Yakult Biochemical, Nishinomiya, Japan
Pectinase	Sigma Chemical Co., St. Louis, Missouri
Hemicellulase HP-150 rhozyme	Rohm and Haas Co., Philadelphia, Pennsylvania

Note: The listing of a company does not signify any commercial preference. Other suppliers are available.

that have wide-ranging wall degradation properties and varying degrees of purity. Table 2.1 lists some of the enzymes that are available. The purity of the enzyme can be a critical factor. For example, if the enzyme preparation contains impurities of proteases or lipases, they will act on the plasma membrane of the protoplast and will result in bursting. It is important to remember that the enzyme digestion mixture should always include the osmotic agent, usually mannitol. After the enzyme is dissolved in the standard culture medium (Murashige and Skoog 1962), the normal method for sterilizing enzyme mixtures is by ultrafiltration through a 0.45-μm filter.

Method of protoplast isolation

The basic technique for isolation has been described by many workers (Cocking 1972; Dodds and Roberts 1982; Gamborg et al. 1981; Yeoman and Reinert 1983). Figure 2.1 is a diagrammatic representation of the typical procedure that is followed. After surface sterilizing (not necessary for axenic shoot material, which is highly recommended), the leaves are rinsed and plasmolyzed for 1 hr in 13% mannitol. The lower epidermis is then carefully teased away to allow penetration of the enzyme mixture. If there are technical problems with removal of the lower epidermis, penetration can be achieved simply by scratching the leaf with a needle or scalpel to break the epidermal and cuticular surfaces.

After an appropriate period of incubation, the leaf will appear to dissolve. The time period involved will vary greatly, depending on the type and physiological state of the material; it can vary from 30 min to

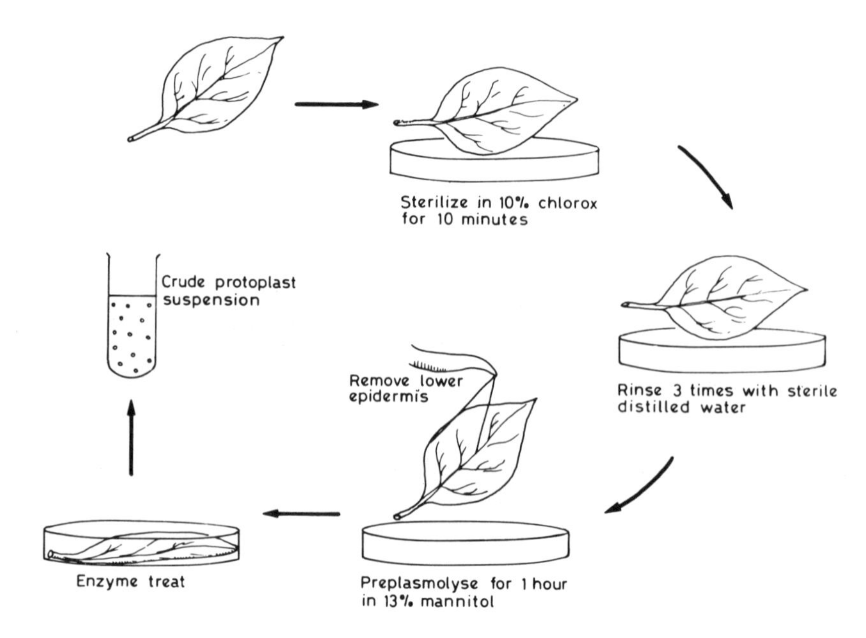

Figure 2.1. Diagrammatic scheme for isolation of crude protoplast preparation.

overnight incubation. In some cases it helps to gently agitate the petri dish during this incubation period at a rate of 40 strokes/min.

After the incubation is over, the protoplasts are released, leaving a crude protoplast suspension floating in a medium containing enzymes. The next step is to remove the enzyme mixture and purify the protoplast preparation.

Purification of protoplast preparation

Several possible methods are available for purifying a crude protoplast preparation. Figure 2.2 shows a centrifuge pelleting technique (Dodds and Roberts 1982), although flotation methods are also available (Gamborg et al. 1981). The pelleting technique simply involves *gently* pelleting the protoplasts to the bottom of a conical-tipped centrifuge tube and discarding the supernatant. At this stage the supernatant will be a mixture of enzymes and cell wall debris. The protoplasts are resuspended at each stage with fresh, osmotically buffered medium without enzymes. After an appropriate num-

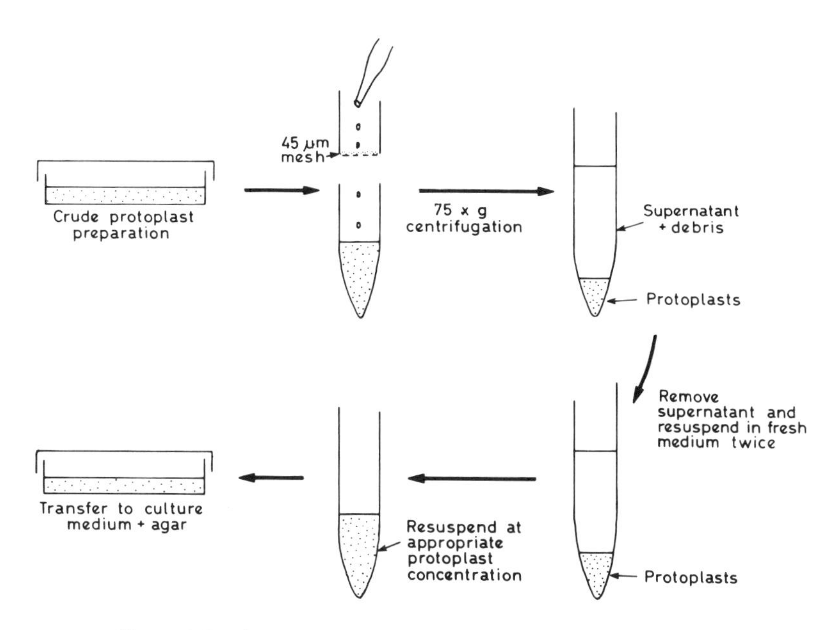

Figure 2.2. Diagrammatic scheme for purification of crude protoplast preparation.

ber of washing steps, normally two or three, purified protoplasts similar to those in Figure 2.3 are obtained.

Planting density and viability determination

The purified suspension will probably contain a mixture of viable and nonviable (damaged, but alive) protoplasts. It is important to determine what percentage of the cells are going to live in order to be able to work out how many protoplasts to inoculate in order to achieve minimum planting density (m.p.d.), as will be described later. The two methods normally used are exclusion of Evans blue stain (Dodds and Roberts 1982) and FDA (fluorescein diacetate) staining. These two staining methods will give an indication whether or not the plasma membrane is still intact (Widholm 1972).

In order to obtain successful growth of the isolated protoplasts, it is important that a sufficiently large number (concentration) of them be obtained. For tobacco, for example, the m.p.d. is $1 \times 10^4/cm^3$. If the protoplasts are cultured at a lower concentration than this, wall regeneration and cell division will not take place.

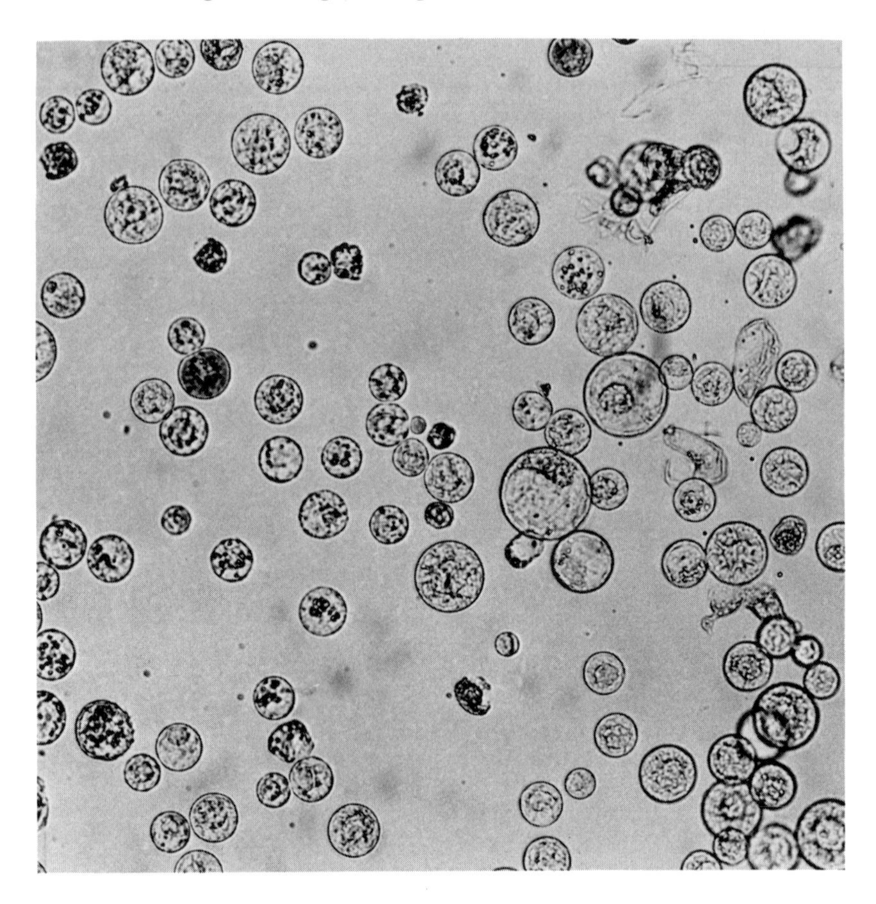

Figure 2.3. Isolated potato mesophyll protoplasts.

Wall regeneration and cell division

After a few hours in culture the isolated protoplasts will begin to develop a new cell wall (Cocking 1970; Davey et al. 1974; Pojnar et al. 1967; Willison and Cocking 1972). This wall formation can be detected by ultraviolet microscopy and the use of calcaflor stain. In a few days a perfectly normal cell wall will have formed. Given the appropriate culture conditions, the cells will then begin to divide and will eventually give rise to a small callus colony (Figure 2.4). It is from these small callus colonies that new intact plants will be regenerated.

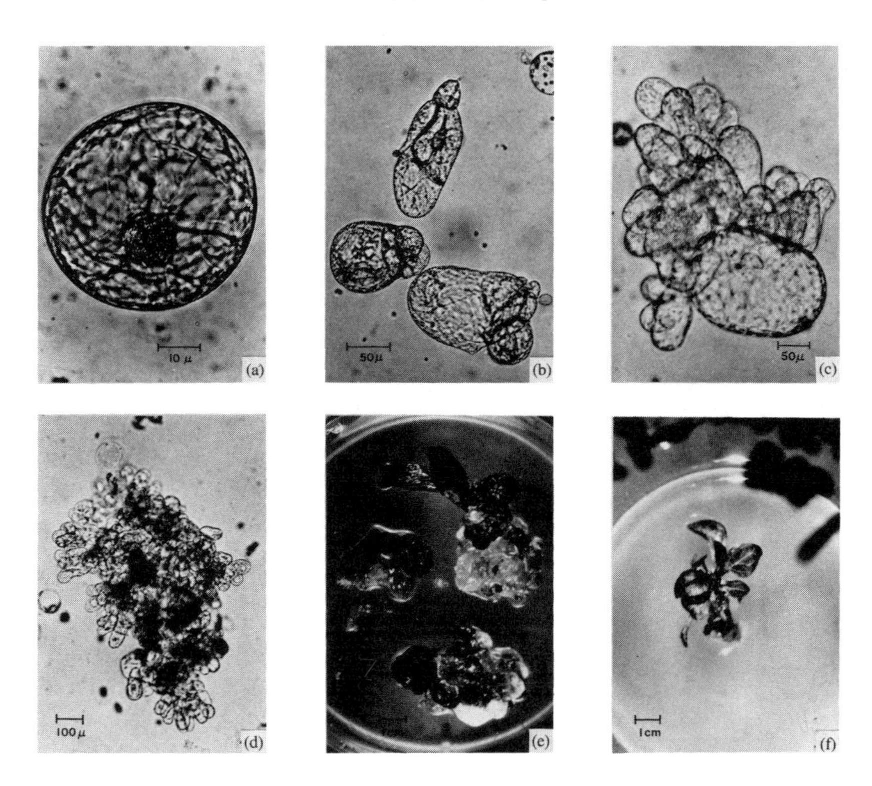

Figure 2.4. Photographic sequence of whole-plant regeneration from single isolated protoplast: (a) freshly isolated protoplast; (b) cell division; (c) small clumps of cells formed; (d) formation of small callus colonies; (e) shoot formation; (f) plantlet. (Courtesy H. Lörz.)

Organogenesis and whole-plant regeneration

Demonstration of the ability to induce organ formation on callus clumps by altering the composition of the culture medium is one of the classic studies in plant tissue culture. Figure 2.5 shows the effects of various hormone combinations on callus segments to promote the inducation of either shoots or roots (Skoog and Miller 1957). By manipulation of the culture medium it is possible to induce shoot formation on protoplast-derived callus (Gamborg et al. 1981; Raveh and Galun 1975).). The regenerated shoots can then be excised and rooted and transferred to sterile soil, thus returning to the normal whole-plant state. The stage of rooting and transfer to

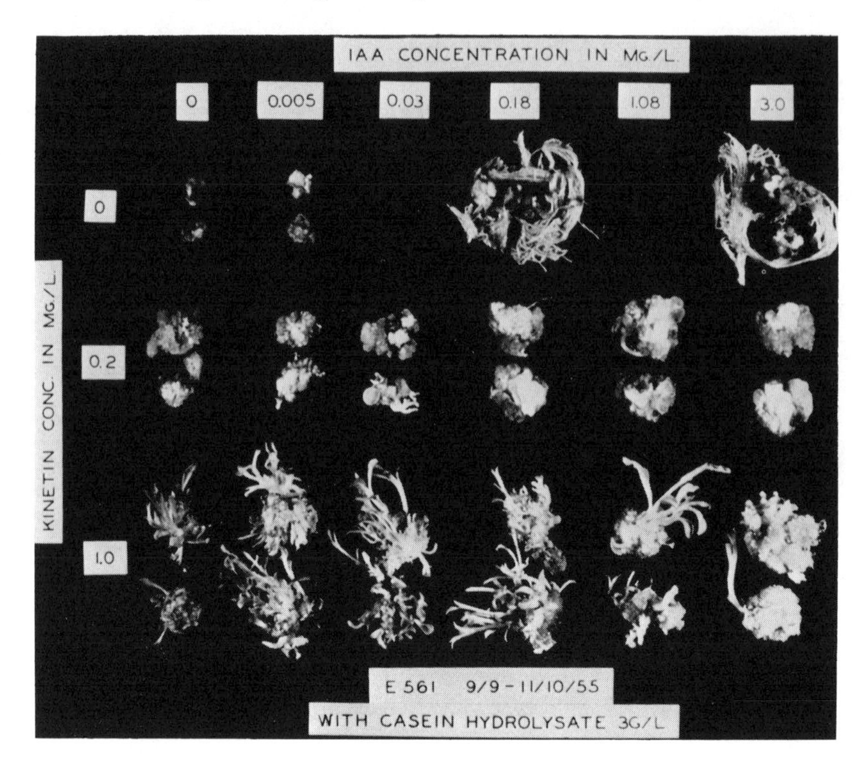

Figure 2.5. Induction of roots and shoots on callus pieces of tobacco. (Courtesy F. Skoog.)

soil is extremely delicate; the plant is moving from a highly protected sterile environment to the nonsterile and more harsh soil environment. After root induction, the small plantlets should be moved to sterile "jiffy" pots and maintained in a growth room with controlled humidity. The in vitro plantlet has a very underdeveloped cuticle and needs the humidity control to prevent water-loss wilting. The humidity should be gradually reduced to slowly harden the plants until they are ready to be transferred to pots and later to the field.

Problems of genetic variations

In the early days of protoplast and single-cell studies it was hoped that regeneration of whole intact plants from this material would yield clonal plants in a highly economic form. However, it has become clear in recent years that the mitoses and cell divisions of callus and single-cell systems are not stable (D'Amato 1978, 1983). The result is that material propagated from isolated protoplasts is highly variable and is not a clone of the original parent plant. A good example of this variability is provided by a study of regeneration of plants from potato protoplasts conducted by Thomas (1981). Plants propagated by meristem budding maintained phenotypic characters and chromosome figures compatible with the original parent plant. However, the protoplast-derived material was highly variable, with changes in leaf form, tuber yield, size, and color. Although on the one hand this has been bad news (i.e., protoplasts cannot be used for clonal propagation), on the other hand it provides an interesting method for induction of genetic variability and possible selection of variants with desired traits.

Selection of stress-tolerant lines

One of the classic methods for selection of resistance to a given stress (e.g., high-salt or heavy-metal concentrations) is to screen very large numbers of seedlings and select from the survivors. This method requires a large amount of space and manpower and is thus extremely expensive. The use of protoplast/single-cell systems offers a new approach. It is easily possible to screen several million cells simply by including the stress agent in a petri dish of culture medium. Those protoplasts with resistance to the stress will divide and form callus colonies. Plants can be regenerated from these callus colonies and the resulting plants screened for resistance to the stress (Boulware and Campter 1972; Carlson 1973). This type of study may provide a more cost-effective method for resistant-plant selection.

Conclusions

This chapter has described the techniques involved in isolation and culture of plant protoplasts. It should be emphasized that these techniques have not been proven to work with all plants and that it is desirable to check the more specialized literature before embarking on work in this area.

The removal of the outer cell wall makes the isolated protoplast amenable to a variety of techniques not possible while the wall is intact; these include uptake into the plant cell of foreign material (see Chapter 4) and the ability to fuse together dissimilar cells. This potential to fuse plant cells opens up totally new horizons in plant breeding and is the subject of the following chapter.

References

Borsby, T., and Shepard, J. F. (1983). Variation between plants regenerated from individual calli produced from separated potato stem callus cells. *Plant Sci. Lett.* **31,** 101–5.

Boulware, M. A., and Campter, N. D. (1972). Effects of selected herbicides on plant protoplasts. *Physiol. Plant.* **26,** 313–17.

Carlson, P. S. (1973). The use of protoplasts for genetic research. *Proc. Natl. Acad. Sci. U.S.A.* **70,** 598–602.

Cocking, E. C. (1960). A method for the isolation of plant protoplasts and vacuoles. *Nature* **187,** 962–3.

(1970). Virus uptake, cell wall regeneration and virus multiplication in isolated plant protoplasts. *Int. Rev. Cytol.* **28,** 89–124.

(1972). Plant cell protoplasts, isolation and development. *Annu. Rev. Plant. Physiol.* **23,** 29–50.

(1973). Plant cell modification, problems and perspectives, in *Protoplastes et fusion de cellules somatiques vegetal,* pp. 327–41. National Center for Scientific Research, Paris.

D'Amato, F. (1978). *Proceedings of the 4th International Tissue Culture Congress;* Calgary, Canada, 1978.

(1983). *Proceedings of the 5th International Tissue Culture Congress,* Tokyo, Japan, 1982.

Davey, M. R., Frearson, E. M., Withers, L. A., and Power, J. B. (1974). Observations on the morphology, ultrastructure and regeneration of tobacco leaf epidermal protoplasts. *Plant Sci. Lett.* **2,** 23–7.

Dodds, J. H., and Roberts, L. W. (1982). *Experiments in Plant Tissue Culture.* Cambridge University Press.

Gamborg, O. L., Shyluk, J. P., and Shahin, E. A. (1981). Isolation, fusion and culture of plant protoplasts, in *Plant Tissue Culture Methods and Application in Agriculture,* ed. T. A. Thorpe, pp. 115–55. Academic Press, New York.

Haberlandt, G. (1902). Kulturversuche mit isolarten Pflanzenzellen. *Sber. Akad. Wiss. Wein* **111,** 69–92.

Melchers, G. (1978). *Proceedings of the 4th International Tissue Culture Congress,* Calgary, Canada, 1978.

(1982). Plenary session address, in *Proceedings of the 5th International Plant Tissue Culture Congress,* Tokyo, Japan, 1982.

Melchers, G., Sacristan, M. D., and Holder, A. A. (1978). Somatic hybrid plants of potato and tomato regenerated from fused protoplasts. *Carlsberg Res. Comm.* **43,** 203–18.

Murashige, T., and Skoog, F. (1962). A revised medium for rapid growth and bioassays with tobacco tissue culture. *Physiol. Plant.* **15,** 473–97.

Pojnar, E., Willison, J. H. M., and Cocking, E. C. (1967). Cell wall regeneration by isolated tomato fruit protoplasts. *Protoplasma* **64,** 460–75.

Raveh, D., and Galun, E. (1975). Rapid regeneration of plants from tobacco protoplasts plated at low densities. *Z. Pflanzenphysiol.* **76,** 76–9.

Rechinger, C. (1893). Untersuchungen über die Grezen der Teilbarketi in Pflanzenzenreich. *Abh. Zool. Bot. Gaz.* **43,** 310–34.

Shepard, J. F. (1982). Cultivar dependent cultural refinements in potato protoplast regeneration. *Plant Sci. Lett.* **26,** 127–32.

Skoog, F., and Miller, C. O. (1957). Chemical regulation of growth and organ formation in plant tissues cultured *in vitro. Symp. Sci. Exp. Biol.* **11,** 118–30.

Takebe, I., Labib, G., and Melchers, G. (1971). Regeneration of whole plants from isolated mesophyll protoplasts of tobacco. *Naturwissenschaften* **58,** 318–20.

Thomas, E. (1981). Plant regeneration from shoot culture derived protoplasts to tetraploid potato. *Plant Sci. Lett.* **23,** 81–8.

Vasil, I. K. (1982). *Proceedings of the 5th International Plant Tissue Culture Congress,* Tokyo, Japan, 1982.

Widholm, J. M. (1972). Anthranilate synthetase from 5 methyltryptophan susceptible and resistant culture *Daucus carota* cells. *Biochem. Biophys. Acta* **279,** 397–403.

Willison, J. H. M, and Cocking, E. C. (1972). The production of microfibrils at the surface of isolated tomato fruit protoplasts. *Protoplasma* **75,** 397–403.

Yeoman, M. M., and Reinert, J. P. S. (1983). *Plant Tissue Culture.* Springer-Verlag, Berlin.

3 Fusion of plant protoplasts

J. H. DODDS

Once the outer cell wall has been removed, liberating protoplasts, one possibility that is opened up is the fusion of different protoplasts together to form somatic hybrids. Fusion of plant protoplasts is not a particularly new discovery. In 1909, Kuster described spontaneous fusion of mechanically isolated protoplasts. It is now known that spontaneous fusion of protoplasts is a strictly intraspecific event and is more likely to occur with protoplasts isolated from young leaves rather than mature leaves.

Animal-cell biologists, who do not have the same cell wall restraints as plant-cell biologists, began to use cell fusion studies to look at a range of developmental and differentiation problems in the 1960s (Barski et al. 1960; Ephrussi and Weiss 1965). It was not until the early 1970s that plant-cell biologists began to study seriously the possibilities of using protoplasts to produce new hybrid plants.

Methods

Protoplasts first have to be isolated, as described in the previous chapter. Then there are several experimental fusion methods available to form hybrids.

Sodium nitrate treatment

By developing further the initial studies of Kuster (1909), Power and associates (1970) were able to study both intraspecific fusion and interspecific fusion. In principle, this uses 5.5% sodium nitrate in 10% sucrose as an aggregation mixture to hold the protoplasts in a pellet while fusion takes place.

Ca^{2+} and pH treatment

The action of $CaCl_2$ (0.05 M) at pH 10.5 for 3 min is sufficient to induce fusion of protoplasts when they are subjected to gentle centrifugation (50 × g for 3 min) (Keller and Melchers 1973).

Polyethylene glycol treatment

In the method described by Wallin and associates (1974), 1 cm^3 of protoplast suspension is added to 1 cm^3 of 60% polyethylene glycol (PEG), and the mixture is agitated for 5 min. In the subsequent 10 min, protoplast fusion occurs. After that period, the PEG is washed away, and the resulting fusion products are plated in culture.

Zimmerman electric method

This method involves the positioning of two protoplasts between two electrodes. A square-wave shock is then used to fuse the two protoplasts (Zimmerman and Schevrich 1981).

Use of other additives

Various other materials have also been used to induce fusion of protoplasts. These compounds include poly-D-lysine, poly-L-ornithine, concanavalin A, and cytochalasin B. These compounds affect the membrane properties of the plasmalemma, thus facilitating fusion (Grout and Coutts 1974).

Possible hybrid products

In the early years of protoplast fusion studies, imagination and speculation were allowed to take over from logical scientific theory, and numerous rather optimistic ideas were put forward. It was proposed that hybrids might be made between root crops (e.g., carrots and parsnips) and top-vegetable crops (e.g., cauliflower and cabbage), the idea being to produce a plant with a cabbage top that after harvesting would be dug up to give a second root-vegetable harvest. To date, none of these ambitious ideas has led to success.

A number of successful somatic crosses have been made with tobacco, the model plant used in tissue culture studies (Carlson et al. 1972; Smith et al. 1976); crosses in the family Dauca have also been

Figure 3.1. Hybrid plant of tomato/potato fusion product. (Courtesy Professor G. Melchers.)

made (Dudits et al. 1977). However, the system that has received the most attention in recent years has been the hybrid produced between tomato and potato (Figure 3.1) (Melchers et al. 1978; Schiller et al. 1982; Zenkteler and Melchers 1978).

This chapter describes the techniques used and the problems that are encountered, but as in the previous chapter, it should be emphasized that what may work for one plant system may not work for another. Much more fundamental research will be required before these can be considered routine techniques.

Stability of the fusion product

One of the main reasons for the interest in the area of somatic hybridization is that it may have use for crossing plants that have conventional incompatibility systems (Lewis 1983). It is evident that some somatic crosses also exhibit a form of cellular incompatibility in that it is impossible for the two nuclei to exist intact in the same cytoplasmic environment. As seen in Figure 3.2, often there is selective elimination of some of the chromosomes, and in extreme cases there is complete elimination of one nucleus. It is important to

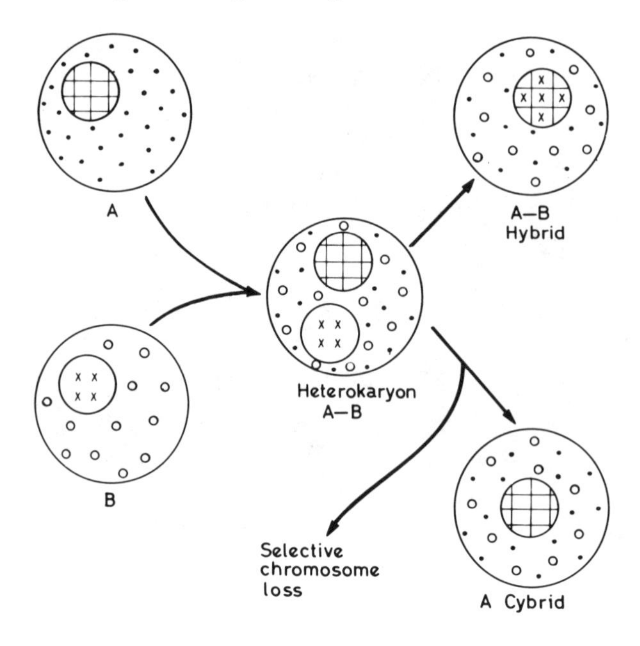

Figure 3.2. Formation of a cytoplasmic hybrid "cybrid."

keep in mind, however, that if one nucleus is completely removed, the two cytoplasms have still been fused. This fusion product is then known as a cytoplasmic hybrid or cybrid. For transfer of extrachromosomal inheritance factors this has significant implications.

Selection of somatic hybrid

There is a method for fusion mentioned previously that allows the possibility of fusing two individually isolated protoplasts: the Zimmerman method. The other methods, however, involve the mixing of large numbers of protoplasts and allowing random fusion of the mixture. It can be seen in Figure 3.3 that various fusion products are possible. It is very common for similar protoplasts to fuse, and this can result in aggregates of two to eight cells forming a large polynucleate fusion production. In most cases the desired hybrid is a 1:1 fusion.

At the present time, one of the most difficult problems in this area is to identify and obtain growth of the desired fusion product. The following selection methods are used as indicated.

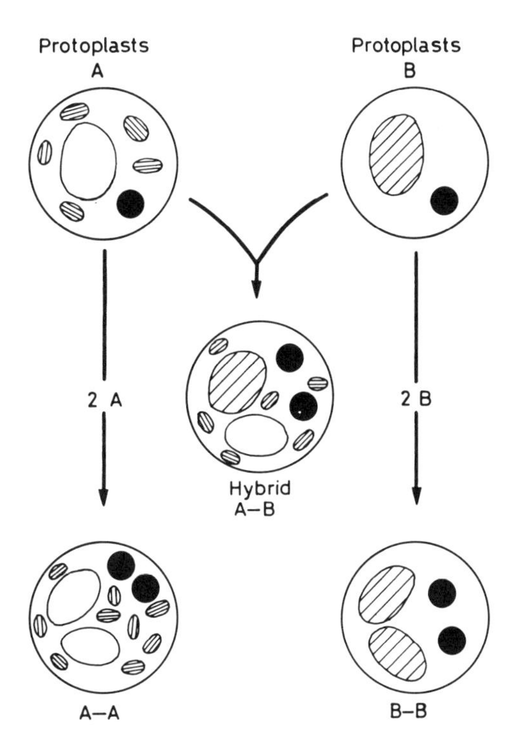

Figure 3.3. Diagram showing various types of fusion products.

Visual selection

The selection of fusion products on the basis of visual selection has on the whole been restricted to those systems with characteristic microscopic markers, such as colorless protoplasts with chloroplast-containing mesophyll protoplasts (Kao et al. 1974) (Figure 3.4). A form of this technique was successfully applied by Patnaik and associates (1982), who labeled cell suspension protoplasts with fluorescein isothiocyanate and then fused them with chloroplast-containing mesophyll protoplasts. When observed by fluorescence microscopy techniques, the chloroplasts fluoresce red because of the chlorophyll, and the isothiocyanate cells exhibit a highly characteristic apple green cytoplasmic fluorescence. Patnaik and associates used micromanipulative methods to separate out green-fluorescing cytoplasm containing red-fluorescing chloroplasts.

Nutritional selection

It was Carlson and associates (1972) who demonstrated the use of biochemical selection systems based on the nutritional re-

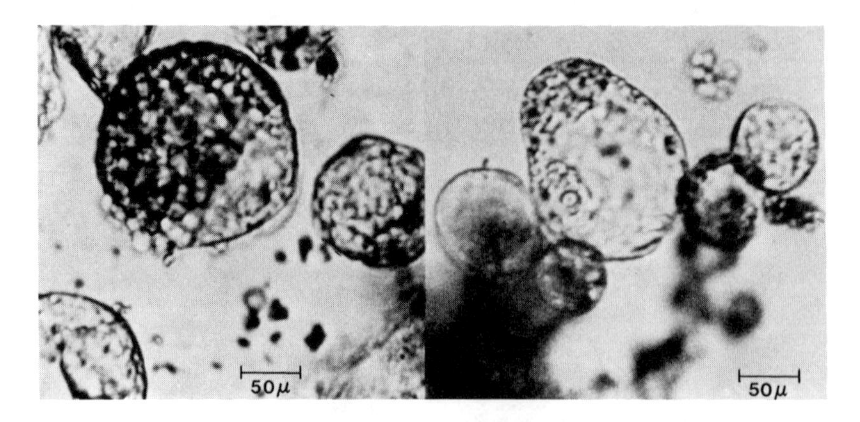

Figure 3.4. Fusion of colorless and chloroplast-containing protoplasts. (Courtesy Professor G. Melchers.)

quirements of a tumorous hybrid between two *Nicotiana* species. It is unfortunate that this is a rather specialized case and is not applicable in other situations. It serves well, however, as a model system to show how a specific selection pressure can be applied. Figure 3.5 outlines the system diagrammatically. If a cross is made between *Nicotiana glauca* and *N. langsdorffii,* the resulting hybrid exhibits a tumorous nature. The two parental protoplast populations are unable to grow on a medium lacking in hormones. The somatic hybrid, however, because of its tumorous nature, produces its own auxins and cytokinins; these allow growth, division, and callus colony formation of the hybrid.

Another method now widely used is complementation with an albino mutant (Dudits et al. 1979).

Fluorescent cell sorting

Originally designed for separation of cancer cells and non-cancer cells, a fluorescence flow cytometry and cell sorting system is now being applied to plant protoplasts (Redenbaugh et al. 1982). The technical operation and construction of this machine are extremely complex; however, the principle is simple. The two parent protoplast populations are labeled with different fluorescent tags, one red and one green; the fusion products will contain both fluorescent markers at the same time. When a mixture of parental protoplasts and fusion products is introduced into the machine, it prepares droplets of an appropriate size so as to allow only a single

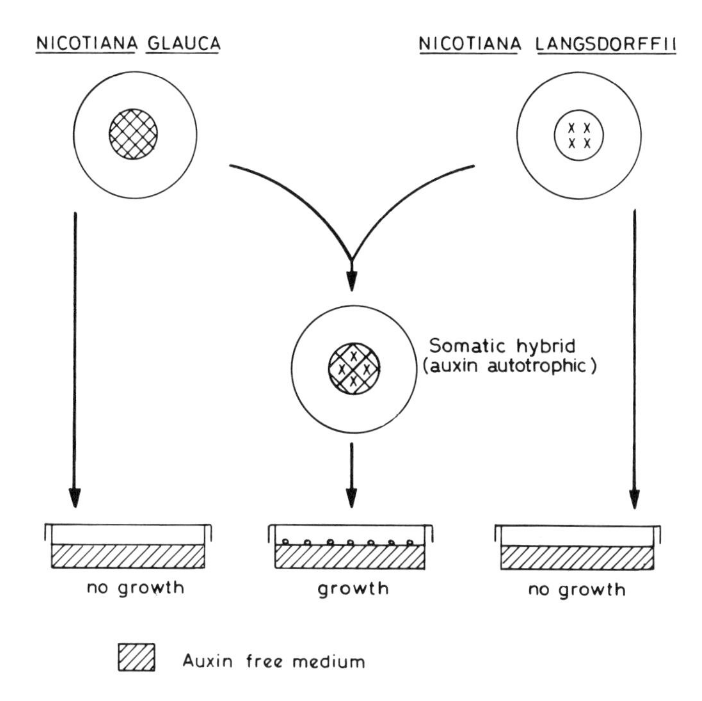

Figure 3.5. Schematic representation of selection scheme for tumorous somatic hybrids.

protoplast or fusion product per droplet. The droplet is then electronically scanned to determine its fluorescence properties. It will be parental type 1, type 2, or fusion hybrid. The droplet is then released between two electric plates that can have a positive or negative charge or be neutral. The charge on the plates will deflect the droplets, allowing separation of the three cell types. It follows, therefore, that after running the mixture through the machine, we will have three tubes, one of which will contain fusion products for further analysis. This technique is still very much in the experimental stage, but it does suggest great potential for the future.

Type of material to hybridize

As was indicated earlier, in the first few years of this work some rather extreme examples were proposed as possible fusion products. In the last few years researchers have concentrated very heavily on the potato/tomato system as a model, and there is now significant interest in the possibility of fusing wild and cultivated potato species as a way of transferring resistance to diseases and viruses from wild to cultivated species (Shepard 1982). I believe that if the correct starting material is chosen, there is a good chance of success in this area.

References

Barski, G., Sorieul, S., and Cornefert, F. (1960). Production dans des cultures in vitro de deux souches cellulaires en association de cellules de caractére "hybride." *Compt. Rend. Acad. Sci.* **251,** 1825–33.

Carlson, P. S., Smith, H. H., and Dearing, R. D. (1972) Parasexual interspecific plant hybridization. *Proc. Natl. Acad. Sci. U.S.A.* **69,** 2292–4.

Dudits, D., Hadlaczky, G. Y., Levi, E., Fejer, O., and Lazar, G. (1977). Somatic hybridisation of *Daucus carota* and *D. capillifolius* by protoplast fusion. *Theor. Appl. Genet.* **51,** 127–32.

Dudits, D., Hadlaczky, G. Y., Basjszar, G., Kones, C. S., Lazar, G., and Horvath, G. (1979). Somatic hybridisation of *Daucus carota* and *D. capillifolius* by protoplast fusion. *Plant Sci. Lett.* **15,** 101–6.

Ephrussi, B., and Weiss, M. C. (1965). Interspecific hybridisation of somatic cells. *Proc. Natl. Acad. Sci. U.S.A.* **53,** 1040–2.

Grout, B. W. W., and Coutts, R. H. A. (1974). Additives for the enhancement of fusion and endocytosis in higher plant protoplasts: an electrophoretic study. *Plant Sci. Lett.* **2,** 397–403.

Kao, K. N., Constabel, F., Michayluk, M. R., and Gamborg, O. L. (1974). Plant protoplast fusion and growth of intergeneric hybrid cells. *Planta* **120,** 215–27.

Keller, W. A., and Melchers, G. (1973). The effect of high pH and calcium on tobacco leaf protoplast fusion. *Z. Naturforsch.* **280,** 737–41.

Kuster, E. (1909). Ueber die Verschmelzung nackter Protoplasten. *Ber. Deut. Botan. Ges.* **27,** 580–98.

Lewis, D. (1983). *Sexual Incompatibility in Plants.* Edward Arnold, London.

Melchers, G., Sacristan, M. D., and Holder, A. A. (1978). Somatic hybrid plants of potato and tomato regenerated from fused protoplasts. *Carlsberg Res. Comm.* **43,** 203–18.

Patnaik, G., Cocking, E. C., Hamill, J., and Pental, D. (1982). A simple procedure for the manual isolation and identification of plant hetereokaryons. *Plant Sci. Lett.* **24,** 105–10.

Power, J. B., Cummins, S. F., and Cocking, E. C. (1970). Fusion of isolated plant protoplasts. *Nature* **225,** 1016–18.

Redenbaugh, K., Ruzing, S., Bartholomew, J., and Bassham, J. A. (1982). Characterisation and separation of plant protoplasts via flow cytometry and cell sorting. *Z. Pflanzenphysiol.* **107,** 65–80.

Schiller, B., Herrmann, R. E., and Melchers, G. (1982). Restriction endonuclease analysis of plastid DNA from tomato, potato and some of their somatic hybrids. *Mol. Gen. Genet.* **186,** 453–9.

Shepard, J. F. (1982). Cultivar dependent cultural refinements in potato protoplast regeneration. *Plant Sci. Lett.* **26,** 127–32.

Smith, H. H., Kao, K. N., and Combatti, N. C. (1976). Interspecific hybridisation by protoplast fusion in *Nicotiana*. Confirmation and extension. *J. Hered.* **67,** 123–7.

Wallin, A., Glinelius, K., and Eriksson, T. (1974). The induction of aggregation and fusion of *Daucus carota* protoplasts by polyethylene glycol. *Z. Pflanzenphysiol.* **74,** 64–80.

Zenkteler, M., and Melchers, G. (1978). In vitro hybridization by sexual methods and by fusion of somatic protoplasts. *Theor. Appl. Gen.* **52,** 81–90.

Zimmerman, U., and Schevrich, P. (1981). High frequency fusion of plant protoplasts by electric fields. *Planta* **151,** 26–32.

4 Isolated cell organelles and subprotoplasts: their roles in somatic cell genetics

H. LÖRZ

Plant protoplast fusion and somatic hybridization offered the potential for combining different genomes and producing new hybrid plants (Cocking 1983; Harms 1983*b*). Fusion of two different plant protoplasts results in a rather complex structure consisting of two nuclear genomes and a mixed cytoplasm with cell organelles from both fusion partners. The eventual fusion of nuclei or degradation of one of the nuclei, degradation and gradual loss of chloroplasts from one fusion partner, and possible segregation, rearrangement, and/or recombination of the mitochondrial genome generate numerous constructs, all of them different with respect to genetic constitution (Lázár 1983; Lörz and Izhar 1983). Application of external selection pressure inhibiting or favoring the development of the constituents of one of the fusion partners offers a possibility to direct the development of the fusion product, but, in principle, the sorting out and recombination are occurring randomly.

Since the inception of somatic plant cell genetics, the transfer of isolated organelles has been discussed as a possibility to circumvent the complex situation of a protoplast fusion product (Carlson 1973; Cocking 1972). With organelle transplantation, only one specific type of cell organelle is transferred into a receptor protoplast, and cells of the following constitution may be produced:

1. after uptake of a nucleus: two nuclear genomes in an unmodified cytoplasmic background
2. after uptake of chromosomes: original nuclear genome and additional chromosomes in an unmodified cytoplasmic background
3. after uptake of chloroplasts (plastids): mixed population of chloroplasts with the receptor cell nuclear genome and mitochondria
4. after uptake of mitochondria: mixed population of mitochondria, but receptor cell nuclear and chloroplast genomes
5. after uptake of other cell organelles or cell fragments (vacuoles, vesicles, ribosomes, membranes, etc.): heterologous situation in re-

spect to a specific type of cell organelle or cell fragment in an otherwise nonmodified receptor cell

It would obviously be easier to study the interactions of nuclear genome, plastidome, and chondriome by means of organelle transfer than by somatic hybridization. Genetic analysis of cytoplasmic male sterility has already been approached by somatic cell genetics (Galun and Aviv 1983). Transfer of isolated cell organelles could provide direct evidence as to whether a specific sterility or resistance (or sensitivity) to plant pathogens, toxins, or herbicides is determined by nuclear genes or by cytoplasmic genes. The study of incompatibility phenomena also could benefit from transfer of discrete parts of the genetic information into plant cells (de Nettancourt 1977). There seems to be no major barrier to the fusion of protoplasts from very different origins; however, during prolonged cell culture and especially with respect to differentiation and morphogenesis, incompatibility reactions often prevent the development of "new" hybrid plants (Harms 1983*a*).

Heterosis is a major topic of interest in applied plant cell biology (Frankel 1983). Although the effects of this pheomenon have been exploited by plant breeders for many years, only limited knowledge of the cellular basis of heterosis has been accumulated. In all these cases, a defined combination of plant cell fragments by subprotoplast fusion or transfer of specific cell organelles would assist the analysis of various phenomena at the level of cultured cells and with modified plants regenerated from in vitro culture.

Referring only to the few examples mentioned earlier, the transfer of cell organelles and subprotoplast fusion appear to be integral parts of the spectrum of possibilities achievable by somatic cell genetics. Organelle transfer has always been discussed and reviewed together with somatic hybridization by protoplast fusion and gene transfer by DNA vectors. Both of the latter fields of research have progressed rapidly (see Chapters 3 and 5), in contrast to the area of organelle uptake into protoplasts. This point of view is supported also by the very few contributions on this topic during the recent protoplast symposium (Potrykus et al. 1983). The intention of this chapter is not to create unrealistic optimism about the possibilities of organelle transfer (as often was done in the past) but rather to summarize our present knowledge and experiences with organelle uptake and to evaluate critically the procedures for preparation of subprotoplasts by protoplast fragmentation and the usefulness of these structures in fusion experiments.

Transfer of isolated cell organelles

Cell organelles carrying genetic information (nuclei, plastids, and mitochondria) are of course the most interesting ones for the topic of plant genetic engineering and will be addressed in this review. Other organelles or cellular particles such as vacuoles, vesicles, ribosomes, and membranes have been discussed previously in the context of organelle transfer and are of interest for physiological studies (Galun 1981; Potrykus 1975).

A basic prerequisite for studies of organelle uptake is preparation of intact and viable organelles. Conventional procedures for organelle isolation often involve mechanical processes such as cutting or grinding for the rupture of plant cells. These methods tend to damage a large proportion of the cell organelles (Hull and Moore 1983). Gentle lysis of protoplasts isolated from various tissues has been used as a new method to isolate intact cell organelles.

Isolation of cell organelles from protoplasts

Nuclei. Protoplasts have been used successfully to prepare large quantities of nuclei. In principle, the isolation procedure involves lysis of protoplasts in hypotonic buffer, treatment with a detergent (e.g., Triton X-100), and subsequent separation of the nuclei from other cell organelles and fragments by gradient centrifugation (Lörz and Potrykus 1978; Ohyama et al. 1977). Nuclei isolated in this way, and also those isolated without the use of a detergent (Tallman and Reeck 1980), have exhibited active transcription (Blascheck et al. 1974) and intact ultrastructure (Hughes et al. 1977; Schel et al. 1983).

Chromosomes. Methods have also been developed for isolation of chromosomes from plant protoplasts of both mitotic and meiotic cells (Griesbach et al. 1982; Malmberg and Griesbach 1980). For this purpose protoplasts are isolated from naturally synchronized cells (meiocytes) or from synthetically synchronized cell cultures in which high percentages of cells are arrested in mitosis. A new method found suitable for mass isolation of chromosomes from monocotyledonous and dicotyledonous species employs a glycine-hexylene glycol buffer system (Hadlaczky et al. 1983). Flow cytometry based on the different DNA contents of the chromosomes has been used for separation of acrocentric and metacentric chromosomes (De Laat and Blaas 1984), and after further refinement of this technique, separation of individual chromosomes may be possible.

Chloroplasts and mitochondria. Protoplasts from numerous plant species, both dicotyledonous and monocotyledonous, have been used for preparation of chloroplasts and mitochondria. Studies on chloroplast isolation have been predominantly biochemical or physiological, and verification of the structural integrity and functionality has been mostly indirect, based on biochemical data (Edwards et al. 1979; Nishimura and Beevers 1978). In a rapid isolation procedure, the protoplasts are forced to move through a nylon net or sieve by centrifugation, and thereby they are ruptured. The resulting homogenate is then immediately (in the same tube) fractioned on a density gradient formed by layers of various constituents, such as silicone oil and sucrose (Hampp 1980; Robinson and Walker 1979). This method allows rapid separation of the plastidal, mitochondrial, and cytoplasmic fractions from protoplasts. Preparations of both chloroplasts and mitochondria obtained via isolated protoplasts have shown only minimal contamination by other organelles and therefore seem to be most suitable for organelle uptake experiments.

Uptake of cell organelles: principles and procedures

Several methods have been developed over the past 15 years to induce uptake of isolated organelles into protoplasts. In the beginning, most efforts were directed toward achieving an efficient uptake rate (i.e., a high percentage of protoplasts with transferred organelles). Only recently has more attention been given to conditions that minimize detrimental effects for the isolated organelles and support viability of the receptor protoplasts.

Endocytosis and/or fusion. Early studies with polystyrene latex spheres showed that uptake of particles into protoplasts occurs mostly by endocytosis (Mayo and Cocking 1969; Willison et al. 1971), a process that requires a supply of energy through oxidative path-

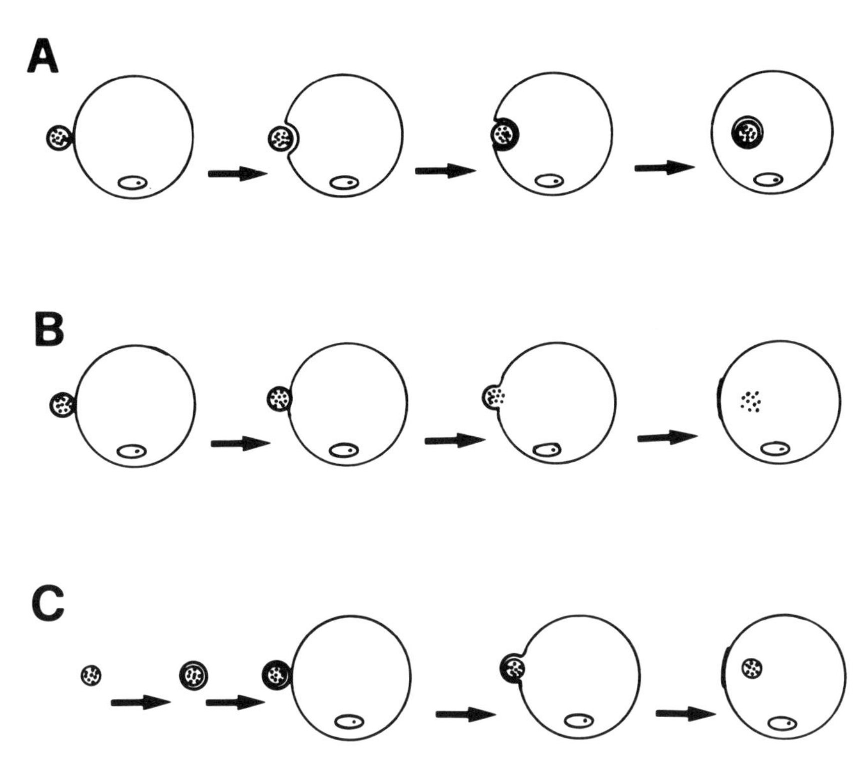

Figure 4.1. Diagram illustrating possible uptake mechanisms of isolated organelles into plant protoplasts. (A) Uptake by endocytosis is initiated by adhesion of the organelle to the protoplast membrane, followed by invagination of the plasma membrane and subsequent release of a membrane-coated organelle into the cytoplasm. (B) Organelle uptake by a fusion process releases the organellar content freely into the receptor cytoplasm. (C) Coating of organelles within a liposome prior to uptake via fusion leads to an intact, transferred organelle, surrounded by the original organellar envelope.

ways (Suzuki et al. 1977). The uptake is initiated by adhesion of the particle to the protoplast membrane, followed by invagination of the plasma membrane, and finally release of the particle into the cytoplasm of the receptor cell (Fowke and Gamborg 1980). Organelle uptake via an endocytotic process leads to a transferred organelle that is surrounded by an additional membrane (Figure 4.1). However, transferred chloroplasts have also been found without an additional membrane coating and also without the outer chloroplast envelope (Davey et al. 1976). In this case the organelles most likely were taken up by a fusion process involving the outer chloroplast envelope and the protoplast membrane, with subsequent release of the organelle content into the cytoplasm. Additionally, there exists the possibility that endocytotic vesicles are degraded and the organelles are released into the cytoplasm of the receptor protoplast.

Sandwich technique. One of the first methods specifically developed to achieve organelle uptake was the so-called sandwich technique (Potrykus 1973, 1975; Potrykus and Hoffmann 1973). Isolated protoplasts are pelleted gently by low-speed centrifugation, and thereafter organelles (e.g., nuclei or chloroplasts) are pelleted onto the protoplast layer. This sandwiching is repeated several times. The conditions most favorable for induction of organelle uptake are such that the buffer solution for the organelles is hypotonic compared with the protoplast medium and contains 0.03% lysozome. Under these conditions the reported uptake rate (i.e., percentage of protoplasts that have taken up one or several foreign organelles) has been 0.1–1%. Repeated centrifugation, the presence of the lysozyme, and several changes of plasmolysis to deplasmolysis and vice versa are certainly extremely stressful for both the protoplasts and the isolated organelles.

Uptake induced by fusogens. Most of the other methods used for organelle uptake were not designed specifically for this purpose but were originally developed for protoplast fusion. Inactivated Sendai virus, a fusion agent, developed for mammalian cells, effected the fusion of human HeLa cells and *Haplopappus* protoplasts (Lima de Faria et al. 1977). It is assumed that isolated organelles also can be fused in this way. Other treatments include incubation of protoplasts and organelles with calcium nitrate (Schieder 1975) or calcium chloride at high pH (Keller and Melchers 1973) and treatment with polyethylene glycol (Kao and Michayluk 1974; Wallin et al. 1974). Endocytotic uptake of individual particles or aggregates of polystyrene spheres can be induced by poly-L-ornithine (Willison et al. 1971). However, these methods are not ideal, for there is a negative correlation between a high efficiency for physical uptake on one hand and biological activity of isolated organelles and protoplast viability on the other hand (Lörz 1977; Lörz and Potrykus 1978).

Polyethylene glycol (PEG) is now routinely used for protoplast fusion. Protoplasts of very different origins have been fused by PEG, and uptake of cell organelles and microorganisms also has been achieved. Uptake can be accomplished by mixing organelles and protoplasts on a microscope slide and adding PEG solution (final concentration 5–20%) for several minutes. The subsequent washing is done with medium containing $CaCl_2$ at high pH or similar treatments as normally used for fusion experiments. The first step, agglutination, is a tight contact between organelle and protoplast membrane. The actual fusion of two adjacent membranes occurs subsequently, during the washing process, and leads to release of the organelle content into the receptor protoplast (Fowke and Gamborg 1980).

It is feasible to conclude that all conditions favorable for protoplast fusion also may enhance organelle uptake into protoplasts. The use of polyvinyl alcohol (PVA) has recently been presented as a novel cell fusion method for protoplasts (Nagata 1978). This compound has not yet been applied for organelle uptake studies. Both PVA and PEG are nonionic weak surfactants, and it has been suggested that nonionic surfactants in general have cell adhesion-fusion activity, and additional compounds of this type may be found.

Vesicles and liposomes. Intact organelle transfer is prevented by uptake procedures based on membrane fusion of the plasmalemma and the original organellar envelope (Figure 4.1). This problem can be circumvented by encapsulating the organelles in synthetically produced additional membranes prior to fusion.

A simple procedure for creating a protective microenvironment was developed for *Cyanobacteria* by Bradley and Leith (1979). Ten percent (by volume) mineral oil or olive oil was added to the suspension of the unicellular *Cyanobacteria* and then vigorously shaken for a few seconds. Oil droplets with diameters of 3–20 μm were formed containing one or more cyanobacterial cells. Isolated protoplasts and oil emulsion were mixed, and oil drops were taken up spontaneously, or the uptake was stimulated by PEG.

Charged lipid vesicles or liposomes constitute a more sophisticated carrier system suitable for fusion with the negatively charged plasmalemma of protoplasts (Cassels 1978). A broad range of unilamellar and multilamellar vesicles that readily fuse with protoplast membranes has been produced from substances such as phosphatidyl serine, cholesterol, phosphatidyl choline, and stearylamine (Papahadjopoulos et al. 1980). Up to now the liposome technique has been applied mainly to encapsulate DNA, RNA, or viruses (Rollo 1983). Loading capacity of the liposomes, fusion ability, and toxicity for the receptor protoplasts are parameters that must be evaluated for the individual systems. Liposomes, especially large unilamellar vesicles, seem to be suitable also to coat and "protect" the integrity of isolated organelles during fusion (Giles 1978).

Electric-field-induced uptake. A rather new technique for fusion of protoplasts is electric-field-induced cell fusion (Vienken et al. 1981; Zimmermann 1982). This method leads to a very high yield of fused cells, and the fusion process occurs synchronously with many protoplasts. A protoplast suspension placed between two electrodes is exposed to a nonuniform alternating electric field. Because of generation of dipoles within the cells, the cells are pulled into the region of highest field intensity and form "chains" along the electric field lines. The length of these typical protoplast chains or the number of cells attached to one another is influenced by the density of the protoplast suspension and the electric field conditions. Fusion of neighboring protoplasts is initiated by injection of a field pulse of high intensity and nanosecond to microsecond duration only. It is worth mentioning that only small areas of the membranes in the contact area between two adjacent cells are subjected to electrical breakdown. The method seems to be most efficient when protoplasts of the same size are used (H. Kohn, personal communication), but protoplasts of very different origins and also isolated vacuoles and protoplasts have been fused successfully (Vienken et al. 1981). A further refinement of this technique has recently been achieved by electric-pulse-induced fusion of two single protoplasts of tobacco in a 100-μl droplet, thus allowing the handling and manipulation of defined single cells (Koop et al. 1983). Electric-field-induced fusion, in combination with microscale culture, has great potential for organelle transfer studies, both for direct fusion of isolated organelles with protoplasts and for fusion of synthetically coated organelles.

Microinjection. Microinjection of isolated organelles into plant cells would be the most direct approach to organelle transfer. Capillary microinjection of biological material such as DNA, RNA, or protein into animal cells has been developed into a routine procedure (Celis et al. 1980). A comparable system for plant cells and protoplasts, described only recently (Steinbiss and Stabel 1983), has demonstrated that protoplast-derived tobacco cells can survive capillary microinjection of a fluorescent dye. The glass needles for this purpose, and also for injection of DNA (P. Stabel, personal communication), have a tip diameter of only about 0.3 μm. Whether or not isolated cells, protoplast-derived cultured cells, or protoplasts are capable of surviving the injury caused by larger needles is not known. However, it has to be expected that the proportion of surviving cells wold be drastically reduced. Discovery of wound-healing substances or conditions supporting rapid cell wound healing could lead to wider application of this technique. At present, microinjection of isolated chromosomes should be considered as a possible approach in specific cases, when, for example, giant cells or large fusion products are used as receptor cells in an oocyte-like system (Torriani and Potrykus 1983).

Micropipettes with openings of about 5 μm were used by Bradley (1979) to inject *Glaucocystis nostochinearum* cyanelles into onion epidermal cells. To minimize the damaging effect of the micropipette, the cells had to be plasmolyzed prior to injection. The approach then was to inject the organelles into the cavity formed between the cell wall and the plasmalemma during plasmolysis. Subsequent deplasmolysis led to a tight contact of injected organelles and the protoplast membrane inside the cell wall. However, uneven expansion of the plasmalemma and "budding" of the membrane were frequently observed, and many of the cells burst.

Transfer of nuclei and chromosomes

The failure of formation of somatic hybrid plants in various fusion-induced combinations (Dudits et al. 1980; Zenkteler and Melchers 1978) by loss of chromosomes from synkaryons or sorting out of chloroplasts of one parental type (Scowcroft and Larkin 1981) indicates the expression of somatic incompatibility at different stages of development. Somatic incompatibility reactions can be reduced by transferring only a part of the plant cell (nucleus, chloroplast, or mitochondria), or only a small part of the nuclear genome (isolated chromosomes). Transplantation of nuclei into amphibian cells has been used in very elegant and successful experiments to study cell differentiation and pluripotentiality of cell nuclei (Gurdon 1963, 1976). The development of protoplast technology and efficient procedures for organelle isolation encouraged such activities also with higher plants.

Potrykus and Hoffmann (1973) first reported the transplantation of nuclei into plant protoplasts. They used the sandwich technique in combination with plasmolysis/deplasmolysis and lysozyme described earlier and found about 0.5% of the treated protoplasts containing one extra nucleus or, in some cases, up to five additional nuclei. The donor nuclei were stained with the fluorescent dye ethidium bromide, and the treated protoplasts were evaluated with a fluorescence microscope that allowed unambiguous distinction of donor (fluorescent) and receptor (nonfluorescent) nuclei.

To investigate cellular incompatibility between two species, Binding (1976) fused protoplasts of tobacco, *Petunia,* and tomato and induced transfer of *Petunia* nuclei into tobacco protoplasts. The nuclei uptake was enhanced by treatment with PEG and $Ca(NO_3)_2$ at pH 9. Light microscopic analysis showed that numerous nuclei entered the protoplasts after agglutination, but *Petunia* chromosomes were not detected in mitotic figures of 100 tobacco regenerants analyzed after five days in culture.

Several transplantation methods were compared in an attempt to improve the uptake procedure (Lörz 1977). These included the use of modified fusion conditions, such as $Ca(NO_3)_2$, $CaCl_2$, high pH, and PEG, or combinations of these, and up to 5% uptake was achieved under conditions that did not interfere with viability and subsequent culture of receptor protoplasts. In attempts to demonstrate biological proof of integration and replication of transferred nuclear genes, two complementing, chlorophyll-deficient, light-sensitive mutants of tobacco were used as sources of nuclei and receptor protoplasts. These mutants had been used earlier by Melchers and Labib (1974) for somatic hybridization by protoplast fusion. Although complementing hybrids could clearly be discriminated against the parental-type cell colonies, no nuclear hybrids were detected out of 5.5×10^7 receptor protoplasts that were cultured following nucleus transplantation experiments (Lörz and Potrykus 1976, 1978). Nitrate-reductase-deficient mutants as receptor protoplasts and wild-type donor nuclei also appeared to be a suitable system for nucleus transplantation experiments; however, in this combination, too, no stable restoration of nitrate reductase activity was found in numerous experiments (O. Schieder and L. Willmitzer, personal communication). The failure to transfer biologically active nuclei into protoplasts was primarily a result of the "stress condition" to which the isolated organelles were exposed during the uptake procedure. It was found that treatment of isolated nuclei for 10 min with PEG under conditions used to stimulate uptake into protoplasts reduced RNA synthesis to about 23% of the control level (Lörz and Potrykus 1978).

A consequent step toward transfer of more defined parts of the plant genome is chromosome transplantation. Purified metaphase chromosomes are extensively used as vectors for transfer of genetic information in mammalian cell genetics (Ruddle 1980). Efficient procedures for synchronization in plant cell cultures and for chromosome isolation, as described earlier, are basic prerequisites for similar experiments with plants. Szabados and associates (1981) provided convincing cytological evidence for incorporation of foreign chromosomes into plant protoplasts. According to light microscopic examinations of cells fixed 3 hr after PEG treatment, the chromosomes introduced into protoplasts preserved their structural and morphological integrity.

Recent findings of J. King and associates (personal communication) indicate that transfer of isolated nuclei of *Vicia hajastana* into *Datura innoxia* protoplasts results in restoration of the pantothenate-requiring mutant cells of *Datura.* In summary, there are strong indications from nucleus and chromosome transfer studies that foreign organelles are taken up into protoplasts; however, conclusive evidence for stable integration and long-term biological activity of transferred organelles is still lacking.

Transfer of chloroplasts

Protoplasts isolated from albino mutants or chlorophyll-free cultured cells as receptor cells and green, wild-type chloroplasts as donor organelles are the most commonly used experimental system for chloroplast uptake studies. Potrykus (1973) described transplantation of wild-type chloroplasts into albino protoplasts of *Petunia.* The uptake was achieved by the sandwich technique described earlier or by incubation of chloroplasts and protoplasts with sodium nitrate. Uptake of green chloroplasts, 1–20 per cell, was found in about 0.1–0.5% of the albino protoplasts, and the transferred organelles remained visible by light microscopy up to six days.

The studies with chloroplasts were stimulated by an early report of Carlson (1973). When protoplasts of an albino, cytoplasmic tobacco mutant were coincubated in a medium containing wild-type chloroplasts, these chloroplasts were taken up into the cytoplasm of the albino protoplasts. It was reported that these organelles were able to replicate and function in the receptor cell cytoplasmic environment. This conclusion was based mainly on the regeneration of a green plant from the treated albino protoplasts. However, some essential control experiments were not described in that report, and the interpretation of the results has not been accepted without criticism (Cocking 1977; Potrykus 1973). A variegated plant also derived from these experiments was analyzed by electrofocusing of fraction-1 protein. In that case, chloroplast DNA and also nuclear DNA of both organelle donor and protoplast receptor-type tobacco were found (Kung et al. 1975). This indicates that nuclei and chloroplasts were transferred simultaneously, and separate transfer of chloroplasts had failed. These experiments have so far not been repeated by other laboratories.

The use of PEG as uptake-inducing agent significantly improved the efficiency of chloroplast uptake. Bonnett and Eriksson (1974) reported that up to 16% of viable *Daucus carota* protoplasts contained green chloroplasts isolated from the algae *Vaucheria dichotoma.* It is worth mentioning that in contrast to the finding of Potrykus and Hoffmann (1973), chloroplasts were found always located in the cytoplasm, but never in the vacuole.

The uptake of *Vaucheria* chloroplasts into protoplasts from cultured carrot cells (Figure 4.2) was examined also by electron microscopy (Bonnett 1976). Chloroplasts were found in uninucleated and multinucleated protoplasts with equal frequency, indicating that chloroplast uptake does not depend only on protoplast fusion. However, greater aggregation of protoplasts induced by PEG also enhanced the frequency of chloroplast uptake. Most significant in respect to a possible functionality of the transferred organelles is the finding that "there is no carrot plamalemma nor other membrane enveloping the chloroplast, the chloroplast is not within a pinocytotic vesicle and is free to interact with the host cytoplasm. The intracellular localization of this foreign organelle has not resulted in any apparent cytoplasmic disorder, as judged by the structural preservation of adjacent cytoplasmic constituents of the carrot protoplast" (Bonnett 1976).

Protoplasts from suspension cells of *Parthenocissus tricuspidata* and chloroplasts of *Petunia hybrida* were used in the study of Davey and associates (1976). Following PEG treatment, intact chloroplasts readily fused with the protoplast plasmalemma to establish continuity between the chloroplast stroma and the protoplast cytoplasm. Broken chloroplasts became localized in membrane-bounded vesicles

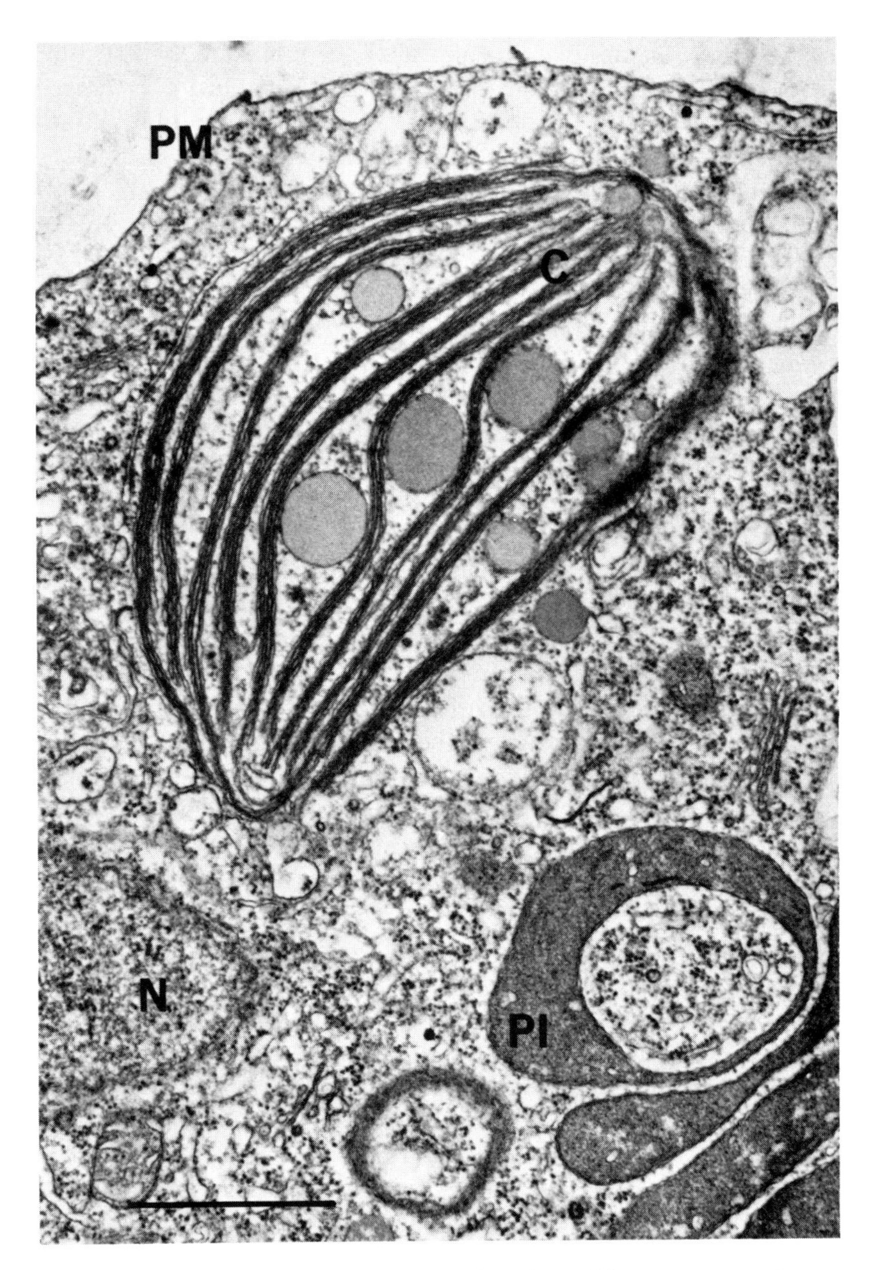

Figure 4.2. Transferred *Vaucheria* chloroplast within a carrot protoplast (×32,000). Uptake was induced by PEG. Bar represents 1 μm; C, *Vaucheria* chloroplasts; PM, plasma membrane; PI, carrot plastid. (Electron micrograph kindly supplied by H. T. Bonnett.)

in the cytoplasm. The behavior of isolated chloroplasts during and after PEG treatment obviously appears to be related to the structure and integrity of the organelle used for transplantation.

The feasibility of transplantation of isolated chloroplasts into protoplasts seems to be even less attractive after two reports published in 1979 (Landgren and Bonnett 1979; Uchimiya and Wildman 1979). In the first case, chloroplasts of *Nicotiana gossei* were transferred into *Nicotiana tabacum* protoplasts, a system that is compatible also by conventional hybridization. No evidence was obtained to suggest that genetic information coded for by the isolated donor organelles was being translated into polypeptides in regenerated shoots derived from treated protoplasts. Landgren and Bonnett (1979) used green chloroplasts of *Nicotiana excelsior* and albino leaf protoplasts of *N. tabacum* as the receptor. Plantlets were obtained from callus-derived albino protoplast populations in which chloroplast incorporation was observed. Although some of these plantlets were pale green, none was capable of autotrophic growth.

The situation described here for chloroplast transfer is very similar to that described earlier for nuclei uptake. Convincing evidence, based on reproducible experiments, for functional integrity and biological activity of transplanted organelles is not yet available. Dramatic changes in chloroplast number and structural changes of plastids have been found during the early stages of mesophyll protoplast culture (Thomas and Rose 1983). Fully developed chloroplasts therefore might not be the ideal organelles for transplantation in protoplasts, and more attention should be given to proplastids.

Transfer of other organelles

In addition to nuclei and chloroplasts, the mitochondrion is a third plant cell organelle carrying genetic information. Transfer of mitochondria into yeast protoplasts has been achieved by fusion of isolated mitochondria as well as by miniprotoplast fusion (Fukuda and Kimura 1980; Gunge and Sakaguchi 1979). No reports have been published thus far dealing with transfer of isolated mitochondria into higher plant cells. Some possibilities to study mitochondrial genetics are provided by protoplast fusion and molecular analysis of somatic hybrid cell lines (Belliard et al. 1979; Nagy et al. 1981). In contrast to the situation with yeast, suitable selective markers are still lacking for mitochondria of higher plants.

Research work directed toward producing new cellular symbiotic associations has included transfer of higher plant organelles into fungal protoplasts (Vasil and Giles 1975) and transfer of microorganisms into plant protoplasts (Burgoon and Bottino 1976; Meeks et al. 1978). Uptake of *Cyanobacteria* and cyanelles isolated from *Glaucocystis nostochinearum* and from *Cyanophora paradoxa* into different receptor-type protoplasts has also been reported (Bradley 1979; Hughes et al. 1978). However, no additional evidence has been presented indicating function and stability of these synthetic symbiotic associations. To circumvent some of the obvious limitations of isolated organelles in transfer experiments, intact protoplasts of *Chlamydomonas reinhardtii* were fused with carrot protoplasts to achieve organelle transfer (Fowke et al. 1979). Cultured fusion products regenerated cell walls and divided. Most of the *Chlamydomonas* organelles degenerated during culture, but transferred chloroplasts were still recognizable in the carrot cytoplasm after 10 days.

In spite of the numerous experiments designed and elaborated during the last decade to transfer isolated cell organelles into higher plant protoplasts, progress has been rather limited. Frequent uptake of different types of organelles is achieved routinely using fusion-inducing conditions such as PEG treatment. Intensive efforts to culture the modified cells have been made only in a few cases. Progress obviously has been impeded by lack of markers for nuclear as well as plastidal genes suitable for stringent selection in very early steps of culture. New mutants isolated specifically for their suitability in somatic cell genetics (Maliga 1980) should facilitate the design of new experiments with a better chance to find the genetically modified cells. Thus far it has not been demonstrated convincingly that new stable associations can be created after uptake of isolated organelles. The major reason for this failure is seen in the negative effects of the treatment used for organelle transplantation. Detrimental effects on the function and structural integrity of isolated nuclei and mitochondria are found after PEG treatment (Table 4.1) (Bendradis and de Virville 1982; Lörz and Potrykus 1978). Better "protection" of the isolated organelles seems to be essential. This could be achieved by synthetic coating with lipid vesicles or liposomes. Although this has been successfully applied to entrap DNA, RNA, and viruses, the sizes of organelles such as nuclei, chloroplasts, and mitochondria have thus far hampered the use of the liposome technique. An alternative approach to overcome the present limitations is the use of subprotoplasts instead of isolated organelles to achieve transfer of defined parts of the plant cell.

Table 4.1. *Uptake of isolated organelles into higher plant protoplasts*

Isolated organelles	Plant species (donor/receptor)	Uptake method	References
Nuclei	*Petunia/Petunia, Nicotiana, Zea*	Sandwich technique	Potrykus and Hoffman (1973)
Nuclei	*Petunia/Nicotiana*	PEG, $Ca(NO_3)_2$	Binding (1976)
Nuclei	*Nicotiana/Nicotiana*	Sandwich technique	Lörz and Potrykus (1978)
	Hordeum/Zea, Triticum	Ca $(NO_3)_2$, PEG, $CaCl_2$ + high pH	
Chromosomes	*Petroselium/Petroselium*	PEG	Szabados et al. (1981)
	Triticum/Triticum, Zea		
Chloroplasts	*Nicotiana/Nicotiana*	Coincubation	Carlson (1973)
Chloroplasts (+ nuclei)	*Nicotiana/Nicotiana*	Low-speed centrifugation, p-L-ornithine	Kung et al. (1975)
Chloroplasts	*Petunia/Petunia*	Sandwich technique, $NaNO_3$	Potrykus (1973)
Chloroplasts	*Vaucheria/Daucus*	PEG	Bonnett and Eriksson (1974)
Chloroplasts	*Vaucheria/Daucus*	PEG	Bonnett (1976)
Chloroplasts	*Petunia/Parthenocissus*	PEG	Davey et al. (1976)
Chloroplasts	*Nicotiana/Nicotiana*	PEG, $CaCl_2$	Uchimiya and Wildman (1979)
Chloroplasts	*Nicotiana/Nicotiana*	PEG	Landgren and Bonnett (1979)
Chloroplasts	*Zea/Daucus*	PEG	Chen et al. (1980)
Cyanelles	*Cyanophora/Hordeum, Nicotiana*	PEG	Hughes et al. (1978)
Cyanelles	*Glaucocystis/Allium, Daucus*	Microinjection, PEG	Bradley (1979)

Subprotoplasts and their role in somatic cell genetics

As outlined in the previous section, separate transfer of nuclear, plastidal, or mitochondrial genomes has not been successfully accomplished by uptake of isolated cell organelles. In mammalian cell genetics the transfer of cytoplasmic characters has been performed at high frequencies by fusing cells with enucleated cell fragments. Nucleocytoplasmatic interrelationships have been studied in *Acetabularia* and other Dasycladaceae by using nucleated and enucleated cell fragments (Kersey 1980; Schweiger and Berger 1979). A similar approach is now being used for higher plant cells. Fragmentation of plant cells into subcellular fractions is possible only after removal of the rigid plant cell wall. Cell-wall-free protoplasts are most suitable for fragmentation, and several techniques for generating subprotoplasts are now available (Binding 1979; Bradley 1983).

Subprotoplasts are fragments derived from protoplasts and by definition do not contain the entire content of plant cells. *Cytoplasts* are subprotoplasts containing most or fractions only of the original cytoplasmic material, but lacking the nucleus. Subprotoplasts containing the nucleus are called *karyoplasts* or *miniprotoplasts.* In this case the nucleus is surrounded by a small but still considerable amount of cytoplasm and the plasma membrane. The term *microprotoplast* was suggested for subprotoplasts containing not all but a few chromosomes (Bradley 1983). The term *microplast* was used to describe protoplast fragments containing only minor fractions of the cytoplasmic material surrounded by an inner membrane of the cell (Bilkey et al. 1982). In contrast to microplasts, subprotoplasts normally are surrounded by the original outer membrane, the plasma membrane. Major emphasis will be given in the following text to enucleated cytoplasts and nucleated miniprotoplasts.

Isolation of subprotoplasts

Functional subprotoplasts. Functional subprotoplasts are produced after chemical or physical treatment of protoplasts to facilitate elimination of some of their functions. These morphologically normal-looking protoplasts are subprotoplasts only in respect to specific genetic or physiological capacities. Zelcer and associates (1978) prepared "functional cytoplasts" by X-ray treatment of freshly isolated protoplasts and thereby inactivated the nuclear genome. Using these treated protoplasts in a fusion experiment they were able to transfer cytoplasmic male sterility independently from the nuclear genome from one tobacco species to another. This technique facilitating the inactivation of one of the fusion partners prior to protoplast fusion has been improved, and application of X-ray and iodoacetate inactivation has led to a high frequency of cybrids (Menczel et al. 1982). Laser-beam microsurgery may be another method to inactivate specific cellular components (Hahne and Hoffmann 1984). However, inactivation of the nuclear genome by physical or chemical means does not exclude absolutely that minor nuclear genomic fragments are "transferred" following protoplast fusion. In fact, there is evidence for limited intergeneric gene transfer after intensive X-ray treatment of the "donor" fusion partner (Dudits et al. 1980; Gupta et al. 1982). DNA hybridization experiments will permit assessment of the degree of limited gene transfer more accurately when line-specific probes are available.

Spontaneously formed subprotoplasts. Protoplasts and subprotoplasts are formed naturally in the pericarps of ripening fruits of some solanaceous species (Binding and Kollmann 1976; Cocking and Gregory 1963; De and Swain 1983). The sap of ripening tomato fruits (6–7 weeks old) also contains, in addition to "normal" nucleated and vacuolated protoplasts, cytoplasts without a nucleus showing different degrees of vacuolization and free vacuoles. All these units obviously are lacking any cell wall. These subprotoplasts are easily isolated by filtration of the juice from the ripening fruit through sieves of different pore sizes and subsequent sedimentation by centrifugation. The sizes of the protoplast fragments range from 2 to 50 μm in diameter, and they can be fractionated further on density gradients. Theoretically, the very small subprotoplasts could represent fragments containing only mitochondria or plastids, but thus far no detailed separation and evaluation have been attempted.

A commonly observed phenomenon during protoplast culture is the so-called budding that gives rise to enucleated subcellular fragments. It appears that during culture the volume of the protoplasts increases, and in some cases nonhomogeneous cell wall formation results in instability or weakness at the cell surface and leads to the formation of buds. Budding of cultured protoplasts is seen most frequently in conditions that are not favorable for cell division and sustained development. Budding was induced efficiently in *Zea mays* internode protoplasts when cultures were kept for two to three days in a medium with a relatively high osmotic value of more than 1,000 mOsm/kg H_2O (Lörz and Potrykus 1980). Multiple formation of cytoplasts from giant protoplasts has been reported by Hoffmann (1981). When cultured protoplasts, isolated from *Nicotiana plumbaginifolia* callus, reached a critical size (about 100 μm in diameter) after several weeks in culture without forming a cell wall, the protoplasts started to release cytoplasts into the medium. Under these conditions the separation of the subprotoplasts from the mother protoplast is spontaneous and complete, whereas during protoplast budding there is no complete separation. Budded subprotoplasts are easily separated from nucleated protoplasts by shaking.

So-called microplasts, small subprotoplasts surrounded by an inner membrane of the cell, are released in large numbers when highly vacuolated thin-walled callus cells are ruptured. Watery friable callus with highly vacuolated cells can be induced on auxin-containing medium from several plant species (Bilkey et al. 1982). Subprotoplasts obtained from ruptured callus cells or after budding of cultivated cells are of very different sizes, and enrichment of specific types of subprotoplasts can be accomplished by subsequent fractionation in density gradients (Harms and Potrykus 1978).

Plasmolytically induced subprotoplasts. Plasmolysis of elongated cells frequently causes the shrinking protoplasts to separate into two or more fragments or subprotoplasts. One of these fragments is a nucleated miniprotoplast, and the others are enucleated cytoplasts. Subsequent treatment of the plasmolyzed tissue with cell-wall-degrading enzymes produces a mixed population of protoplasts, miniprotoplasts, and cytoplasts (Figure 4.3).

Because of physical parameters, elongated cells form subprotoplasts with higher probability than more isodiametric cells. Bradley (1978) has suggested that subprotoplasts will be formed during plasmolysis when the cell length is greater than π times the cell diameter. Observations with onion epidermal cells support this physical model, where the protoplast is taken as a fluid drop. This simplified model is surely incomplete, because subprotoplasts are not formed in all plasmolyzed cells with length greater than π times the cell diameter (Archer et al. 1982).

Vatsya and Bhaskaran (1981) isolated subprotoplasts from precul-

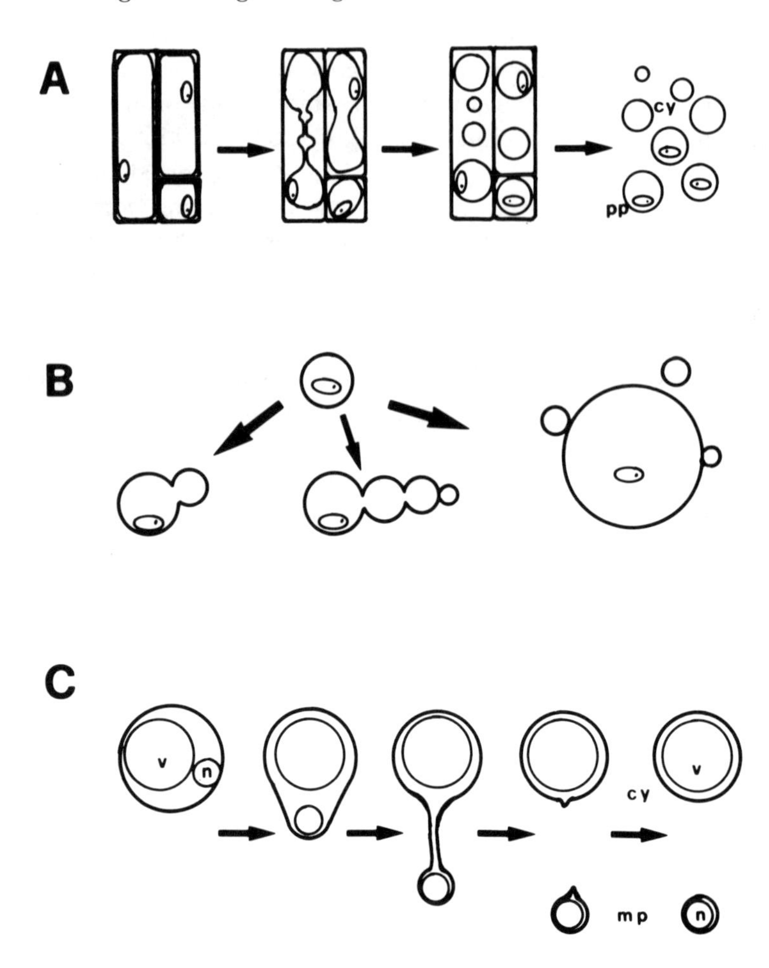

Figure 4.3. Formation of subprotoplasts. (A) Diagram illustrating the production of subprotoplasts induced by plasmolysis. (B) Budding of protoplasts giving rise to enucleated subprotoplasts. The formation of single or multiple buds and multiple subprotoplast formation are induced during protoplast culture. (C) Fragmentation of protoplasts during centrifugation. The illustration is drawn after microscropic observations of protoplasts from cultured cells centrifuged in Percoll gradients for periods from 10 to 60 min at 40,000 × *g;* cy, cytoplast; mp, miniprotoplast; pp, protoplast; n, nucleus; v, vacuole.

tured cotyledonary leaves of *Brassica oleracea* and showed that the yield of subprotoplasts was dependent on the osmolarity of the enzyme mixture used for protoplast isolation. Greater hypertonicity of the enzyme mixture caused more subprotoplast formation. These observations can be explained also in mathematical terms and are related to the increase in the ratio of surface area to volume in hypertonic medium (Vatsya and Bhaskaran 1983). Changing the mannitol concentration from 0.44 M to 0.77 M in the enzyme mixture caused an increase in the percentage of enucleated subprotoplasts from *B. oleracea* cells from 5% to about 35%. Ten to 15% of cytoplasts were found in isolates obtained from treatment of *Nicotiana* (deb)tbc leaves with cellulase and pectinase (Archer et al. 1982). Further enrichment of cytoplasts was achieved by separation of protoplasts and subprotoplasts on a two-step Percoll density gradient, leading to preparations with more than 60% cytoplasts.

Incubation of pollen grains in suitable medium induces the growth of pollen tubes (pollen germination) that are extremely elongated cells. Subprotoplasts have been released from enzymatically treated, germinating pollen tubes of *Hyoscyamus muticus* and *Nicotiana tabacum* (Lörz and Potrykus 1980). Of special interest with this type of material is the status of haploidy, the possibility to isolate karyoplasts containing either the vegetative nucleus or the generative nucleus or both, and the possibility to isolate cytoplasts containing totally dedifferentiated plastids. Plasmolytically induced subprotoplasts thus can be isolated from tissue and cells of very different origins, provided that not only isodiametric cells but also elongated cells are available.

Subprotoplast preparation by centrifugation. The most general approach to preparation of subprotoplasts is fragmentation of isolated protoplasts by centrifugal force (Bradley 1983; Lörz 1984). The different cellular components (e.g., nucleus, plastids, mitochondria, and cytoplasmic material) exhibit different specific densities and consequently can be separated in density gradients. Because of the centrifugal forces during centrifugation, cellular material of high density (e.g., nucleus) is oriented toward the bottom of the centrifuge tube, whereas less dense material (e.g., vacuoles) is oriented toward the top of the tube. During the initial phase of centrifugation, protoplasts are dramatically deformed and drawn out. Prolonged centrifugation leads to separation of nuclei-containing subprotoplasts (miniprotoplasts) and enucleated cryptoplasts (Lörz et al. 1981). After removal from the gradient and transfer to osmotically stabilized medium, subprotoplasts again become spherical within a few minutes.

Exposure of isolated protoplasts to the fungal metabolite cytochalasin B in combination with centrifugation was found to support protoplast fragmentation (Bracha and Sher 1981; Wallin et al. 1978; Zhou-Ping and Cheng 1981). Cytochalasins induce a wide range of effects in both animal and plant cells and usually are associated with cellular movements and activities of microfilaments. One of the effects observed is enucleation of mammalian cells growing in monolayer culture. A combination of centrifugation and cytochalasin B treatment increases the yield of enucleated cells (Ringertz and Savage 1976). These observations have encouraged researchers to use cytochalasin B also for plant cells. However, because unknown effects of the fungal metabolite on plant cells cannot be excluded, and protoplast fragmentation is possible without cytochalasin, it is no longer included in gradients (T. Eriksson and L. Laser, personal communication).

Many different types of protoplasts have been used for gradient fragmentation; however, the composition of the gradient, speed of centrifugation, time, and temperature have to be worked out individually for the specific type of protoplast used for enucleation (Table 4.2). Suitable components for establishing gradients for protoplast centrifugation are inorganic salts, sugars, and modified silica gels such as Percoll. After centrifugation, enucleated cytoplasts normally are located in the top fraction of the gradient, and nucleated miniprotoplasts form a band between the more dense fractions in the lower part of the centrifuge tube.

Success in enucleation of protoplasts is greatly influenced by the "quality" of the protoplast preparation (homogeneity of the protoplasts, stability, etc.). The yield of miniprotoplasts can be over 90% of the number of protoplasts placed onto a gradient (Bracha and Sher 1981; Lörz et al. 1981). Enucleated cytoplasts are more fragile, and the recovery rate has been lower. In microscopic analysis of the cytoplast fractions it was found that the cytoplast preparations were highly pure, and only minimal contamination with nucleated protoplasts (less than 5%) was obtained.

Thus, cytoplasts produced by centrifugation are suitable material for fusion experiments aiming for exclusive transfer of cytoplasmic genetic information. The reverse task of separating the nuclear genome from cytoplasmic material can be solved only partially. Fragmentation of protoplasts from cultured cells leads to very small miniprotoplasts or karyoplasts, but electron microscopy has shown clearly that all the nucleated subprotoplasts still contain a considerable

Table 4.2. *Production of subprotoplasts by gradient centrifugation*

Source of protoplasts	Species	Gradient composition	Subprotoplasts	References
Mesophyll, petals	Several	$CaCl_2$, mannitol, sucrose	Evacuolated miniprotoplasts, vacuoplasts	Lörz et al. (1976)
Cell suspension	*Nicotiana tabacum*	Cytochalasin B, sorbitol, sucrose	Cytoplasts, miniprotoplasts	Wallin et al. (1978)
Mesophyll, stem parenchyma	*Hordeum vulgare* *Zea mays*		Evacuolated miniprotoplasts, cytoplasts	Lörz and Potrykus (1980)
Cell suspension	*Hyoscyamus muticus* *Nicotiana tabacum* *Zea mays*	Percoll, $CaCl_2$, mannitol	Cytoplasts, miniprotoplasts	Lörz et al. (1981)
Epidermis, mesophyll	*Allium cepa*	Cytochalasin B, mannitol, sucrose	Cytoplasts, nucleoplasts	Bracha and Sher (1981)
Callus	*Allium sativum*	Cytochalasin B, mannitol, sucrose	Enucleated miniprotoplasts	Zhou-Ping and Cheng (1981)
Mesophyll	*Petunia parodii*	Percoll, $CaCl_2$, mannitol	Evacuolated miniprotoplasts	Griesbach and Sink (1983)

amount of cytoplasmic material (Lörz et al. 1981). This observation is supported by the fact that some of the miniprotoplasts are still able to recover and develop normally.

In general, protoplasts isolated from cultured cells are more suitable for enucleation than mesophyll protoplasts containing fully developed, green chloroplasts. Because of their similar specific densities, chloroplasts and nuclei are not separated,. and fragmentation often results in a "light" fragment consisting of the large vacuole and some residual cytoplasm and a "heavy" subprotoplast containing the nucleus and most of the chloroplasts (Griesbach and Sink 1983; Lörz et al. 1976). Miniprotoplasts without a large central vacuole and with reduced cytoplasm can be interesting material for somatic hybrid selection systems or for microinjection studies. Highly vacuolated cells are easily damaged during injection, whereas nonvacuolated miniprotoplasts are more stable for this type of manipulation, and the smaller cells additionally provide a better chance to aim directly into the nucleus.

Fusion studies with subprotoplasts

The outer surfaces of subprotoplasts and protoplasts, the plasma membranes, are of identical or very similar constitutions, and all methods developed for protoplast fusion are also applicable to subprotoplasts. Subprotoplasts have been fused most frequently by a modified PEG method.

Bearing in mind the limitation mentioned with respect to isolation of "ideal karyoplasts," miniprotoplast fusion is essentially protoplast fusion using small evacuolated protoplasts characterized by a dense cytoplasm. Wallin and associates (1979) isolated miniprotoplasts of two nitrate-reductase-deficient tobacco mutants and fused miniprotoplasts with miniprotoplasts or with protoplasts from the other line. Hybrids were selected by their ability for growth in medium containing nitrate as the sole nitrogen source. This result indicates that genetic information from both parental nuclei was expressed.

Fusion of enucleated cytoplasts and nucleated miniprotoplasts for direct production of *Allium cepa* cybrids was reported by Bracha and Sher (1981); however, no culture experiments were described. Incompatibility after fusion has been investigated by intergeneric fusion experiments in the family Solanacea. Binding (1976) attempted the following combinations: tobacco protoplasts + *Petunia* protoplasts, tobacco protoplasts + *Petunia* nuclei, tobacco protoplasts + tomato subprotoplasts, and *Petunia* protoplasts + tomato subprotoplasts. On an average, protoplasts and subprotoplasts fused about 10% to form heteroplasmic products. Cell division and segregation of plastids were observed in tobacco + *Petunia* protoplast fusion products. No mitotic activity was found with combinations of tobacco protoplast + tomato subprotoplast, even if only a small portion of tomato cytoplasm was involved in the symplasm formation. The same results were obtained with fusion products of *Petunia* protoplasts + tomato subprotoplasts. Although only limited positive results are available, it has been possible to investigate the influences of nucleus and cytoplasm separately by using subprotoplasts as carriers of individual types of organelles.

Cytoplast-protoplast fusion was successfully applied for interspecific chloroplast transfer in *Nicotiana* (Maliga et al. 1982). Protoplasts of *Nicotiana tabacum,* line SR 1, carrying a maternally inherited streptomycin resistance were enucleated by centrifugation through an isoosmotic Percoll gradient (Lörz et al. 1981). Resulting cytoplasts containing resistant plastids were fused with sensitive *Nicotiana plumbaginifolia* protoplasts. Resistant clones recovered after fusion were presumed to be derived from interspecific cytoplast-protoplast fusion. Several plants were regenerated from streptomycin-resistant clones, and plastid transfer was confirmed by the Eco-RI restriction pattern of the chloroplast DNA. However, resistant plants also were found resulting from a fusion of *N. plumbaginifolia* protoplasts with nucleated protoplasts present in the *N. tabacum* cytoplast preparation as contaminants. Even with this drawback, cytoplast-protoplast fusion can be envisaged as an alternative method for organelle transfer. Further improvement of the technique will make it feasible to use cytoplast-fusion-mediated organelle transfer also for a wider range of experiments.

Conclusion

Uptake of isolated cell organelles into protoplasts and subprotoplast fusion are promising fields, and their roles in somatic cell genetics are expected to be more important in the future than up to now. A great deal of basic research is still required. The mechanism of organelle uptake has been studied in some detail, but it is still neither predictable nor experimentally accessible whether the actual transplanted organelles retain an intact outer membrane or the organellar and protoplast membranes fuse during uptake to release the organellar content into the cytoplasm. The methodologic parameters used thus far for organelle uptake have to be considered as major reasons for the failure to transfer biologically active organelles. Further modifications are needed to isolate and transfer organelles more gently.

A new aspect is the manipulation of individual, defined single cells cultured in very small amounts of medium. Organelle transfer can be approached by applying electric-field-induced uptake in single-cell microdroplet cultures or by using capillary microinjection. The success of organelle microinjection may be limited because of cellular damage caused by the rather wide-open needles necessary for organelles.

Application of the liposome technique to protect and to coat organelles synthetically is advancing. However, the usefulness of liposomes for organelle transfer depends on the technical possibilities to coat rather large structures with artificial membranes. The method should be considered especially for smaller organelles such as mitochondria or isolated chromosomes.

Another promising approach to discrete transfer of cytoplasmic and nuclear genetic information is the use of subprotoplasts. Spontaneously formed subprotoplasts are available from various sources, or they can be prepared by centrifugal fragmentation of protoplasts and fused by standard procedures. In one instance, evidence is available that a chloroplast-coded resistance was transferred independently from the nuclear genome by cytoplast-protoplast fusion.

Somatic hybridization by protoplast fusion is more advanced than genetic manipulation by organelle transfer. Application of defined selection schemes and careful analyses of fusion products have revealed very different types of "somatically constructed" cells. Transfer of isolated organelles or fusion with subprotoplasts could avoid time-consuming processes of selection and sorting out of organelles. Besides gene transfer with isolated DNA, organelle uptake and subprotoplast fusion are the most direct approaches to selective transfer of genetic information into plant cells.

Acknowledgment

The author is grateful to Drs. Peggy Ozias-Akins, Elke Göbel, and Tage Eriksson for their helpful discussions during preparation of this manuscript and to Mrs. M. Pasemann for typing the manuscript.

References

Archer, E. K., Landgren, C., and Bonnett, H. T. (1982). Cytoplast formation and enrichment from mesophyll protoplast populations of *Nicotiana* spp. *Plant Sci. Lett.* **25,** 175–85.

Belliard, G., Vedel, F., and Pelletier, G. (1979). Mitochondrial recombination in cytoplasmic hybrids of *Nicotiana tabacum* by protoplast fusion. *Nature* **281,** 401–3.

Bendradis, A., and de Virville, J. D. (1982). Effects of polyethylene glycol treatment used for protoplast fusion and organelle transplantation on the functional and structural integrity of mitochondria isolated from spinach leaves. *Plant Sci. Lett.* **26,** 257–64.

Bilkey, P. C., Davey, M. R., and Cocking, E. C. (1982). Isolation, origin and properties of enucleated plant microplasts. *Protoplasma* **110,** 147–51.

Binding, H. (1976). Somatic hybridization experiments in solanaceous species. *Mol. Gen. Genet.* **144,** 171–6.

(1979). Subprotoplasts and organelle transplantation, in *Plant Cell and Tissue Culture – Principles and Applications,* ed. W. R. Sharp, P. O. Larsen, E. F. Paddock, and V. Raghavan, pp. 789–805. Ohio State University Press, Columbus.

Binding, H., and Kollmann, R. (1976). The use of subprotoplasts for organelle transplantation, in *Cell Genetics in Higher Plants,* ed. D. Dudits, G. L. Farkas, and P. Maliga, pp. 191–206. Akademiai Kiadó, Budapest.

Blaschek, W., Hess, D., and Hoffmann, F. (1974). Transcription in nuclei prepared from isolated protoplasts of *Nicotiana* and *Petunia. Z. Pflanzenphysiol.* **72,** 262–71.

Bonnett, H. T. (1976). On the mechanism of the uptake of *Vaucheria* chloroplasts by carrot protoplasts treated with polyethylene glycol. *Planta* **131,** 229–33.

Bonnett, H. T., and Eriksson, T. (1974). Transfer of algal chloroplasts into protoplasts of higher plants. *Planta* **120,** 71–9.

Bracha, M., and Sher, N. (1981). Fusion of enucleated protoplasts with nucleated miniprotoplasts in onion (*Allium cepa* L.). *Plant Sci. Lett.* **23,** 95–101.

Bradley, P. M. (1978). Production of enucleated plant protoplasts of *Allium cepa. Plant Sci. Lett.* **13,** 287–90.

(1979). Micromanipulation of cyanelles of a cyanobacterium into higher plant cells. *Physiol. Plant.* **46,** 293–8.

(1983). The production of higher plant subprotoplasts. *Plant Mol. Biol. Reporter* **1,** 117–23.

Bradley, P. M., and Leith, A. (1979). Uptake of *Cyanobacteria* contained in oil drops by plant protoplasts. *Naturwissenschaften* **66,** 111–12.

Burgoon, A. C., and Bottino, P. J. (1976). Uptake of the nitrogen fixing blue-green algae *Gloeocapsa* into protoplasts of tobacco and maize. *J. Hered.* **67,** 223–6.

Carlson, P. S. (1973). The use of protoplasts for genetic research. *Proc. Natl. Acad. Sci. U.S.A.* **70,** 598–602.

Cassels, A. C. (1978). Uptake of charged lipid vesicles by isolated tomato protoplasts. *Nature* **275,** 760.

Celis, J. E., Graessmann, A., and Loyter, A. (editors) (1980). *Transfer of Cell Constituents into Eukaryotic Cells.* Plenum Press, New York.

Chaleff, R. S., and Carlson, P. S. (1974). Somatic cell genetics in higher plants. *Annu. Rev. Genet.* **8,** 267–78.

Chen, M., Lin, Z., Zhao, Y., and Liu, H. (1980). Transfer of corn chloroplasts into protoplasts of carrot. *Academia Botanica Sinica* **23,** 27–30.

Cocking, E. C. (1972). Plant cell protoplasts – isolation and development. *Annu. Rev. Plant Physiol.* **23,** 20–50.

(1977). Uptake of foreign genetic material by plant protoplasts. *Int. Rev. Cytol.* **48,** 323–43.

(1983). Plant genetic manipulation from plant somatic cell genetics, in *Genetic Engineering in Eukaryotes,* ed. P. F. Lurquin and A. Kleinhofs, pp. 187–94. Plenum Press, New York.

Cocking, E. C., and Gregory, D. W. (1963). Organized protoplasmic units of plant cells: their occurrence, origin and structure. *J. Exp. Bot.* **14,** 504–11.

Davey, M. R., Frearson, E. M., and Power, J. B. (1976). Polyethylene glycol-induced transplantation of chloroplasts into protoplasts: an ultrastructural assessment. *Plant Sci. Lett.* **7,** 7–16.

De, D. N., and Swain, D. (1983). Protoplast, cytoplast and subprotoplast from ripening tomato fruits: their nature and fusion properties, in *Plant Cell Culture in Crop Improvement,* ed. S. K. Sen and K. L. Giles, pp. 201–8. Plenum Press, New York.

Dudits, D., Fejér, O., Hadlaczky, G., Koncz, C., Lázár, G. B., and Horváth, G. (1980). Intergeneric gene transfer mediated by plant protoplast fusion. *Mol. Gen. Genet.* **179,** 283–8.

Edwards, G. E., McLilley, R., Craig, S., and Hatch, M. D. (1979). Isolation of intact and functional chloroplasts from mesophyll and bundle sheath protoplasts of the C_4 plant *Panicum miliaceum. Plant Physiol.* **63,** 821–7.

Ferenczy, L., and Maráz, A. (1977). Transfer of mitochondria by protoplast fusion in *Saccharomyces cerevisiae. Nature* **268,** 524–5.

Fowke, L. C., Gresshoff, P. M., and Marchant, H. J. (1979). Transfer of organelles of the algae *Chlamydomonas reinhardtii* into carrot cells by protoplast fusion. *Planta* **144,** 341–8.

Fowke, L. C., and Gamborg, O. L. (1980). Application of protoplasts to the study of plant cells. *Int. Rev. Cytol.* **68,** 9–51.

Frankel, R. (1983). *Heterosis. Monographs on Theoretical and Applied Genetics. Vol. 6,* Springer-Verlag, Berlin.

Fukuda, H., and Kimura, A. (1980). Transfer of mitochondria into protoplasts of *Saccharomyces cerevisiae* by mini-protoplast fusion. *FEBS Lett.* **113,** 58–61.

Galun, E. (1981). Plant protoplasts as physiological tools. *Annu. Rev. Plant Physiol.* **32,** 237–66.

Galun, E., and Aviv, D. (1983). Cytoplasmic hybridization: genetic and breeding applications, in *Handbook of Plant Cell Culture,* ed. D. A. Evans et al., pp. 358–92. Macmillan, New York.

Giles, K. L. (1978). The uptake of organelles and microorganisms by plant protoplasts: old ideas but new horizons, in *Frontiers of Plant Tissue Culture 1978,* ed. T. A. Thorpe, pp. 67–74. University of Calgary, Alberta.

Griesbach, R. J., Malmberg, R. L., and Carlson, P. S. (1982). An improved technique for the isolation of higher plant chromosomes. *Plant Sci. Lett.* **24,** 55–60.

Griesbach, R. J., and Sink, K. C. (1983). Evacuolation of mesophyll protoplasts. *Plant Sci. Lett.* **30,** 297–301.

Gunge, N., and Sakaguchi, K. (1979). Fusion of mitochondria with protoplasts in *Saccharomyces cerevisiae. Mol. Gen. Genet.* **170,** 243–8.

Gupta, P. P., Gupta, M., and Schieder, O. (1982). Correction of nitrate reductase defect in auxotrophic plant cells through protoplast-mediated intergeneric gene transfer. *Mol. Gen. Genet.* **188,** 378–83.

Gurdon, J. B. (1963). Nuclear transplantation in amphibia and the importance of stable nuclear changes in promoting cellular differentiation. *Q. Rev. Biol.* **38,** 54–78.

(1976). The pluripotentiality of cell nuclei, in *The Developmental Biology of Plants and Animals,* ed. C. F. Graham and P. F. Wareing, pp. 55–63. Blackwell, Oxford.

Hadlaczky, G., Bisztray, G., Praznovsky, T., and Dudits, D. (1983). Mass isolation of plant chromosomes and nuclei. *Planta* **157,** 278–85.

Hahne, G., and Hoffmann, F. (1984). The effect of laser microsurgery on cytoplasmic strands and cytoplasmic streaming in isolated plant protoplasts. *Eur. J. Cell Biol.* **33,** 175–9.

Hampp, R. (1980). Rapid separation of the plastid, mitochondrial and cytoplasmic fractions from intact leaf protoplasts of *Avena. Planta* **150,** 291–8.

Harms, C. T. (1983*a*). Somatic incompatibility in the development of higher plant somatic hybrids. *Q. Rev. Biol.* **58,** 325–53.

(1983*b*). Somatic hybridization by plant protoplast fusion, in *Protoplasts 1983,* ed. I. Potrykus et al., pp. 69–84. Birkhäuser, Basel.

Harms, C. T., and Potrykus, I. (1978). Fractionation of plant protoplast types by iso-osmotic density gradient centrifugation. *Theor. Appl. Genet.* **53,** 57–63.

Hoffman, F. (1981). Formation of cytoplasts from giant protoplasts in culture. *Protoplasma* **107,** 387–91.

Hughes, B. G., Hess, W. M., and Smith, M. A. (1977). Ultrastructure of nuclei isolated from plant protoplasts. *Protoplasma* **93,** 267–74.

Hughes, B. G., White, F. G., Vernon, L. P., and Smith, M. A. (1978). Uptake of cyanelles and blue-green algae into barley and tobacco protoplasts. *Plant Physiol. Suppl.* **61,** 54.

Hull, J. L., and Moore, A. L. (1983). *Isolation of Membranes and Organelles from Plant Cells.* Academic Press, New York.

Kao, K. N., and Michayluk, M. R. (1974). A method for high-frequency intergeneric fusion of plant protoplasts. *Planta* **115,** 355–69.

Keller, W. A., and Melchers, G. (1973). The effect of high pH and calcium on tobacco leaf protoplast fusion. *Z. Naturforsch.* **C28,** 737–41.

Kersey, Y. (1980). Cytoplasts from coenocytic algal cells, in *Handbook of Physiological Methods–Developmental and Cytological Methods,* ed. E. Gantt, pp. 171–7. Cambridge University Press.

Koop, H. W., Dirk, J., Wolff, D., and Schweiger, H. G. (1983). Somatic hybridization of two selected single cells. *Cell Biology Int. Rep.* **7,** 1123–8.

Kung, S. D., Gray, J. C., Wildman, S. G., and Carlson, P. S. (1975). Polypeptide composition of fraction 1 protein from parasexual hybrid plants in the genus *Nicotiana. Science* **187,** 353–5.

Laat, A. M. M., de, and Blaas, J. (1984). Flow-cytometric characterization and sorting of plant chromosomes. *Theor. Appl. Genet.* **67,** 463–7.

Landgren, C. R., and Bonnett, H. T. (1979). The culture of albino tobacco protoplasts treated with PEG to induce chloroplast incorporation. *Plant Sci. Lett.* **16,** 15–22.

Lázár, G. B. (1983). Recent developments in plant protoplast fusion and selection technology, in *Protoplasts 1983,* ed. I. Potrykus et al., pp. 61–67, Birkhäuser, Basel.

Lesney, M. S., Callon, P. C., and Sink, K. C. (1983). A simplified method for bulk production of cytoplasts from suspension culture-derived protoplasts of *Solanum nigrum,* in *Protoplasts 1983,* ed. I. Potrykus et al., pp. 116–17. Birkhäuser, Basel.

Lima de Faria, A., Eriksson, T., and Kjellen, L. (1977). Fusion of human cells with *Haplopappus* protoplasts by means of *Sendai* virus. *Hereditas* **87,** 57–66.

Lörz, H. (1977). Protoplasten höherer Pflanzen als Objekte der genetischen Manipulation. Thesis, Universität Heidelberg.

(1984). Enucleation of protoplasts: preparation of cytoplasts and miniprotoplasts, in *Cell Culture and Somatic Cell Genetics of Plants, Vol. 1,* ed. I. K. Vasil, pp. 448–53. Academic Press, New York.

Lörz, H., Harms, C. T., and Potrykus, O. (1976). Isolation of "vacuoplasts" from protoplasts of higher plants. *Biochem. Physiol. Pflanzen* **169,** 617–20.

Lörz, H., and Izhar, S. (1983). Organelle transfer, sorting out, recombination, in *Protoplasts 1983,* ed. I. Potrykus et al., p. 129. Birkhäuser, Basel.

Lörz, H., Paszkowski, J., Dierks-Ventling, C., and Potrykus, I. (1981). Isolation and characterization of cytoplasts and miniprotoplasts derived from protoplasts of cultured cells. *Physiol. Plant.* **53,** 385–91.

Lörz, H., and Potrykus, I. (1976). Uptake of nuclei into higher plant protoplasts, in *Cell Genetics in Higher Plants,* ed. D. Dudits et al., pp. 239–44. Académiai Kiadó, Budapest.

(1978) Investigations on the transfer of isolated nuclei into plant protoplasts. *Theor. Appl. Genet.* **53,** 251–6.

(1980). Isolation of subprotoplasts for genetic manipulation studies, in *Advances in Protoplast Research,* ed. L. Ferenczy, G. L. Farkas, and G. Lázár, pp. 377–82. Akadémiai Kiadó, Budapest.

Maliga, P. (1980). Isolation, characterization, and utilization of mutant cell lines in higher plants. *Int. Rev. Cytol. Suppl.* **11A,** 225–50.

Maliga, P., Lörz, H., Lázár, G., and Nagy, F. (1982). Cytoplast-protoplast fusion for interspecific chloroplast transfer in *Nicotiana. Mol. Gen. Genet.* **185,** 211–15.

Malmberg, R. L., and Griesbach, R. J. (1980). The isolation of mitotic and meiotic chromosomes from plant protoplasts. *Plant Sci. Lett.* **17,** 141–7.

Mayo, M. A., and Cocking, E. C. (1969). Pinocytic uptake of polystyrene latex particles by isolated tomato fruit protoplasts. *Protoplasma* **68,** 223–30.

Meeks, J. C., Malmberg, R. L., and Wolk, C. P. (1978). Uptake of auxotrophic cells of a heterocyst-forming cyanobacterium by tobacco protoplasts, and the fate of their associations. *Planta* **139,** 55–60.

Melchers, G., and Labib, G. (1974). Somatic hybridization of plants by fusion of protoplasts. I. Selection of light resistant hybrids of "haploid" light sensitive varieties of tobacco. *Mol. Gen. Genet.* **135,** 277–94.

Menczel, L., Galiba, G., Nagy, F., and Maliga, P. (1982). Effect of radiation dosage on efficiency of chloroplast transfer by protoplast fusion in *Nicotiana. Genetics* **100,** 487–95.

Nagata, T. (1978). A novel cell-fusion method of protoplasts by polyvinylalcohol. *Naturwissenschaften* **65,** 263–4.

Nagy, F., Török, I., and Maliga, P. (1981). Extensive rearrangements in the mitochondrial DNA in somatic hybrids of *Nicotiana tabacum* and *N. knightiana. Mol. Gen. Genet.* **183,** 437–9.

de Nettancourt, D. (editor) (1977). *Incompatibility in Angiosperms. Monographs on Theoretical and Applied Genetics, Vol. 3.* Springer-Verlag, Berlin.

Nishimura, M., and Beevers, H. (1978). Isolation of intact plastids from protoplasts from castor bean endosperm. *Plant Physiol.* **62,** 40–3.

Ohyama, K., Pelcher, L. E., and Horn, D. (1977). A rapid, simple method for nuclei isolation from protoplasts. *Plant Physiol.* **60,** 179–81.

Papahadjopoulos, D., Wilson, T., and Taber, R. (1980). Liposomes as macromolecular carriers for the introduction of RNA and DNA into cells, in *Transfer of Cell Constituents into Eukaryotic Cells,* ed. J. E. Celis, A. Graessmann, and A. Loyter, pp. 155–72. Plenum Press, New York.

Potrykus, I. (1973). Transplantation of chloroplasts into protoplasts of *Petunia. Z. Pflanzenphysiol.* **70,** 364–6.

(1975). Uptake of cell organelles into isolated protoplasts, in *Modification of the Information Content of Plant Cells,* ed. R. Markham et al., pp. 169–79. North Holland, Amsterdam.

Potrykus, I., Harms, C. T., Hinnen, A., Hütter, R., King, P. J., and Shillito, R. D. (editors) (1983). *Protoplasts 1983. Experientia Supplementum, Vol. 46.* Birkhäuser, Basel.

Potrykus, I., and Hoffmann, F. (1973). Transplantation of nuclei into protoplasts of higher plants. *Z. Pflanzenphysiol.* **69,** 287–9.

Potrykus, I., and Lörz, H. (1976). Organelle transfer into isolated protoplasts, in *Cell Genetics in Higher Plants,* ed. D. Dudits et al., pp. 183–90. Akadémiai Kiadó, Budapest.

Ringertz, N. R., and Savage, R. E. (1976). *Cell Hybrids.* Academic Press, New York.

Robinson, S. P., and Walker, D. A. (1979). Rapid separation of the chloroplast and cytoplasmic fractions from intact leaf protoplasts. *Arch. Biochem. Biophys.* **196,** 319–402.

Rollo, F. (1983). Liposomes as a tool for introducing biologically active viral nucleic acids into plant protoplasts, in *Genetic Engineering in Eukaryotes,* ed. P. F. Lurquin and A. Kleinhofs, pp. 179–86, Vol. 61, NATO ISI series. Plenum, New York.

Ruddle, F. H. (1980). Gene transfer in eukaryotes, in *Transfer of Cell Constituents into Eukaryotic Cells,* ed. J. E. Celis, A. Graessmann, and A. Loyter, pp. 295–309. Plenum Press, New York.

Schel, J. H. W., van Lammeren, A. A. M., and Poelma, W. M. J. (1983). Isolation of nuclei from endosperm protoplasts of *Zea mays* L., in *Fertilization and Embryogenesis of Ovulated Plants,* pp. 323–7. VEDA, Bratislava.

Schieder, O. (1975). Selektion einer somatischen Hybride nach Fusion von Protoplasten auxotropher Mutanten von *Sphaerocarpos donnellii* Aust. *Z. Pflanzenphysiol.* **74,** 357–65.

Schweiger, H.-G., and Berger, S. (1979). Nucleocytoplasmic interrelationships in *Acetabularia* and some other Dasycladaceae. *Int. Rev. Cytol. Suppl.* **9,** 12–44.

Scowcroft, W. R., and Larkin, P. J. (1981). Chloroplast DNA assorts randomly in intraspecific somatic hybrids of *Nicotiana debneyi. Theor. Appl. Genet.* **60,** 179–84.

Steinbiss, H. H., and Broughton, W. J. (1983). Methods and mechanism of gene uptake in protoplasts. *Int. Rev. Cytol. Suppl.* **16,** 191–208.

Steinbiss, H. H., and Stabel, P. (1983). Protoplast derived tobacco cells can survive capillary microinjection of the fluorescent dye lucifer yellow. *Protoplasma* **116,** 223–7.

Suzuki, M., Takebe, I., Kajita, S., Honda, Y., and Matsui, C. (1977). Endocytosis of polystyrene spheres by tobacco leaf protoplasts. *Exp. Cell Res.* **105,** 127–35.

Szabados, L., Hadlaczky, G., and Dudits, D. (1981). Uptake of isolated plant chromosomes by plant protoplasts. *Planta* **151,** 141–5.

Tallman, G., and Reeck, G. R. (1980). Isolation of nuclei from plant protoplasts without the use of a detergent. *Plant Sci. Lett.* **18,** 271–5.

Thomas, M. R., and Rose, R. J. (1983). Plastid number and plastid structural changes associated with tobacco mesophyll protoplast culture and plant regeneration. *Planta* **158,** 329–38.

Torriani, U., and Potrykus, I. (1983). Attempts to develop a plant analogue to the oocyte system, in *Protoplasts 1983,* ed. I. Potrykus et al., pp. 268–9. Birkhäuser, Basel.

Uchimiya, H., and Wildman, S. G. (1979). Nontranslation of foreign genetic information for fraction 1 protein under circumstances favorable for direct transfer of *Nicotiana gossei* isolated chloroplasts into *N. tabacum* protoplasts. *In Vitro* **15,** 463–8.

Vasil, I. K., and Giles, K. L. (1975). Induced transfer of higher plant chloroplasts into fungal protoplasts. *Science* **190,** 680.

Vatsya, B., and Bhaskaran, S. (1981). Production of subprotoplasts in *Brassica oleracea* var. capitata – a function of osmolarity of the media. *Plant Sci. Lett.* **23,** 277–82.

(1983). Factors responsible for the production of subprotoplasts in *Brassica oleracea* var. capitata, in *Plant Cell Culture in Crop Improvement,* ed. S. K. Sen and K. L. Giles, pp. 485–9. Plenum Press, New York.

Vienken, J., Ganser, R., Hampp, R., and Zimmermann, U. (1981). Electric field induced fusion of isolated vacuoles and protoplasts of different developmental and metabolic provenience. *Physiol. Plant.* **53,** 64–70.

Wallin, A., Glimelius, K., and Eriksson, T. (1974). The induction of aggregation and fusion of *Daucus carota* protoplasts by polyethylene glycol. *Z. Pflanzenphysiol.* **74,** 64–80.

(1978). Enucleation of plant protoplasts by cytochalasin B. *Z. Pflanzenphysiol.* **87,** 333–40.

(1979). Formation of hybrid cells by transfer of nuclei via fusion of miniprotoplasts from cell lines of nitrate reductase deficient tobacco. *Z. Pflanzenphysiol.* **91,** 89–94.

Willison, J. H. M., Grout, B. W. W., and Cocking, E. C. (1971). A mechanism for the pinocytosis of latex spheres by tomato fruit protoplasts. *Bioenergetics* **2,** 371–82.

Zelcer, A., Aviv, D., and Galun, E. (1978). Interspecific transfer of cytoplasmic male sterility by fusion between protoplasts of normal *Nicotiana sylvestris* and X-ray irradiated protoplasts of male-sterile *N. tabacum. Z. Pflanzenphysiol.* **90,** 397–407.

Zenkteler, M., and Melchers, G. (1978). In vitro hybridization by sexual methods and by fusion of somatic protoplasts. *Theor. Appl. Genet.* **52,** 81–90.

Zhou-Ping, L., and Cheng, M. (1981). Enucleation of garlic *Allium sativum* protoplasts by cytochalasin B and centrifugation, in *International Cell Biology 1980–1981,* ed. H. G. Schweiger, p. 879. Springer-Verlag, Berlin.

Zimmermann, U. (1982). Electric field-mediated fusion and related electrical phenomena. *Biochim. Biophys. Acta* **694,** 227–77.

5 *Agrobacterium* as a vector system for the introduction of genes into plants

L. HERRERA-ESTRELLA, M. DE BLOCK,
P. ZAMBRYSKI, M. VAN MONTAGU,
AND J. SCHELL

Plant pathologists have long been interested in the soil bacterium *Agrobacterium tumefaciens,* because some strains of this organism have the capacity to infect almost all dicotyledonous plants after they have been wounded (Braun 1978, 1982). As a result of the infection, the wound tissue begins to proliferate as a neoplastic growth, commonly referred to as a crown gall tumor. Once induced, the tumor no longer requires the presence of the bacteria to grow and can be cultivated in vitro as an axenic culture (Braun 1943). Two main properties characterize crown galls: their ability to grow in vitro without the supplement of hormones required by normal plant cells and their ability to synthesize a set of new metabolites termed opines, which are not present in normal cells. Opines are amino acid or sugar derivatives that can be used as carbon and nitrogen sources by the bacteria responsible for inciting the tumor (Tempé and Petit 1982).

Agrobacterium first attracted the attention of molecular biologists when it was discovered that a large plasmid, the tumor-inducing or Ti plasmid, was responsible for the oncogenic capacity of the bacterium (Ven Larebeke et al. 1974; Watson et al. 1975; Zaenen et al. 1974) and that tumor formation is the direct result of the transfer and stable integration of a defined segment of the Ti plasmid (T-DNA) into the nuclear genome of plant cells (Chilton et al. 1977, 1980; Willmitzer et al. 1980).

Recently, further interest in *Agrobacterium* has arisen because the long-envisaged possibility of using the Ti plasmid as a vector for transfer and expression of foreign genes in plants has become a reality and has opened the door for studies of problems of gene regulation in plants. Such studies were previously hampered by the lack of methods to introduce DNA into the genome of plant cells.

In this chapter we shall discuss some of the most recent advances in our understanding at the molecular level of how *Agrobacterium* trans-

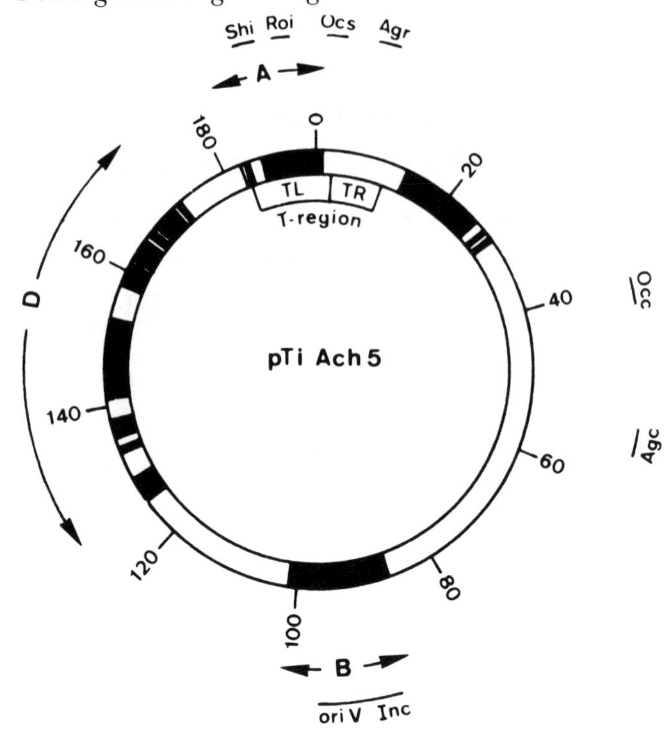
Shi
Roi
Ocs
Agr
A
0
20
40
60
80
100
120
140
160
180
TL
TR
T-region
pTi Ach5
D
B
ori V
Inc
Occ
Agc

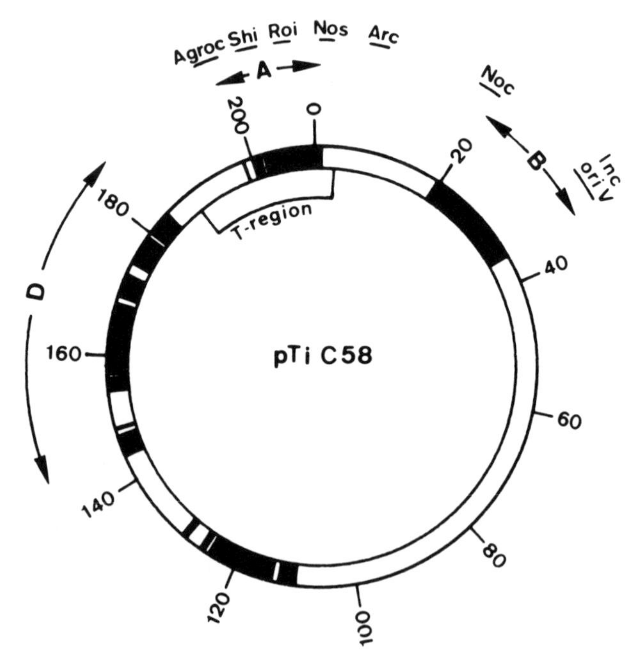
Agroc
Shi
Roi
Nos
Arc
A
Noc
B
Inc
ori V
0
20
40
60
80
100
120
140
160
180
200
T-region
pTi C58
D

fers DNA to plant cells and how this understanding allowed the systematic design of Ti plasmid derivatives efficient at introducing foreign DNA sequences to plant cells without altering their normal capacity to differentiate into whole plants. Also, some of the first examples of expression of foreign genes in plant tissue culture systems and in whole plants will be described, including an example of the expression of a bacterial gene under the control of a light-inducible plant promoter.

Genetic and functional organization of the Ti plasmid

Following the discovery of the plasmid nature of the tumor-inducing principle of *Agrobacterium,* the Ti plasmid became the main target of studies aiming to identify and analyze the functions of genes responsible for tumor formation. Ti plasmids can be classified according to the types of opines they induce the plant to produce. To date, most studies have concentrated on the Ti plasmids that induce tumors that produce nopaline or octopine. Crown gall induced by the nopaline-type Ti plasmids also produces agrocinopine A and B (Ellis and Murphy 1981; Firmin and Fenwick 1978), whereas octopine-induced tumors also produce agropine and mannopine (Tate et al. 1982).

A detailed picture of the genetic organization of the octopine and nopaline Ti plasmids has been obtained, initially by means of transposon-insertion mutagenesis and later using extensive deletion mutants. It was demonstrated that not only the genes responsible for opine synthesis are located on the Ti plasmid but also those genes that enable *Agrobacterium* to utilize opines as carbon and nitrogen sources. Moreover, these studies defined two regions of the Ti plasmid that are essential for tumor formation. These results are diagrammed in Figure 5.1. Mutants in one of these regions, called the

Figure 5.1. Functional organization of the octopine pTiAch5 and the nopaline pTiC58 plasmids. The numbers indicate the sizes of the plasmids in kilobases (kb). The black bars represent regions of homology between the two plasmids, as described by Engler et al. (1981); *shi* and *roi* indicate the positions of the loci involved in shoot and root inhibition, respectively. The genes responsible for opine synthesis are indicated as *ocs* (octopine synthase), *agroc* (agrocinopine synthase), *agr* (agropine synthase), and *nos* (nopaline synthase). The regions containing the genes involved in octopine and nopaline catabolism are indicated *occ* and *noc,* respectively. The D or Vir region has large stretches of homology between both plasmids and is the region essential for the virulence of *Agrobacterium.* More detailed maps and descriptions of these plasmids are given by De Greve et al. (1981) and Holsters et al. (1980).

Vir region, have a nononcogenic phenotype, whereas mutants in the other region of the Ti plasmid important for tumor induction (the T region) either have an attenuated oncogenic phenotype or induce tumors with altered morphology (De Greve et al. 1981; Holsters et al. 1980).

Cot and Southern blot hybridization analysis demonstrated that the T region contains all of the Ti plasmid sequences that are transferred and integrated into the nuclear genome in tumor lines (Chilton et al. 1977, 1980; Lemmers et al. 1980; Thomashow et al. 1980). In some octopine plasmids (e.g., pTiA6NC, pTiAch5, and pTiB6S3) the T region is divided into two adjacent DNA segments, one of 13 kb (left T-DNA) and one of approximately 7 kb (right T-DNA) (De Beuckeleer et al. 1981; Thomashow et al. 1980). These two DNA segments can be transferred to the plant genome either independent of each other or else as a continuous stretch of DNA collinear with the T region present in the Ti plasmid. Nopaline plasmids, like pTiC58 and pTiT37, always transfer a single DNA segment of approximately 23 kb (Hepburn et al. 1983*a;* Lemmers et al. 1980; Ursic et al. 1983) (Figure 5.1).

Octopine and nopaline T-DNAs are expressed in plant cells, encoding several polyadenylated transcripts, which are transcribed by the RNA polymerase II of the plant (Bevan and Chilton 1982; Willmitzer et al. 1981, 1982, 1983). The positions of these transcripts were mapped by hybridization with different T-region probes, and the directions of transcription were identified by hybridization to both strands of the corresponding T-region fragment. Six of the transcripts (5, 2, 1, 4, 6a, and 6b) are homologous to both T-DNAs and are encoded by a 9-kb DNA segment common to both octopine and nopaline T-DNAs (Chilton et al. 1978; Depicker et al. 1978; Engler et al. 1981; Willmitzer et al. 1983). Compared with other cellular transcripts, the T-DNA transcripts contribute no more than 10^{-5}–10^{-6} of the total poly(A)$^{+}$ RNA population; nevertheless, the expression of the T-DNA genes has a dramatic effect on the physiology and phenotype of the transformed tissue. At least three of these transcripts seem to be directly responsible for the tumorous mode of growth of transformed cells.

An analysis of plant cells transformed by transposon insertion and deletion mutants of the T-DNA region of Ti plasmids established a correlation between the phenotype of the tumor and the activities of some of the genes encoding the different T-DNA transcripts. Normal crown gall tumors grow as unorganized tissue. Mutations in genes *1* and *2* induce tumors that produce an abundance of shoots, indicating that these genes encode functions that suppress shoot formation in wild-type tumors (*shi,* Figure 5.2). Ti plasmids with mutations in gene *4* induce tumors that characteristically show vast root proliferation, indicating that gene *4* encodes a function that blocks the normal ability of plant cells to undergo differentiation to form roots (*roi,* Figure 5.2) (Garfinkel et al. 1981; Joos et al. 1983*a;* Leemans et al. 1982).

Other T-DNA genes have been shown to encode the enzymes responsible for opine synthesis. Such is the case for the nopaline (Holsters et al. 1980) and agrocinopine (Joos et al. 1983*a*) synthases in pTiC58 and octopine and agropine synthases in pTiAch5 and pTiB6S3 (De Greve et al. 1981; Salomon et al. 1984). Although there are several other genes that are transcribed, it has not been possible to correlate their mutants with any easily detectable change in the phenotype of the corresponding tumor. However, it is quite possible that some of them do influence the tumor phenotype. For instance, mutants in gene *1* or *2* in octopine plasmids induce tumors that sprout normal shoots derived from untransformed cells, in contrast to similar mutants in nopaline plasmids that induce tumors with mostly abnormal shoots containing opines (Joos et al. 1983*a;* Leemans et al. 1982). These differences could possibly be accounted for by the presence of transcripts a–e in nopaline T-DNAs that are absent in octopine T-DNAs, or by differences in the levels of transcription of the genes homologous to both plasmids that have been correlated with tumor formation.

One of the main conclusions from studies using deletion mutants is that none of the genes encoded by T-DNA are essential for transfer or integration of T-DNA. T-DNA transfer occurred with all mutants tested, including those that were unable to induce tumor formation. In the latter case, using Ti plasmid mutants with large deletions removing most of the internal part of the T-DNA and containing only one of the genes encoding opine synthesis and the T-DNA borders, it was shown that cells at the wound site were still able to synthesize opines, indicating that T-DNA transfer and integration had occurred (Joos et al. 1983*a;* Leemans et al. 1982; Zambryski et al. 1983).

By analogy to what is known about plant growth regulators, the function of gene *4* can be correlated to a cytokinin-like effect.

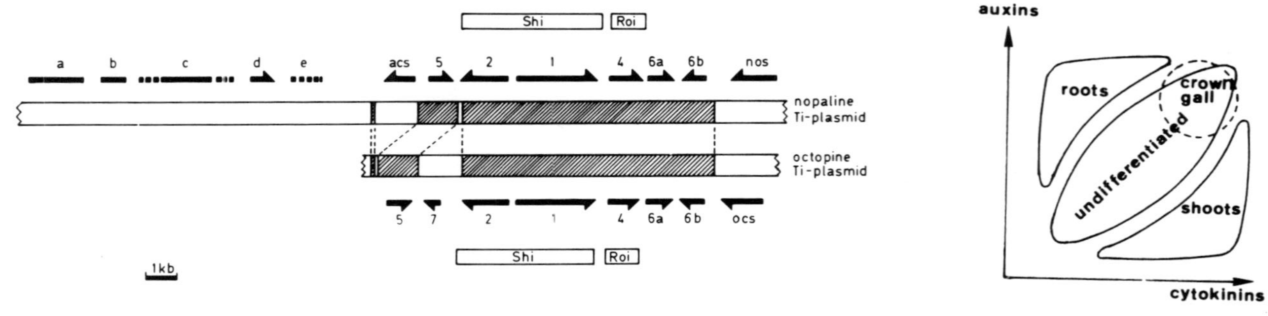

Figure 5.2. Transcriptional map of the octopine and nopaline T-DNAs. The regions of homology between the two T-DNAs (common region) are shown as cross-hatched areas. The positions, lengths, and directions of the T-DNA transcripts are as described by Willmitzer et al. (1982, 1983). Transcripts 1, 2, 3, 4, 5, 6a, and 6b are common to both T-DNAs; *shi* indicates transcripts that control shoot inhibition; *roi* indicates the transcript that controls root inhibition; *nos*, nopaline synthase; *ocs*, octopine synthase, *acs*, agrocinopine synthase. The auxin-cytokinin diagram is a schematic representation of how the hormone balance can influence callus morphology, as first described by Skoog and Miller (1957).

Therefore, a mutant in gene *4* would result in a low level of cytokinins, whereas the level of auxins would be elevated by the expression of genes *1* and *2*, and hence roots would form and proliferate. Similarly, the combined effects of genes *1* and *2* can be correlated to an auxin-like effect, such that mutants in these genes result in shoot formation because of the expression of gene *4* alone (Skoog and Miller 1957) (Figure 5.2).

Recently it has been demonstrated that genes *1* and *2* do, indeed, code for enzymes involved in the metabolic pathway of auxin production. It has been shown that gene *2* is expressed not only in plant cells but also in *Escherichia coli* and *Agrobacterium.* In bacteria this gene codes for a protein of 49,000 daltons, able to convert indole-3-acetamide into indole-3-acetic acid, which is the naturally active auxin present in plants. When crown gall or nontransformed, hormone-independent plant tissues were analyzed for the presence of this enzymatic activity it was found that the conversion of indole-3-acetamide to indole-3-acetic acid is specific for T-DNA-transformed tissue (Schröder et al. 1983, 1984). Further evidence supporting the role of genes *1* and *2* in auxin production comes from experiments showing that external addition of the synthetic auxin α-naphthale acetic acid restores the wild-type phenotype of tumors induced by mutants in gene *1* or *2,* probably by restoring the high level of auxins normally present in crown galls. It was also found that α-naphthale acetamide (a chemical without any hormone activity) could restore the phenotype of mutants in gene *1,* but not mutants in gene *2* indicating that the protein encoded by gene *2,* participates in the conversion of α-naphthale acetamide to a biologically active auxin, presumably naphthalene acetic acid (NAA). The fact that NAA, but not α-naphthale acetamide, can also restore the phenotype of mutants in gene *2* suggests that gene *1* is involved in an earlier step in the pathway of auxin production present in crown galls (Inzé et al. 1984). Because the proposed substrate (indole-3-acetamide) for the enzyme encoded by gene *2* is not normally present in plant cells (Schröder et al. 1984), gene *1* is a reasonable candidate for the production of such a substrate. Taken together, all these results suggest that gene *2* codes for an amino hydrolase directly involved in auxin biosynthesis. Remarkably, this gene appears to be expressed both in *Agrobacterium* and in plant cells. This is the first indication that T-DNA induces tumor formation by introducing a novel enzymatic pathway in plant cells that probably cannot be controlled by the regulatory mechanism of the host and that results in the formation of growth-stimulating factors (Schröder et al. 1984). Interestingly, an enzymatic activity similar to that encoded by gene *2* has been described in *Pseudomonas savastanoi* (Comai and Kosuge 1982), a bacterium that can also induce tumorous growth on some plants.

T-DNA integration

The mechanism by which *Agrobacterium* transfers T-DNA to the genome of plant cells is still unknown. It is probable that the interaction of *Agrobacterium* with the competent plant cells initiates a series of events, perhaps activating specific genes in both the bacteria and the host. To understand how T-DNA transfer occurs, one must first locate precisely the regions of the Ti plasmid that are integrated into the plant genome, and then determine those that are essential for the transfer and integration of this region.

Four nopaline Ti-plasmid-induced tumor lines have been analyzed in detail (Zambryski et al., 1982). The data obtained from this analysis show that there is little variation in the T-DNAs integrated into the genome in the different tumor lines analyzed. Some of these tumor lines appear to have a single copy of the T-DNA; others have multiple copies, some of which are organized in tandem arrays. Molecular cloning of the ends of the T-DNA from the genomes of transformed plant cells has shown that some DNA fragments contain sequences derived only from the right or left end of the T-DNA, whereas others contain sequences derived from both, confirming the presence of tandem copies of the T-DNA in transformed cells (Holsters et al. 1983; Yadav et al. 1982; Zambryski et al. 1980, 1982).

Determination of the nucleotide sequences of the junction fragments has revealed the precise ends of the T-DNA present in the genome. In all cases the homology between the sequences in the tumor DNA and those present in the Ti plasmid end within or close to a 25-bp sequence that flanks the T-region in the Ti plasmid as direct repeats (Holsters et al. 1983; Simpson et al. 1982; Yadav et al. 1982; Zambryski et al. 1982). Within the 25 bp of these repeats, 12 bp show perfect homology, and the rest imperfect homology. These terminal repeated sequences are present not only at the ends of the T region of nopaline plasmids but also at the ends of the TL and TR regions of the octopine plasmid (Barker et al. 1983; Holsters et al. 1983). Figure 5.3 shows the nucleotide sequences of the 25-bp sequences present in nopaline and octopine T-DNAs. Analysis of octopine TL borders, recloned from transformed cells, has shown that the TL integration is quite analogous to that found for nopaline T-DNA, including the presence of multiple T-DNA copies in tandem array. Detailed analysis of one of the junctions between two copies of the T-DNA in tandem showed that the junction consisted of a 40-bp unit of plant origin repeated six times. The presence of plant DNA between the tandem T-DNA copies suggests that the generation of the tandem in this particular tumor took place during or after insertion of an original copy of the T-DNA in the plant genome (Holsters et al. 1983).

A search in two nucleotide sequence data banks has revealed that the 12 bp of perfect homology of the 25-bp T-DNA flanking repeats is present only in the Ti plasmid flanking the T-DNA, not in any other DNA sequence known to date (Barker et al. 1983). It has been found that the right end of the T-DNA in the plant occurs either at the first base of the 25-bp repeat or one base before in at least three different tumor lines. The left border of the integrated T-DNA seems to be more variable than the right border, but nonetheless it occurs within 100 bp from the left 25-bp repeat (Zambryski et al. 1982).

Although the 25-bp repeat is present at both ends of the T-DNA, genetic analysis using deletion mutants has shown that the T-DNA ends are functionally different. Deletions of the right ends of the nopaline T-DNA make the Ti plasmid virtually avirulent on most plant species; in contrast, however, deletions of the left end have little or no effect on the transfer and integration of the T-DNA (Joos et al. 1983*a*).

Recently, a series of experiments has shown that the right end of the nopaline T-DNA is sufficient to direct the transfer and integration of bacterial sequences in the plant genome. It has been shown that the nopaline synthase gene that is linked to the right end of the T-DNA can still be efficiently transferred when inserted in a different part of the Ti plasmid, or when inserted in an independent replicon (A. Caplan, personal communication). A more detailed analysis of the DNA sequences required for DNA transfer and integration has shown (1) that the right 25-bp repeat is sufficient to promote DNA transfer and (2) that transfer occurs in a directional fashion. In other words, transfer is taking place only when the 25-hp

G C T G G	T G G C A G G A T A T A T T G	T G	G T G T A A A C	A A A T T	Nopaline L
G T G T T	T G A C A G G A T A T A T T G	G C	G G G T A A A C	C T A A G	Nopaline R
A G C G G	C G G C A G G A T A T A T T C	A A	T T G T A A A T	G G C T T	Octopine A
C T G A C	T G G C A G G A T A T A T A C	C G	T T G T A A T T	T G A G C	Octopine B
A A A G G	T G G C A G G A T A T A T C G	A G	G T G T A A A A	T A T C A	Octopine C
A C T G A	T G G C A G G A T A T A T G C	G G	T T G T A A T T	C A T T T	Octopine D

Figure 5.3. Comparison of the 25-bp terminal sequences flanking the T-DNAs of octopine and nopaline Ti plasmids. The boxes indicate the regions with higher degrees of homology between these terminal repeats. The two base pairs in between the boxes are not conserved. Nopaline L (left) and R (right) are the repeats flanking the T-DNA of nopaline-type plasmids pTiC58 or pTiT37. Octopine A/B and C/D (left/right) represent the terminal repeats flanking the TL and TR regions, respectively, present in the octopine-type Ti plasmids pTiAch5 and pTi15955. Data from Zambryski et al. (1982), Yadav et al. (1982), Simpson et al. (1982), Holsters et al. (1983), and Barker et al. (1983).

box is oriented in the same direction with respect to the DNA to be transferred as that found with respect to the T-DNA in the Ti plasmid (Wang et al. 1984).

As will be described later (Figure 5.6), dominant selectable markers linked to the right end of the T-DNA were introduced between the normal T-DNA ends. This produced a T-DNA with the two external borders in the normal orientation but with the right border attached to the selectable marker, inverted with respect to its normal orientation. In these experiments it was observed that the selectable marker alone was integrated into the plant genome at a high frequency without being linked to the rest of the T-DNA (De Block et al. 1984). In this engineered construction the selectable marker was located to the right of the inverted internal right border. In the normal orientation this would be toward the left and internal to the T-DNA. This confirms that the right border can direct DNA transfer independent of its orientation and that transfer and/or integration are directional, such that in relation to the normal orientation of the right border, sequences to the left of it will always be transferred and integrated. These experiments, in conjunction with the differences observed between the left and right borders, suggests that the sequence recognized as the right end of the T-DNA also directs DNA transfer from the bacteria to the plant or acts as the initiation point of a replicative integration process.

It has also been shown that in order to be transferred, the T-DNA does not have to be physically linked to the Vir region. Tumor formation by *Agrobacterium* harboring two compatible plasmids, one containing the Vir region and the other the T-DNA, is indistinguishable from tumor formation by *Agrobacterium* harboring the wild-type Ti plasmid. However, *Agrobacterium* containing any of these plasmids alone is not oncogenic. This demonstrates that the Vir region is trans-acting in relation to the necessary functions for T-DNA transfer (de Framond et al. 1983; Hoekema et al. 1983). The finding that the octopine Vir region can complement the nopaline T-DNA, and vice versa, demonstrates that there are no plasmid-specific virulence functions.

There are several hypotheses to explain how the actual process of T-DNA transfer occurs. One possibility is that the T-region is excised from the Ti plasmid in agrobacteria during the process of infection; in this hypothesis the T region would be cleaved out by enzymes that are probably encoded by the Vir region and that specifically recognize the T-DNA borders. Because none of the T-DNA genes are involved in T-DNA integration, and if only the T-DNA region enters the plant cell, then the host must provide the functions necessary for integration. A second hypothesis suggests that the whole Ti plasmid enters the plant cell, and after the T-DNA has been inserted into a chromosome, the remainder of the Ti plasmid is lost. In this case, either the functions needed for precise excision and integration of the T-DNA are provided by the host plant and specifically recognize the T-DNA border, or part of the Vir region is transiently expressed in the plant to provide the functions needed to accomplish the insertion of the T-DNA into the plant genome.

Because the Vir region and the T-DNA need not be physically linked in the same replicon to allow T-DNA transfer, this suggests that transfer of the whole Ti plasmid to the plant cell is not required for T-DNA integration, unless parts of more than one replicon can enter the host. However, it has been shown that at a certain frequency, segments larger than the T-DNA, if not the whole Ti plasmid, can enter the plant cell (Joos et al. 1983*b*).

One could assume that the right T-DNA border serves as an origin of transfer and that transfer continues until a second signal for termination of transfer is reached, namely the left border. As the left border is less stringently required, it could also be assumed that sometimes the transfer does not stop at the left border and that transfer proceeds, allowing other parts of the Ti plasmid to enter the plant cells.

Most mutations in the Vir region decrease or eliminate the oncogenicity of the Ti plasmid. Complementation studies using cosmid clones overlapping these mutations have shown that the *vir* genes are clustered in 10 complementation groups (Klee et al. 1982, 1983), and if we assume that the cosmid is not transferred to the host, this also suggests that most, if not all, the genes encoded by the Vir region are functional in the bacterium. It is known that when a plant is coinfected on the same wound with bacteria containing two different Ti plasmids, the T-DNAs of both plasmids can integrate in the same cell (Ooms et al. 1982). When different Vir mutants are used to coinfect on the same wound, they are not able to complement each other (Iyer et al. 1982). This suggests, first, that none of the products of the Vir region is readily diffusible between bacteria, and also that if the whole Ti plasmid enters the plant, the Vir region is not expressed there, because the mutations cannot complement each other.

Transfer and expression of foreign sequences in plant cells

Once it was firmly established that the interaction between *Agrobacterium* and wounded plant cells results in the transfer of a segment of Ti plasmid (T-DNA) into the genome in plant cells, the question arose whether or not other DNA sequences, experimentally inserted within the T-DNA, would be efficiently cotransferred. Genetic analysis of the T region by transposon mutagenesis provided Ti plasmid mutants with bacterial sequences inserted in the T region, which made it possible to test this hypothesis. A Tn7 insertion in the nopaline synthase locus produced a Ti plasmid mutant still capable of T-DNA transfer. Analysis of DNA from these tumors showed that the whole T-DNA region, including the inserted Tn7 sequence, had been transferred and integrated as a single 38×10^3-bp segment into the genome in tobacco cells (Hernalsteens et al. 1980).

Several different DNA sequences have since been introduced into various regions of the T region of Ti plasmids. Analysis of the DNA transferred by such mutated Ti plasmids led to the conclusion that probably any DNA sequences inserted within the T-DNA are cotransferred and integrated without any detectable rearrangement. So far, no limit has been found to the size of DNA that can be transferred and integrated when inserted between T-DNA borders. Sequences of up to 50 kb have been introduced into the genome of tobacco as a single DNA segment.

Early attempts to express foreign genes in plants, using the antibiotic resistance genes encoded by the transposons used to mutagenize the T region, showed that prokaryotic genes are not expressed in plant cells, presumably because the prokaryotic transcriptional signals of these genes are not recognized by the transcriptional machinery of the plant. Further attempts to express heterologous genes in plants were made using genes from other eukaryotic organisms, such as the yeast alcohol dehydrogenase gene (Barton et al. 1983), or genes from mammalian cells, such as β-globin, interferon, and genes under control of the early SV40 promoter (L. Herrera-Estrella et al., unpublished results). Analysis of the messenger RNA of tissues containing these genes revealed that none of them were expressed. This suggests that specific factors or signals required for their expression are present only in the cells or specific tissues of their original host and are not present in plant cells. This can be correlated with proteins that may interact with specific segments of DNA in the promoter region, like enhancers (Benoist and Chambon 1981), or sequences that determine differential or tissue-specific expression, as has been demonstrated for some animal genes (Walker et al. 1983).

At that stage it became obvious that an essential requirement for expression of heterologous genes in plants was to use the transcriptional signal of a gene known to be expressed in plants. Most of the plant genes cloned to date are highly regulated and can be expressed only at a particular stage of plant development. For instance, the gene families encoding for the zein or phaseolin storage proteins are expressed only during seed formation, and the leghemoglobins are induced only during interaction between legumes and *Rhizobium* (Brisson and Verma 1982; Geraghty et al. 1981; Pedersen et al. 1982; Wiborg et al. 1982). The best candidates for constitutively expressed genes proved to be those encoded by the T-DNA, particularly the ones encoding nopaline synthase or octopine synthase. These genes have been shown to be normally expressed both in callus tissue and in all tested differentiated tissues of plants regenerated from calli containing the opine genes (De Greve et al. 1982*a;* Wöstemeyer et al. 1983).

The nucleotide sequences of the nopaline and octopine synthase genes, as well as other T-DNA genes, revealed that they have potential transcriptional signals that bear close resemblance to the consensus sequences that have been found important for the start and stop of transcription of other eukaryotic genes. At the 5′ flanking region the genes contain sequences homologous to the TATA or Goldberg-Hogness box, 30 to 40 nucleotides upstream of the start of transcription, and a sequence similar to the CAT box, 60 to 80 nucleotides upstream of the 5′ end of the transcript. In addition, they have sequences similar to the AATAAA between 20 and 100 nucleotides 5′ to the polyadenylation site that strongly resemble the consensus sequence similarly placed in animal genes (Barker et al. 1983; Bevan et al. 1983*a;* De Greve et al. 1982*b;* Depicker et al. 1982; Gielen et al. 1984).

So far, most of our information about the transcriptional signals used by plant cells has been obtained from studies using T-DNA-encoded genes. Thus, it has been shown that for the *ocs* gene, sequences upstream of position −170 from the start of transcription are essential for efficient transcription of this gene. The functional *ocs* promoter is comprised with 294 bp of the 5′ region of the start point of transcription (Koncz et al. 1983).

The *ocs* gene produces transcripts that are polyadenylated at either of two positions in the 3′ untranslated region of this gene, although the longer transcript is found to be more abundant. A mutant of the *ocs* with deletion of the 3′ untranslated region was found to be still active. This deletion terminates 19 bp upstream of the major polyadenylation site of the wild-type transcript, removing the AATAAA signal preceding this site. The *ocs* mutant gene produces transcripts with a polyadenylated tail added at four different positions (Dhaese et al. 1983). The most abundant of these transcripts is polyadenylated at the same position as the minor one in the wild-type gene. All these results suggest that the AATAAA signal in plants is mainly involved in the posttranscriptional processing of the transcripts, in a fashion similar to that previously found in animal cells.

Based on this information, a series of vectors was constructed using the promoter and terminator signals for transcription of the nopaline synthase (*nos*) gene. In between these signals, restriction sites were introduced for the insertion of any desired sequence. The functionality of these vectors was first tested using the coding sequence of another T-DNA gene, the octopine synthase. It was proved that the *nos* promoter was able to direct functional expression of the octopine synthase coding sequence when transferred to plant cells (Herrera-Estrella et al. 1983*b*). The first example of a heterogeneous gene being expressed in plants was the bacterial chloramphenicol acetyltransferase (*cat*) gene from pBR325. When the coding sequence of the *cat* gene under control of the *nos* signals was transferred to tobacco cells, the transformed cells were shown to express this chimeric gene and to produce a functional protein (Herrera-Estrella et al. 1983*b*).

After it was shown that foreign genes can be expressed in plant cells, the next step was expression of genes that can be used as dominant selectable markers. Selectable markers ideally should be expressed in a variety of host cells, and their expression should produce a selectable change in the phenotype of the normal host cells. Selectable markers have been the key for establishment of DNA-mediated transformation systems in bacteria, yeast, *Dictyostelium,* and various animal systems.

Bacterial genes, such as aminoglycoside phosphotransferase APH(3′)II and APH(3′)I carried by transposons Tn*5* and Tn*903*, respectively, inactivate related aminoglycoside antibiotics like neo-

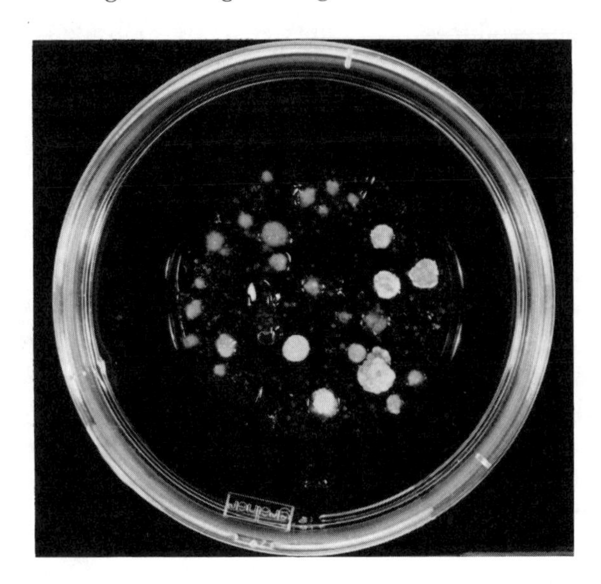

Figure 5.4. Selection of calli expressing the *nos*-Km gene in kanamycin-containing medium. Six-week-old plant cell colonies derived from cocultivation of *N. plumbaginifolia* with *Agrobacterium* containing pGV3850::*nos*-Km were transferred to medium containing kanamycin sulfate (50μg/ml). Two weeks later, as shown on the photograph, the transformed calli can be easily distinguished from the dead sensitive calli.

mycin, kanamycin, and G418. The APH(3′)II from Tn*5* had been shown to provide selectable resistance to G418 in mammalian cells, when expressed under control of the SV40 promoter, and in *Dictyostelium* when expressed from the *Dictyostelium* actin 8 promoter (Hirth et al. 1982; Southern and Berg 1982). Because plant cells are sensitive to several of these aminoglycosides, expression of an APH activity could therefore be expected to determine a resistance phenotype.

Chimeric genes containing the the APH(3′)II and APH(3′)I coding sequences under control of the *nos* transcriptional signals were constructed and inserted into the T-DNA of the Ti plasmids (Bevan et al. 1983*b;* Fraley et al. 1983; Herrera-Estrella et al. 1983*a*). A modification of an in vitro cell transformation system, based on cocultivation of regenerating plant cell protoplasts with *Agrobacterium,* was used to transfer the chimeric gene into plant cells (Márton et al. 1979). When small, fast-growing plant cell colonies derived from protoplasts used for the cocultivation experiments were transferred to selective plates containing kanamycin at 50 μg/ml, resistant colonies were obtained with frequencies ranging from 3% to 15% (Figure 5.4). The resistant colonies were indeed shown to be transformed by Southern blot hybridization and testing for linkage with the hormone independence markers of the T-DNA. Later experiments have shown that the methotrexate-insensitive dihydrofolate reductase encoded by the plasmid R67, as well as chloramphenicol acetyltransferase, can also be used as selectable markers, although the resistance phenotype is not as marked as in the case of the APH(3′)II (De Block et al. 1984; Herrera-Estrella et al. 1983*a*). All these experiments have shown that four different bacterial coding sequences can be properly transcribed and their messenger RNA translated into functional proteins. Thus, the codon usage for the plant translation machinery may allow expression of any other bacterial, fungal, or mammalian gene, including those that could confer a useful trait to plant cells.

The first chimeric *nos*-Km^R gene used for transformation experiments contained an ATG codon in the 5′ untranslated sequence out of frame with the APH coding sequence. It has been shown that in most eukaryotic genes, the first ATG in the messenger RNA is the initiation codon, and that when ATGs other than the initiation codon are present in the 5′ untranslated region, the proper translation of the mRNA is less efficient. To test whether or not the same phenomenon occurs for plant mRNAs, we constructed a chimeric gene in which the extra ATG in the leader of the APH(3′)II gene was deleted. When plant cells transformed with this gene were tested for the level of resistance to kanamycin, they were found to grow normally at a level four times higher than that of the original construction. Because the level of transcription is the same for both genes, this suggests that at least for this gene the messenger RNA is more efficiently translated if the initiation codon is the first ATG present in the transcribed part of the gene (Herrera-Estrella et al. in press).

Regeneration of whole plants containing foreign DNA sequences

Once it was known that foreign genes can be expressed in plant cells, the second most important step toward the study of gene regulation and differentiation, for purposes of genetic engineering, was to obtain transferred cells able to differentiate into normal plants rather than to grow in a tumorous fashion. Knowing which functions encoded by the T-DNA prevent differentiation and which functions enable the Ti plasmid to transfer DNA sequences to plant cells, it was possible to start to systematically design Ti plasmid derivatives capable of efficient transfer of DNA to plant cells, but devoid of the genes preventing normal differentiation. It is also important to have a genetic marker to monitor for the presence of the transferred DNA. Based on the Ti plasmid pTiC58, a deletion was constructed that removed most of the T-DNA but retained the T-DNA border sequences allowing transfer, as well as the nopaline synthase gene as a T-DNA-specific marker. In this Ti plasmid the internal portion of the T-DNA was replaced by the cloning vehicle pBR322. When wounded plants were inoculated with this Ti plasmid derivative (pGV3850), it was found that although there was no tumor formation, the cells of the wound site did produce nopaline (Zambryski et al. 1983). After propagation in vitro, the wound calli were transferred to medium that stimulated shoot regeneration. Plantlets derived from these shoots were screened for nopaline production, and it was found that between 10% and 70% of these plantlets were nopaline-positive (Zambryski et al. 1983). Southern blot analysis showed that the nopaline-positive plantlets were indeed transformed and that they contained one or more copies of the modified pGV3850 T-DNA. These plantlets are able to grow into mature plants that are fertile and produce normal amounts of seeds. It has also been shown that the newly acquired trait is stable through the meiotic process and is transmitted normally to the progeny. It is worth mentioning that some of these plants transmitted the nopaline synthase trait to 100% of the F_1 progeny. This suggests either that these plants contained copies of the T-DNA in several chromosomes or that some abnormal loss and/or duplication of chromosomes occurred. Similar chromosome aberrations have been observed by simply regenerating plants from tissue cultures; this has been termed somaclonal variation (Larkin and Scowcroft 1981).

Because pBR322 is inserted in between the T-DNA borders in pGV3850, this modified Ti plasmid is a versatile acceptor for introducing genes into plants. Any DNA sequence cloned into pBR322, or its derivatives, can be easily introduced into the pGV3850 T-DNA by a single homologous recombination event between the pBR present in pGV3850 and the homologous sequence present in the cloning vehicle (Figure 5.5). pBR322-like plasmids can be easily mobilized from *E. coli* to *Agrobacterium,* if one provides *in trans* the Mob

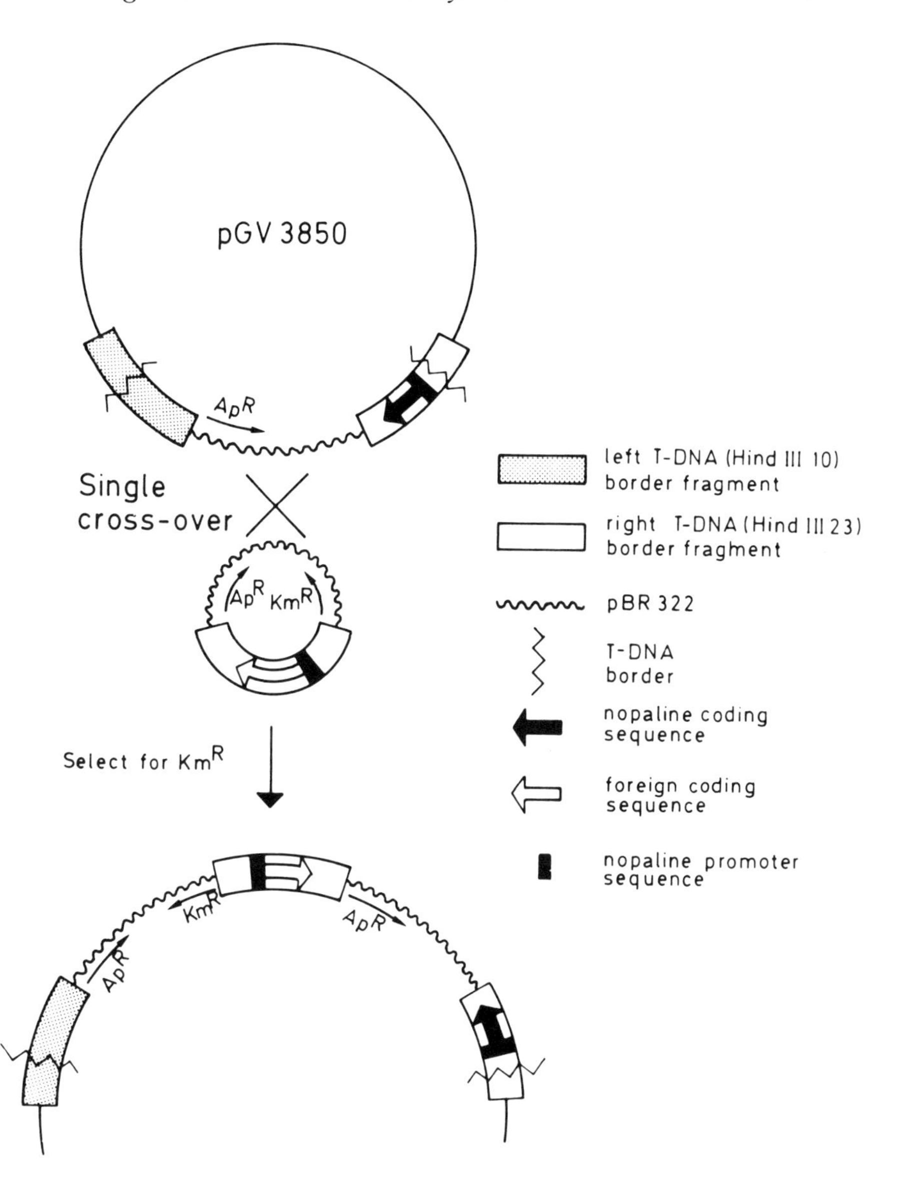

Figure 5.5. Ti-plasmid-derived vector pGV3850 and its use as an acceptor plasmid for the introduction of foreign DNA sequences into plants. The pGV3850 T-DNA contains only the nopaline synthase gene and the border sequences essential for T-DNA transfer. The wavy line represents pBR322 used to replace the internal part of the wild-type T-DNA. As described in the text, a single recombination event between the pBR322 sequences present in the T-DNA and the homologous sequences in the cloning vehicle allows the introduction of foreign sequences into the T-DNA.

functions of ColE1 and the Tra functions of a self-transmissible plasmid-like R64*drd*11 (Van Haute et al. 1983). As the origin of replication of pBR322 is not functional in *Agrobacterium,* the only way that its resistance markers can be maintained after being transferred to *Agrobacterium* is by recombination with the homologous region in the pGV3850 plasmid carrying the genetic markers. The cointegrates can be selected for and maintained in *Agrobacterium* by including a drug resistance marker other than ampicillin in the cloning vehicle. As shown in Figures 5.5 and 5.6, the cointegrates contain a direct duplication of the pBR sequences. In *Agrobacterium,* the cointegrate can be maintained by selective growth in medium containing the appropriate drugs. It is unlikely that these cointegrates will be unstable in plant cells, as the plant genome is composed of much repeated DNA, including, at times, the T-DNA itself, which can be present in tandem copies. So far, we have not found any evidence whatsoever for instability of the T-DNA from these cointegrates in tissue or plants that have been growing for several months. In other experiments it has been found that some tumor tissues grown in vitro for long periods of time do not produce any detectable amount of opines. In other cases, tumors that were thought to be opine-negative produced transformed shoots that regained opine production. In both cases, the loss of opine production has been correlated with methylation of the gene encoding the corresponding opine synthase, and opine production can be restored by adding 5-azacytidine compound, which inhibits DNA methylation (Hepburn et al. 1983*b;* Van Slogteren et al. 1983).

Expression of foreign genes in whole plants

Once it was known that foreign genes could be expressed in plant cells when placed under control of a gene normally expressed in plants and that nononcogenic Ti plasmids were still capable of DNA transfer, the next step was to investigate whether foreign genes could be expressed in the differentiated tissue of regenerated plants.

Using the pGV3850 system described before, cointegrates were constructed between pGV3850 and pBR vectors containing chimeric genes that confer resistance to either kanamycin or methotrexate or chloramphenicol in tissue cultures of plant cells (Figure 5.6). The T-DNA of these cointegrates was transferred to tobacco cells using

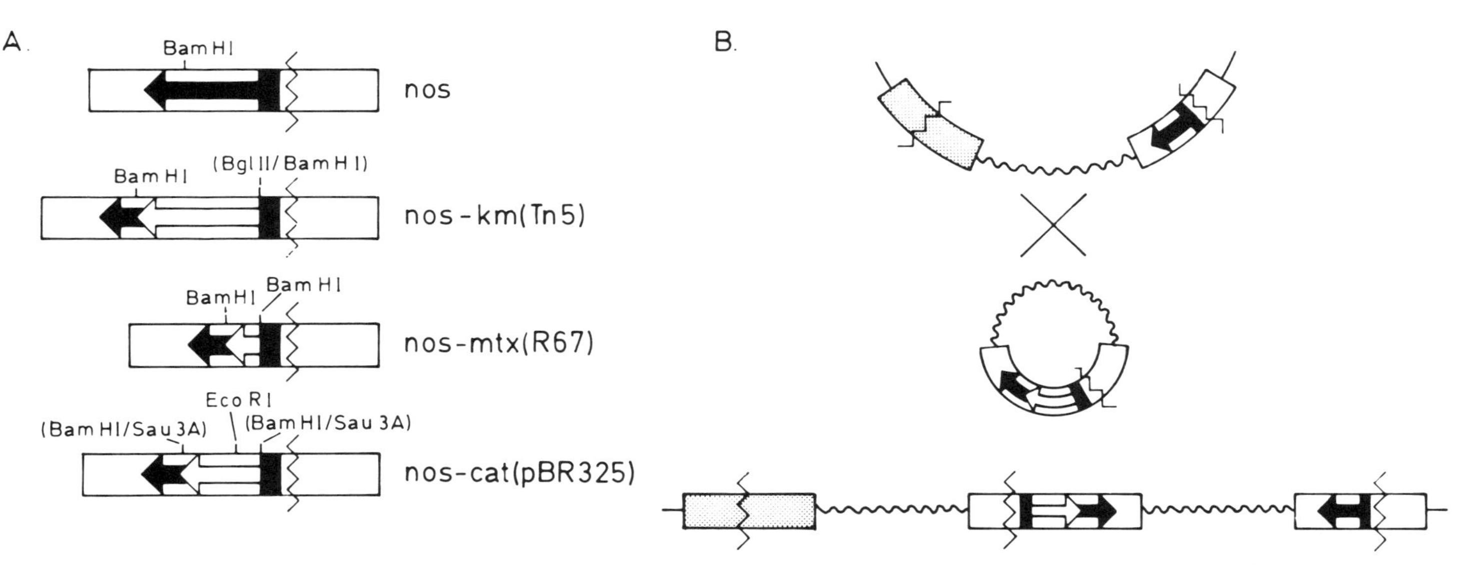

Figure 5.6. Use of pGV3850 as acceptor to introduce dominant selectable markers into plant cells. (A) Schematic representation of the various chimeric genes containing bacterial coding sequences under control of the *nos* promoter. The wild-type *nos* gene is shown compared with the structure of the chimeric gene. (B) Structure of one of the cointegrates formed between one of the chimeric genes shown in part A and pGV3850. Such cointegrates contain an extra right T-DNA border sequence (jagged line).

the cocultivation method (De Block et al. 1984). Plant cell colonies obtained after cocultivation were plated on selective medium containing the appropriate antibiotic. Transformed colonies able to grow in the presence of the antibiotics were obtained at a frequency between 3% and 15% of the original number of protoplasts. The cocultivation method was used to obtain resistance colonies, because these colonies arise from a single protoplast or very few protoplasts, and in most cases represent a single T-DNA transfer event.

After being propagated in vitro, the transformed colonies were transferred to regeneration medium. After a few weeks plantlets were obtained; these plantlets were shown to grow into mature plants and to produce seeds, as expected from cells transformed by the nononcogenic Ti plasmid pGV3850.

Using Southern blot analysis it was shown that these transformed plants indeed contained the chimeric genes in their genomes. It was demonstrated in three different ways that the regenerated plants expressed the chimeric genes: (1) pGV3850::neotransformed shoots and control shoots were transferred to agar medium containing kanamycin at 100 μg/ml; after three weeks the transformed shoots were able to develop roots and maintain normal growth, in contrast to control shoots, which did not form roots and eventually died; (2) seeds obtained from pGV3850::neotransformed and control nontransformed plants were germinated in the presence of kanamycin (100 μg/ml), although both transformed and nontransformed seeds were able to germinate; control seeds became etiolated and died after three weeks, and transformed seeds were able to maintain normal growth after germination despite the presence of the antibiotic; (3) plants transformed with pGV3850::cat were shown to express CAT activity in all their differentiated tissues.

Using the ability of pGV3850::neotransformed seeds to germinate and to grow in kanamycin-containing medium, the genetic transmission pattern of the kanamycin resistance trait was examined by self-fertilizing the regenerated plants. In all cases a ratio close to 3:1 for resistant and sensitive seeds was found (Figure 5.7), suggesting that the mother plant was heterozygous for the resistance trait and that the trait was transmitted as a single dominant Mendelian factor. The linkage between the drug resistance markers and the nopaline synthase gene was determined in transformed plant cell colonies by testing for the synthesis of nopaline or Southern blot analysis. This analysis showed that in most of the cases the resistance

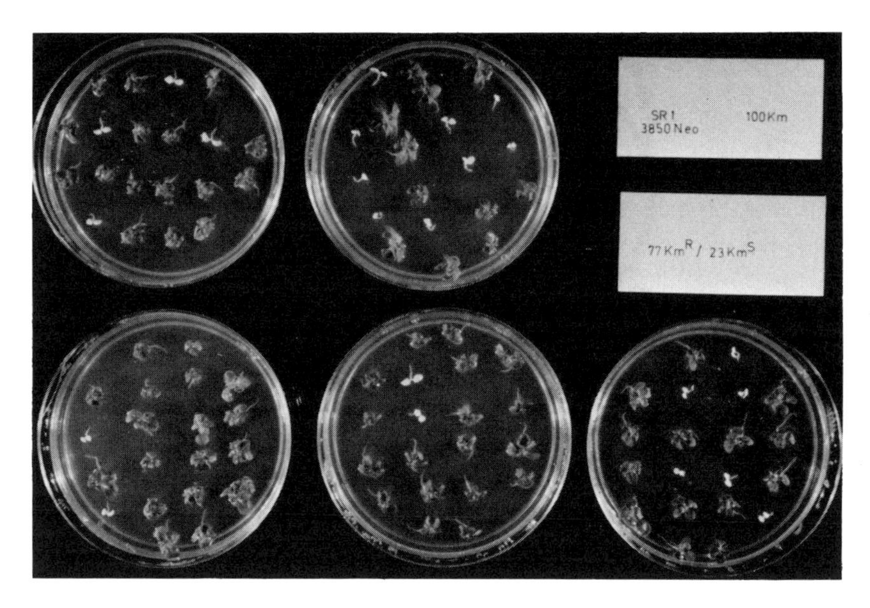

Figure 5.7. Germination of seed obtained by selfing plants containing the *nos*-Km^R chimeric gene in kanamycin-containing medium. Seeds were germinated in solid medium containing Murashige and Skoog medium salts and kanamycin sulfate (100 μg/ml). All seeds germinated, but whereas those harboring the *nos*-Km^R gene developed normally, the sensitive ones etiolated, and eventually died. The genetic transmission pattern of the first generation (S1) from the self-fertilized plant containing the *nos*-Km^R gene is shown in the photograph. From 100 seeds tested, 77 showed resistance to kanamycin, whereas 23 were sensitive. Thus, the Km^R trait was inherited as a dominant Mendelian factor, as deduced from the 3:1 resistant-to-sensitive ratio.

marker had been transferred and integrated by itself, without being linked to the rest of the pGV3850 T-DNA. This fact is not completely unexpected, because together with the chimeric genes used in these experiments, an extra copy of the right T-DNA border was introduced into the T-DNA of the cointegrates. These findings confirm that the only requirement for DNA transfer is the presence of a right T-DNA border. Moreover, this suggests that the transfer is directional and, possibly, that two right borders present next to each other and in opposite orientations interfere with each other for the transfer of DNA (Figure 5.6).

A new set of chimeric genes and expression vectors lacking this extra T-DNA right border has recently been constructed. When these new constructs, together with pGV3850, were used to transform plant cells, it was found that the genetic linkage between the selectable markers and the *nos* gene was reestablished, confirming that the right T-DNA border was responsible for the independent transfer of the selectable marker in earlier experiments (Herrera-Estrella et al. in press).

Regulated expression of foreign genes in plant cells

The results discussed earlier demonstrate that the Ti plasmid can be used to transfer and express genes in plants. Now experiments can be attempted in order to study genes that are expressed only in certain tissues of a plant, or only at certain stages in embryogenesis, or only in genes that are induced or repressed by internal or external factors.

The induction by light of many of the genes involved in the photosynthetic pathway of plants is one of the most exciting areas in plant biology. Several members of the gene families encoding the small subunit of ribulose bisphosphate carboxylase (Rubisco) have been isolated from different plant species (Berry-Lowe et al. 1982; Broglie et al. 1983; Cashmore 1983). Rubisco performs the first step in the Calvin cycle (hydrolytic cleavage of ribulose-1,5-bisphosphate to form two molecules of 3-phosphoglycerate). Rubisco is composed of two subunits, the large subunit that is encoded by the chloroplast genome and the small subunit that is encoded by the nuclear genome. The small subunit (ss) is synthesized on free cytoplasmic ribosomes as a 20,000-dalton precursor. The ss precursor is posttranslationally processed during transport into chloroplasts to yield the mature polypeptide.

The light-inducible expression of the ss gene family has been shown to be regulated, at least partially, at the transcriptional level in green tissue (Gallagher and Ellis 1982). To determine if this type of light-regulated expression is controlled either by the sequences 5′ to the promoter or by sequences located in another region of the gene, a chimeric gene was constructed containing 900 bp of the promoter region of an ss gene isolated from pea, coupled to the CAT coding sequence from Tn9 and the 3′-end sequence of the *nos* gene (Herrera-Estrella et al. 1984). The *cat* gene is an important tool to monitor for the transcription of promoters, because a simple assay allows quantitation of the level of enzymatic activity in cells of different organisms, which usually represents the level of transcription initiated at the promoter under study (Gorman et al. 1982).

This chimeric gene was introduced into tobacco cells, and the influence of light on the CAT activity was determined. It was found that within 900 bp of the 5′-upstream region of the *ss* gene used in this construction, there are signals sufficient to confer light-inducible expression of the *cat* gene. Analysis of the levels of CAT mRNA confirmed that there is a 20-fold increase in the level of expression of the *ss-cat* chimeric gene when green tissue is grown under light, as compared with tissue grown in the dark. It was also observed that in white undifferentiated tissue, the level of expression is extremely low, even when the tissue is grown under light conditions (Herrera-Estrella et al. 1984).

These findings constitute the first evidence that sequences upstream of the start of transcription are sufficient to confer regulated expression of a plant gene and, additionally, that a promoter from one plant species can be functional in cells of another plant species (tobacco).

Prospectives

It has been shown by our group and others that foreign DNA sequences can be transferred to and expressed in plant cells. Now it is also possible to regenerate whole plants from the original transformed cells, which can also express the foreign DNA sequences.

The first attempts to increase our understanding of how plant cells regulate gene expression are beginning to be carried out. Using the system described here, it is hoped that it will be possible to identify the DNA sequences directly responsible for the tissue-specific expression of genes. Because T-DNA seems to integrate randomly in the plant genome, it will be also possible to investigate whether or not the site of insertion influences the regulation of the expression of the gene being studied. Genes from completely unrelated plants can be tested in tobacco or *Petunia,* which are now the model systems, and potentially in any other plant that can be transformed by *Agrobacterium.*

Dominant selectable markers can be used to develop or perfect other DNA-mediated transformation systems, such as direct DNA transformation, microinjection, or pollen transformation. This will also facilitate the testing of other potential vectors, such as cauliflower mosaic virus, or some of the members of the single-stranded DNA viruses (Gemini viruses) whose genomes have recently been cloned and that have been shown to be infective as double-stranded molecules when excised from the cloning vehicle (Bisaro et al. 1982; Stanley 1983; Stanley and Gay 1983).

Because the promoter of the *nos* gene is functional in a broad range of unrelated dicotyledonous plants, it would not be surprising to find that it could also be used to express dominant selectable markers in monocotyledonous plants. These selectable markers or similar ones based on the promoter of a gene expressed in monocotyledonous plants can be used to attempt a microinjection system using the transposable element of maize described much earlier by McClintock (1967); for a review, see Fedoroff (1983); the element has recently been cloned (Fedoroff et al. 1983). This system might function in a fashion similar to the one designed for *Drosophila* using the transposable P elements (Rubin and Spradling 1982; Spradling and Rubin 1982).

The expression of genes that will provide a useful trait to the plant cell can now be attempted: for example, genes encoding for resistance to viral or fungal infection, or those preventing the attack of plants by insects. We hope that these types of analysis will start to provide interesting insights into the basic and applied research in plant biology.

Acknowledgments

We thank Ms June Simpson, Dr. Allan Caplan, and Dr. Janice Sharp for critical reading of the manuscript, and all the members of the Laboratory of Genetics for their helpful discussion and advice during these studies. We thank Ms Martine De Cock, Albert Verstraete, and Karel Spruyt for excellent assistance in the preparation of the manuscript. L.H.E. is indebted to CONACYT Mexico for a Ph.D. fellowship; M.D.B. is a Senior Research Assistant of the National Fund for Scientific Research (Belgium); P.Z. is supported by a long-term EMBO fellowship. This research was supported by grants from the A.S.L.K.-Kankerfonds, from the Instituut tot Aanmoediging van het Wetenschappelijk Onderzoek in Nijverheid en Landbouw (I.W.O.N.L. 3894A), from the Services of the Prime Minister (O.O.A. 12052179), from the Fonds voor Geneeskundig Wetenschappelijk Onderzoek (F.G.W.O. 3.001.82), to M.V.M. and J.S., and was carried out under research contract GVI-4-017-B (R.S.) of the Biomolecular Engineering Programme of the Commission of the European Communities.

References

Barker, R. F., Idler, K. B., Thompson, D. V., and Kemp, J. D. (1983). Nucleotide sequence of the T-DNA region from the *Agrobacterium tumefaciens* octopine Ti plasmid pTi15955. *Plant Mol. Biol.* **2,** 335–50.

Barton, K. A., Binns, A. N., Matzke, A. J. M., and Chilton, M.-D. (1983). Regeneration of intact tobacco plants containing full length copies of genetically engineered T-DNA, and transmission of T-DNA to R1 progeny. *Cell* **32,** 1033–43.

Benoist, C., and Chambon, P. (1981). In vivo sequence requirements of the SV40 early promoter region. *Nature (London)* **290,** 304–10.

Berry-Lowe, S. L., McKnight, T. D., Shah, D. M., and Meagher, R. B. (1982). The nucleotide sequence, expression, and evolution of one member of a multigene family encoding the small subunit of ribulose-1,5-bisphosphate carboxylase in soybean. *J. Mol. Appl. Genet.* **1,** 483–98.

Bevan, M., Barnes, W. M., and Chilton, M.-D. (1983*a*). Structure and transcription of the nopaline synthase gene region of T-DNA. *Nucl. Acids Res.* **11,** 369–85.

Bevan, M. W., and Chilton, M.-D. (1982). Multiple transcripts of T-DNA detected in nopaline crown gall tumors. *J. Mol. Appl. Genet.* **1,** 539–46.

Bevan, M. W., Flavell, R. B., and Chilton, M.-D. (1983*b*). A chimaeric antibiotic resistance gene as a selectable marker for plant cell transformation. *Nature (London)* **304,** 184–7.

Bisaro, D. M., Hamilton, W. D. O., Coutts, R. H. A., and Buck, K. W. (1982). Molecular cloning and characterization of the two DNA components of tomato golden mosaic virus. *Nucl. Acids Res.* **10,** 4913–22.

Braun, A. C. (1943). Studies on tumor inception in crown gall disease. *Am. J. Bot.* **30,** 674–7.

(1978). Plant tumours. *Biochim. Biophys. Acta* **516,** 167–91.

(1982). A history of the crown gall problem, in *Molecular Biology of Plant Tumors,* ed. G. Kahl and J. Schell, pp. 155–210. Academic Press, New York.

Brisson, N., and Verma, D. P. (1982). Soybean leghemoglobin gene family: normal, pseudo, and truncated genes. *Proc. Natl. Acad. Sci. U.S.A.* **79,** 4055–9.

Broglie, R., Coruzzi, G., Lamppa, G., Keith, B., and Chua, N.-H. (1983). Structural analysis of nuclear genes encoding for the precursor to the small subunit of wheat ribulose-1,5-bisphosphate carboxylase. *Bio/technology* **1,** 55–61.

Cashmore, A. R. (1983). Nuclear genes encoding the small subunit of ribulose-1,5-bisphosphate carboxylase, in *Genetic Engineering of Plants–An Agricultural Perspective,* ed. T. Kosuge, C. P. Meredith, and A. Hollaender, p. 29–38. Plenum Press, New York.

Chilton, M.-D., Drummond, M. H., Merlo, D. J., and Sciaky, D. (1978). Highly conserved DNA of Ti-plasmids overlaps T-DNA, maintained in plant tumors. *Nature* (*London*) **275,** 147–9.

Chilton, M.-D., Drummond, M. J., Merlo, D. J., Sciaky, D., Montoya, A. L., Gordon, M. P., and Nester, E. W. (1977). Stable incorporation of plasmid DNA into higher plant cells: the molecular basis of crown gall tumorigenesis. *Cell* **11,** 263–71.

Chilton, M.-D., Saiki, R. K., Yadav, N., Gordon, M. P., and Quetier, F. (1980). T-DNA from *Agrobacterium* Ti plasmid is in the nuclear DNA fraction of crown gall tumor cells. *Proc. Natl. Acad. Sci. U.S.A.* **77,** 4060–4.

Comai, L., and Kosuge, T. (1982). Cloning and characterization of iaaM, a virulence determinant of *Pseudomonas savastanoi. J. Bacteriol.* **149,** 40–6.

De Beuckeleer, M., Lemmers, M., De Vos, G., Willmitzer, L., Van Montagu, M., and Schell, J. (1981). Further insight on the transferred-DNA of octopine crown gall. *Mol. Gen. Genet.* **183,** 283–8.

De Block, M., Herrera-Estrella, L., Van Montagu, M., Schell, J., and Zambryski, P. (1984). Expression of foreign genes in regenerated plants and in their progeny. *EMBO Journal* **3,** 1681–9.

de Framond, A.J., Barton, K. A., and Chilton, M.-D. (1983). Mini-Ti: a new vector strategy for plant genetic engineering. *Bio/technology* **1,** 262–9.

De Greve, H., Decraemer, H., Seurinck, J., Van Montagu, M., and Schell, J. (1981). The functional organization of the octopine *Agrobacterium tumefaciens plasmid pTiB6S3. Plasmid* **6,** 235–48.

De Greve, H., Leemans, J., Hernalsteens, J. P., Thia-Toong, L., De Beuckeleer, M., Willmitzer, L., Otten, L., Van Montagu, M., and Schell, J. (1982*a*). Regeneration of normal and fertile plants that express octopine synthase, from tobacco crown galls after deletion of tumor-controlling functions. *Nature* (*London*) **300,** 752–5.

De Greve, H., Dhaese, P., Seurinck, J., Lemmers, M., Van Montagu, M., and Schell, J. (1982*b*). Nucleotide sequence and transcript map of the *Agrobacterium tumefaciens* Ti plasmid-encoded octopine synthase gene. *J. Mol. Appl. Genet.* **1,** 499–512.

Depicker, A., Stachel, S., Dhaese, P., Zambryski, P., and Goodman, H. M. (1982). Nopaline synthase: transcript mapping and DNA sequence. *J. Mol. Appl. Genet.* **1,** 561–74.

Depicker, A., Van Montagu, M., and Schell, J. (1978). Homologous DNA sequences in different Ti-plasmids are essential for oncogenicity. *Nature* (*London*) **275,** 150–3.

Dhaese, P., De Greve, H., Gielen, J., Seurinck, J., Van Montagu, M., and Schell, J. (1983). Identification of sequences involved in the polyadenylation of higher plant nuclear transcripts using *Agrobacterium* T-DNA genes as models. *EMBO Journal* **2,** 419–26.

Ellis, J. G., and Murphy, P. J. (1981). Four new opines from crown gall tumours–their detection and properties. *Mol. Gen. Genet.* **181,** 36–43.

Engler, G., Depicker, A., Maenhaut, R., Villarroel-Mandiola, R., Van Montagu, M., and Schell, J. (1981). Physical mapping of DNA base sequence homologies between an octopine and a nopaline Ti-plasmid of *Agrobacterium tumefaciens. J. Mol. Biol.* **152,** 183–208.

Fedoroff, N. (1983). Controlling elements in maize, in *Mobile Genetic Elements,* ed. J. A. Shapiro, pp. 1–63. Academic Press, New York.

Fedoroff, N., Wessler, S., and Shure, M. (1983). Isolation of the transposable maize controlling elements *Ac* and *Ds. Cell* **35,** 235–42.

Firmin, J. L., and Fenwick, G. R. (1978). Agropine–a major new plasmid-determined metabolite in crown gall tumours. *Nature* (*London*) **276,** 842–4.

Fraley, R. T., Rogers, S. G., Horsch, R. G., Sanders, P. R., Flick, J. S., Adams, S. P., Bittner, M. L., Brand, L. A., Fink, C. L., Fry, J. S., Galluppi, G. R., Goldberg, S. B., Hoffmann, N. L., and Woo, S. C. (1983). Expression of bacterial genes in plant cells. *Proc. Natl. Acad. Sci. U.S.A.* **80,** 4803–7.

Gallagher, T. F., and Ellis, R. J. (1982). Light-stimulated transcription of genes for two chloroplast polypeptides in isolated pea leaf nuclei. *EMBO Journal* **1,** 1493–8.

Garfinkel, D. J., Simpson, R. B., Ream, L. W., White, F. F., Gordon, M. P., and Nester, E. W. (1981). Genetic analysis of crown gall: fine structure map of the T-DNA by site-directed mutagenesis. *Cell* **27,** 143–53.

Geraghty, D., Peifer, M. A., Rubenstein, I., and Messing, J. (1981). The primary structure of a plant storage protein: zein. *Nucl. Acids Res.* **9,** 5163–74.

Gielen, J., De Beuckeleer, M., Seurinck, J., Deboeck, F., De Greve, H., Lemmers, M., Van Montagu, M., and Schell, J. (1984). The complete nucleotide sequence of the TL-DNA of the *Agrobacterium tumefaciens* plasmid pTiAch5. *EMBO Journal* **3,** 835–46.

Gorman, C. M., Moffat, L. F, and Howard, B. H. (1982). Recombinant genomes which express chloramphenicol acetyltransferase in mammalian cells. *Mol. Cell. Biol.* **2,** 1044–51.

Hepburn, A. G., Clarke, L. E., Blundy, K. S., and White, J. (1983*a*). Nopaline Ti plasmid, pTiT37, T-DNA insertions into a flax genome. *J. Mol. Appl. Genet.* **2,** 211–24.

Hepburn, A. G., Clarke, L. E., Pearson, L., and White, J., (1983*b*). The role of cytosine methylation in the control of nopaline synthase gene expression in a plant tumor. *J. Mol. Appl. Genet.* **2,** 315–29.

Hernalsteens, J. P., Van Vliet, F., De Beuckeleer, M., Depicker, A., Engler, G., Lemmers, M., Holsters, M., Van Montagu, M., and Schell, J. (1980). The *Agrobacterium tumefaciens* Ti plasmid as a host vector system for introducing foreign DNA in plant cells. *Nature* (*London*) **287,** 654–6.

Herrera-Estrella, L., De Block, M., Messens, E., Hernalsteens, J.-P., Van Montagu, M., and Schell, J. (1983*a*). Chimeric genes as dominant selectable markers in plant cells. *EMBO Journal* **2,** 987–95.

Herrera-Estrella, L., Depicker, A., Van Montagu, M., and Schell, J. (1983*b*). Expression of chimaeric genes transferred into plant cells using a Ti-plasmid-derived vector. *Nature* (*London*) **303,** 209–13.

Herrera-Estrella, L., Van den Broeck, G., Maenhaut, R., Van Montagu, M., Schell, J., Timko, M., and Cashmore, A. (1984). Light-inducible and chloroplast-associated expression of a chimaeric gene introduced into *Nicotiana tabacum* using a Ti plasmid vector. *Nature* (*London*) **310,** 115–20.

Herrera-Estrella, L., Zambryski, P., Simpson, J., Van Montagu, M., Schell, J., and Hernalsteens, J. P. (*in press*).

Hirth, K.-P., Edwards, C. A., and Firtel, R. A. (1982). A DNA-mediated transformation system for *Dictyostelium discoideum. Proc. Natl. Acad. Sci. U.S.A.* **79,** 7356–60.

Hoekema, A., Hirsch, P. R., Hooykaas, P. J. J., and Schilperoort, R. A. (1983). A binary plant vector strategy based on separation of *vir*- and T-region of the *Agrobacterium tumefaciens* Ti plasmid. *Nature* (*London*) **303,** 179–81.

Holsters, M., Silva, B., Van Vliet, F., Genetello, C., De Block, M., Dhaese, P., Depicker, A., Inzé, D., Engler, G., Villarroel, R., Van Montagu, M., and Schell, J. (1980). The functional organization of the nopaline *A. tumefaciens* plasmid pTiC58. *Plasmid* **3,** 212–30.

Holsters, M., Villarroel, R., Gielen, J., Seurinck, J., De Greve, H., Van Montagu, M., and Schell, J. (1983). An analysis of the boundaries of the octopine TL-DNA in tumors induced by *Agrobacterium tumefaciens. Mol. Gen. Genet.* **190,** 35–41.

Inzé, D., Follin, A., Van Lijsebettens, M., Simoens, C., Genetello, C., Van Montagu, M., and Schell, J. (1984). Genetic analysis of the individual T-DNA genes of *Agrobacterium tumefaciens;* further evidence that two genes are involved in indole-3-acetic acid synthesis. *Mol. Gen. Genet.* **194,** 265–74.

Iyer, V. N., Klee, H. J., and Nester, E. W. (1982). Units of genetic expression in the virulence region of a plant tumor-inducing plasmid of *Agrobacterium tumefaciens. Mol. Gen. Genet.* **188,** 418–24.

Joos, H., Inzé, D., Caplan, A., Sormann, M., Van Montagu, M., and Schell, J. (1983*a*). Genetic analysis of T-DNA transcripts in nopaline crown galls. *Cell* **32,** 1057–67.

Joos, H., Timmerman, B., Van Montagu, M., and Schell, J. (1983*b*). Genetic analysis of transfer and stabilization of *Agrobacterium* DNA in plant cells. *EMBO Journal* **2,** 2151–60.

Klee, H. J., Gordon, M. P., and Nester, E. W. (1982). Complementation analysis of *Agrobacterium tumefaciens* Ti plasmid mutations affecting oncogenicity. *J. Bacteriol.* **150,** 327–31.

Klee, H. J., White, F. F., Iyer, V. N., Gordon, M. P., and Nester, E. W. (1983). Mutational analysis of the virulence region of an *Agrobacterium tumefaciens* Ti plasmid. *J. Bacteriol.* **153,** 878–83.

Koncz, C., De Greve, H., André, D., Deboeck, F., Van Montagu, M., and Schell, J. (1983). The octopine synthase genes carried by Ti plasmids contain all signals necessary for expression in plants. *EMBO Journal* **2,** 1597–603.

Larkin, P. J., and Scowcroft, W. R. (1981). Somaclonal variation – a novel source of genetic variability from cell cultures for improvement. *Theor. Appl. Genet.* **60,** 197–214.

Leemans, J., Deblaere, R., Willmitzer, L., De Greve, H., Hernalsteens, J. P., Van Montagu, M., and Schell, J. (1982). Genetic identification of functions of TL-DNA transcripts in octopine crown galls. *EMBO Journal* **1,** 147–52.

Lemmers, M., De Beuckeleer, M., Holsters, M., Zambryski, P., Depicker, A., Hernalsteens, J. P., Van Montagu, M., and Schell, J. (1980). Internal organization, boundaries and integration of Ti-plasmid DNA in nopaline crown gall tumours. *J. Mol. Biol.* **144,** 353–76.

McClintock, B. (1967). Genetic systems regulating gene expression during development. *Dev. Biol. Suppl.* **1,** 84–112.

Márton, L., Wullems, G. J., Molendijk, L., and Schilperoort, R. A. (1979). In vitro transformation of cultured cells from *Nicotiana tabacum* by *Agrobacterium tumefaciens. Nature (London)* **277,** 129–30.

Ooms, G., Molendijk, L., and Schilperoort, R. A. (1982). Double infection of tobacco plants by two complementing octopine T-region mutants of *Agrobacterium tumefaciens. Plant Mol. Biol.* **1,** 217–26.

Pedersen, K., Devereux, J., Wilson, D. R., Sheldon, E., and Larkins, B. A. (1982). Cloning and sequence analysis reveal structural variation among related zein genes in maize. *Cell* **29,** 1015–26.

Rubin, G. M., and Spradling, A. C. (1982). Genetic transformation of *Drosophila* with transposable element vectors. *Science* **218,** 348–53.

Salomon, F., Deblaere, R., Leemans, J., Hernalsteens, J.-P. Van Montagu, M., and Schell, J. (1984). Genetic identification of functions of TR-DNA transcripts in octopine crown galls. *EMBO Journal* **3,** 141–6.

Schröder, G., Klipp, W., Hillebrand, A., Ehring, R., Koncz, C., and Schröder, J. (1983). The conserved part of the T-region in Ti-plasmids expresses four proteins in bacteria. *EMBO Journal* **2,** 403–9.

Schröder, G., Waffenschmidt, S., Weiler, E. W., and Schröder, J. (1984). The T-region of Ti plasmids codes for an enzyme synthesizing indole-3-acetic acid. *Eur. J. Biochem.* **138,** 387–91.

Simpson, R. B., O'Hara, P. J., Kwok, W., Montoya, A. L., Lichtenstein, C., Gordon, M. P., and Nester, E. W. (1982). DNA from the A6S/2 crown gall tumors contains scrambled Ti-plasmid sequences near its junctions with the plant DNA. *Cell* **29,** 1005–14.

Skoog, F., and Miller, C. O. (1957). Chemical regulation of growth and origin formation in plant tissues cultured *in vitro. Symp. Soc. Exp. Biol.* **11,** 118–31.

Southern, P. J., and Berg, P. (1982). Transformation of mammalian cells to antibiotic resistance with a bacterial gene under control of the SV40 early region promoter. *J. Mol. Appl. Genet.* **1,** 327–41.

Spradling, A. C., and Rubin, G. M. (1982). Transposition of cloned P elements into *Drosophila* germ line chromosomes. *Science* **218,** 341–7.

Stanley, J. (1983). Infectivity of the cloned gemini virus genome requires sequences from both DNAs. *Nature* (*London*) **305,** 643–5.

Stanley, J., and Gay, R. M. (1983). Nucleotide sequence of cassava latent virus. *Nature* (*London*) **301,** 260–2.

Tate, M. E., Ellis, J. G., Kerr, A., Tempé, J., Murray, K., and Shaw, K. (1982). Agropine: a revised structure. *Carbohyd. Res.* **104,** 105–20.

Tempé, J, and Petit, A. (1982). Opine utilization by *Agrobacterium,* in *Molecular Biology of Plant Tumors,* ed. G. Kahl and J. Schell, pp. 451–9. Academic Press, New York.

Thomashow, M. F., Nutter, R., Montoya, A. L., Gordon, M. P., and Nester, E. W. (1980). Integration and organisation of Ti-plasmid sequences in crown gall tumors. *Cell* **19,** 729–39.

Ursic, D., Slightom, J. L., and Kemp, J. D. (1983). *Agrobacterium tumefaciens* T-DNA integrates into multiple sites of the sunflower crown gall genome. *Mol. Gen. Genet.* **190,** 494–503.

Van Haute, E., Joos, H., Maes, M., Warren, G., Van Montagu, M., and Schell, J. (1983). Intergeneric transfer and exchange recombination of restriction fragments cloned in pBR322: a novel strategy for the reversed genetics of Ti plasmids of *Agrobacterium tumefaciens. EMBO Journal* **2,** 411–18.

Van Larebeke, N., Engler, G., Holsters, M., Van den Elsacker, S., Zaenen, I., Schilperoort, R. A., and Schell, J. (1974). Large plasmid in *Agrobacterium tumefaciens* essential for crown gall-inducing ability. *Nature* (*London*) **252,** 169–70.

Van Slogteren, G. M. S., Hoge, J. H. C., Hooykaas, P. J. J., and Schilperoort, R. A. (1983). Clonal analysis of heterogeneous crown gall tumor tissue induced by wild-type and shooter mutant strains of *A. tumefaciens. Plant Mol. Biol.* **2,** 317–21.

Walker, M. D., Edlund, T., Boulet, A. M., and Rutter, W. J. (1983). Cell-specific expression controlled by the 5′-flanking region of insulin and chymotrypsin genes. *Nature* (*London*) **306,** 557–61.

Wang, K., Herrera-Estrella, L., Van Montagu, M., and Zambryski, P. (1984). Right 25 bp terminus sequence of the nopaline T-DNA is essential for and determines direction of DNA transfer from *Agrobacterium* to the plant genome. *Cell* **38,** 455–62.

Watson, B., Currier, T. C., Gordon, M. P., Chilton, M.-D., and Nester, E. W. (1975). Plasmid required for virulence of *Agrobacterium tumefaciens. J. Bacteriol.* **123,** 255–64.

Wiborg, O., Hyldig-Nielsen, J., Jensen, E., Paludan, K., and Marcker, K. (1982). The nucleotide sequences of two leghemoglobin genes from soybean. *Nucl. Acids Res.* **10,** 3487–94.

Willmitzer, L., De Beuckeleer, M., Lemmers, M., Van Montagu, M., and Schell, J. (1980). DNA from Ti-plasmid is present in the nucleus and absent from plastids of plant crown-gall cells. *Nature* (*London*) **287,** 359–61.

Willmitzer, L., Schmalenbach, W., and Schell, J. (1981). Transcription of T-DNA in octopine and nopaline crown gall tumours is inhibited by low concentrations of α-aminitin. *Nucl. Acids Res.* **9,** 4801–12.

Willmitzer, L., Simons, G., and Schell, J. (1982). The TL-DNA in octopine crown gall tumours codes for seven well-defined polyadenylated transcripts. *EMBO Journal* **1,** 139–46.

Willmitzer, L., Dhaese, P., Schreier, P. H., Schmalenbach, W., Van Montagu, M., and Schell, J. (1983). Size, location, and polarity of T-DNA-encoded transcripts in nopaline crown gall tumors; evidence for common transcripts present in both octopine and nopaline tumors. *Cell* **32,** 1045–56.

Wöstemeyer, A., Otten, L., De Greve, H., Hernalsteens, J. P., Leemans, J., Van Montagu, M., and Schell, J. (1983). Regeneration of plants from crown gall cells, in *Genetic Engineering in Eukaryotes,* NATO ASI Series A, Vol. 61, ed. P. Lurquin and A. Kleinhofs, pp. 137–51. Plenum Press, New York.

Yadav, N. S., Vanderleyden, J., Bennett, D. R., Barnes, W. M., and Chilton, M.-D. (1982). Short direct repeats flank the T-DNA on a nopaline Ti plasmid. *Proc. Natl. Acad. Sci. U.S.A.* **79,** 6322–6.

Zaenen, I., Van Larebeke, N., Teuchy, H., Van Montagu, M., and Schell, J. (1974). Supercoiled circular DNA in crown gall inducing *Agrobacterium* strains. *J. Mol. Biol.* **86,** 109–27.

Zambryski, P., Depicker, A., Kruger, K., and Goodman, H. (1982). Tumor induction by *Agrobacterium tumefaciens:* analysis of the boundaries of T-DNA. *J. Mol. Appl. Genet.* **1,** 361–70.

Zambryski, P., Holsters, M., Kruger, K., Depicker, A., Schell, J., Van Montagu, M., and Goodman, H. M. (1980). Tumor DNA structure in plant cells transformed by *A. tumefaciens. Science* **209,** 1385–91.

Zambryski, P., Joos, H., Genetello, C., Leemans, J., Van Montagu, M., and Schell, J. (1983). Ti plasmid vector for the introduction of DNA into plant cells without alteration of their normal regeneration capacity. *EMBO Journal* **2,** 2143–50.

6 Viruses as vectors for plant genes

R. HULL

Early attempts to introduce "foreign" genes into plant cells showed that the DNA enters the cell but is not replicated or expressed, as reviewed elsewhere (Kleinhofs and Behki 1977; Lurquin and Kado 1979). Those and subsequent observations indicated that for successful replication and expression, defined sequences of DNA are needed that are recognized by the enzymes involved. These sequences can be provided by a vector that can also contain sequences that allow the successfully transformed cells to be identified and selected. Selection markers are an advantage but not a necessity. Plant viruses can be considered as foreign nucleic acids that successfully replicate in and are expressed in plant cells. Thus, they must have the nucleic acid sequences that can be recognized by plant cells to perform these functions. For this reason they have attracted much attention as possible vectors for taking genes into plant cells.

There have been several previous reviews of the subject; see Hull (1978, 1980*b*, 1981, 1983), Howell (1982), Gardner (1983), Hull and Davies (1983), and van Vloten-Doting (1983). In this chapter I do not wish simply to repeat the information given in these previous reviews. I intend to highlight the salient points, to bring the information up to date, and to try to put the subject into its current perspective. For more detailed information, the reader is referred to the sources cited earlier.

Groups of plant viruses

Of the three hundred or so well-characterized plant viruses, about 78% have a genome comprising single-stranded RNA in the plus or messenger sense. About 13% contain single-stranded RNA in the minus sense, and double-stranded RNA, double-stranded DNA, and single-stranded DNA each account for about 3%. However, as far as genetic manipulation is concerned, it is the viruses containing double-stranded DNA that have attracted the greatest attention thus far. There is now awakening interest in examining the genomes of viruses that contain single-stranded DNA and single-stranded "plus" RNA as potential gene vectors.

Double-stranded DNA viruses

The caulimoviruses are the only group of plant viruses whose particles are known to contain double-stranded DNA. The properties of the type member, cauliflower mosaic virus (CaMV), and of the other members of the group have been reviewed extensively (Hull 1979, 1981, 1983, 1984; Hull and Davies 1983; Shepherd 1976, 1977, 1979, 1981). The pertinent facts concerning the molecular biology of CaMV in relation to vector construction are the following:

1. The virion DNA has been sequenced and shown to be 8 kb (Balazs et al. 1982; Franck et al. 1980; Gardner et al. 1981).
2. The encapsidated virion DNA has discontinuities at unique sites, one in one strand and two in the other (one CaMV isolate has a deletion that removes one of the two). These discontinuities have a triple-stranded structure with the 5′ end at a fixed site and the 3′ end overlapping the 5′ end by up to 40 or more residues (Franck et al. 1980; Richards et al. 1981). In some cases there is a ribonucleotide at the 5′ end (Guilley et al. 1983*b*). Other caulimoviruses have either three or four discontinuities, but always one in one strand (Donson and Hull 1983; Hull and Donson 1982; Richins and Shepherd 1983).
3. The sequence confirms previous studies showing that transcription is asymmetrical (Howell and Hull 1978; Hull et al. 1979).
4. From the sequence six or possibly eight regions coding for proteins of more than 10,000 daltons have been recognized (Franck et al. 1980; Hohn et al. 1981). Functions have been ascribed to three of these regions with reasonable certainty. Region II codes for a protein of 18,000 daltons that is associated with aphid transmission of the virus (Armour et al. 1983; Woolston et al. 1983). Region IV codes for the precursor to the coat protein (Daubert et al. 1982; Franck et al. 1980; Hahn and Shepherd 1982). Region VI codes for

the inclusion-body matrix protein (Al Ani et al. 1980; Covey and Hull 1981; Odell and Howell 1980; Xiong et al. 1982). There is a suggestion that the product from region V might be associated with viral replication (Hull and Covey 1983*a;* Toh et al. 1983; Volovitch et al. 1984).

5. Most of the coding regions are clustered, either overlapping or separated by only a few nucleotides. There is a short intergenic region of 60 bp (IR2) between coding regions V and VI and a larger intergenic region (IR1) of about 600 bp between coding regions VI and VII; VII is separated from I also by about 60 bp.

6. Two major CaMV-specific polyadenylated RNA transcripts are found in infected cells. The 19S transcript is the mRNA for gene product VI (Covey and Hull 1981; Odell and Howell 1980; Xiong et al. 1982). The other major RNA, the 35S transcript, is a more than full-length transcript in that it has a terminal direct repeat of about 180 nucleotides (Covey et al. 1981; Dudley et al. 1982; Guilley et al. 1982). Various other CaMV-specific RNAs have been reported (Condit and Meagher, 1983; Condit et al. 1983; Guilley et al. 1982), but there is no information whether or not they are transcribed from individual promoters.

There have been two basic approaches to assessing the potential of CaMV DNA as a gene vector: (1) detailed analysis of its replication and expression and (2) attempts to cut the DNA and insert foreign DNA while still retaining biological activity.

Expression of CaMV

As noted earlier, there are two major virus-specific transcripts in cells infected with CaMV. Mapping shows that the two transcripts share a common 3′ end (Covey et al. 1981; Dudley et al. 1982; Guilley et al. 1982) at nucleotide 7615. Some 25 nucleotides upstream from this is the sequence AATAAA that is recognized as a strong signal for polyadenylation (Proudfoot and Brownlee 1976). However, as discussed by Proudfoot (1984), the signal for termination of transcription in eukaryotes is not yet known.

The 5′ end of the 19S transcript is at nucleotide 5764, giving a leader sequence of 11 nucleotides before the initiation codon of open region VI. Some 29 nucleotides upstream of this is a TATA box (TATTTAAA), and a further 63 nucleotides upstream is a CAT box (CTAATT), sequences that are usually found in the eukaryotic promoter region.

The 5′ end of the 35S transcript is at nucleotide 7435, some 180 nucleotides upstream of the 3′ end, which the RNA polymerase must pass in the first circle of transcription. Nothing is known about the mechanism that determines why the site of the 3′ end is not recognized the first time it is passed but is recognized the second time. Upstream of the 5′ end of the 35S transcript is a TATA box (TATATAA) at -25 and a CAT box (CACAAT) at -60.

Promoter sequences in eukaryotic DNA not only comprise TATA and CAT boxes but often various other regions upstream, as reviewed elsewhere (Breathnach and Chambon 1981). Nothing is known about these in plants at present, but it is likely that the full promoter regions of CaMV DNA cover a hundred or more nucleotides.

Replication of CaMV

It has recently been recognized that several features of CaMV nucleic acid resemble those of retroviruses, and there is now increasing evidence for reverse transcription being involved in the replication of CaMV (Guilley et al. 1983*a*, 1983*b*; Hohn et al. 1983*a*, 1983*b*; Hull and Covey 1983*a*, 1983*c*, 1983*d*; Marco and Howell 1984; Menissier et al. 1983; Pfeiffer and Hohn 1983). Much of the evidence is reviewed by Hull and Covey (1983*c*). The basic replication cycle starts with the virion DNA entering the cell. This DNA must be transported to the nucleus and have the discontinuities sealed to give covalently closed molecules. These form minichromosomes that are the template for transcription, most likely using cellular RNA polymerase II (Guilfoyle 1980; Guilfoyle and Olszewski 1983; Menissier et al. 1982; Olszewski et al. 1982). The 35S transcript is, in turn, the template for formation of minus-strand DNA by reverse transcription. Priming is at the site of G1 (the discontinuity in the strand with the single discontinuity) by $tRNA^{met}_{i}$, and the first stage is the formation of minus-strand DNA up to the 5′ end of the template. This strong-stop DNA, comprising about 600 nucleotides of DNA with the tRNA at the 5′ end, has been characterized by Covey and associates (1983) and Turner and Covey (1984). It is presumed that the template is removed by an RNase H-like activity, leaving the 3′ end of the newly synthesized minus-strand DNA able to hybridize to the terminal repeat portion of the 3′ end of the 35S RNA. Synthesis of the minus-stranded DNA continues, and plus-strand DNA formation is primed at the sites of G2 and G3 (the other two discontinuities). Various forms of unencapsidated CaMV DNA have been reported by Hull and Covey (1983*b*). Some of these are interpreted as being formed by defects in the synthesis of minus-strand DNA between G2 and G3 and between G3 and G1. A second strand switch is effected using the sequence complementary to the tRNA primer at G1, and the circle is completed. When the primers are removed, the 5′ side of each discontinuity is in a fixed position. There is limited strand displacement when the advancing strand meets the priming site, this giving the triple-stranded structure at the discontinuities.

This model for replication explains many of the unusual features of CaMV DNAs, both encapsidated and unencapsidated. The weight of evidence now supports the model, but we await definitive proof of it operating in vitro. The properties of the enzymic activities involved, especially in the reverse transcription step, need to be determined. Pfeiffer and Hohn (1983) reported a replication complex that would incorporate radioactive deoxyribonucleotides into CaMV-specific DNA. It appeared that the template was RNA. Volovitch and associates (1984) have partially characterized an RNA-dependent DNA polymerase activity in CaMV-infected leaves and have shown that it differs in template specificity from the normal γ-like DNA polymerase found in plant cells.

Mutagenesis of CaMV DNA

The literature up to 1982 on the effects of insertions or deletions on the infectivity of CaMV RNA was reviewed by Hull and Davies (1983). Up to that time, all mutations studied were lethal, except the ones in the intergenic region near the unique *Bst*EII site and around the unique *Xho* I site in coding region II. Since then, Daubert and associates (1983) and Dixon and associates (1983) have found several other nonlethal sites that are illustrated in Figure 6.1. It can be seen from Figure 6.1 that coding regions I, III, and V appear to be absolutely essential for viral replication and spread through the plant. Coding regions IV and VI will tolerate some perturbations in their C-terminal regions, but not in the N-terminal

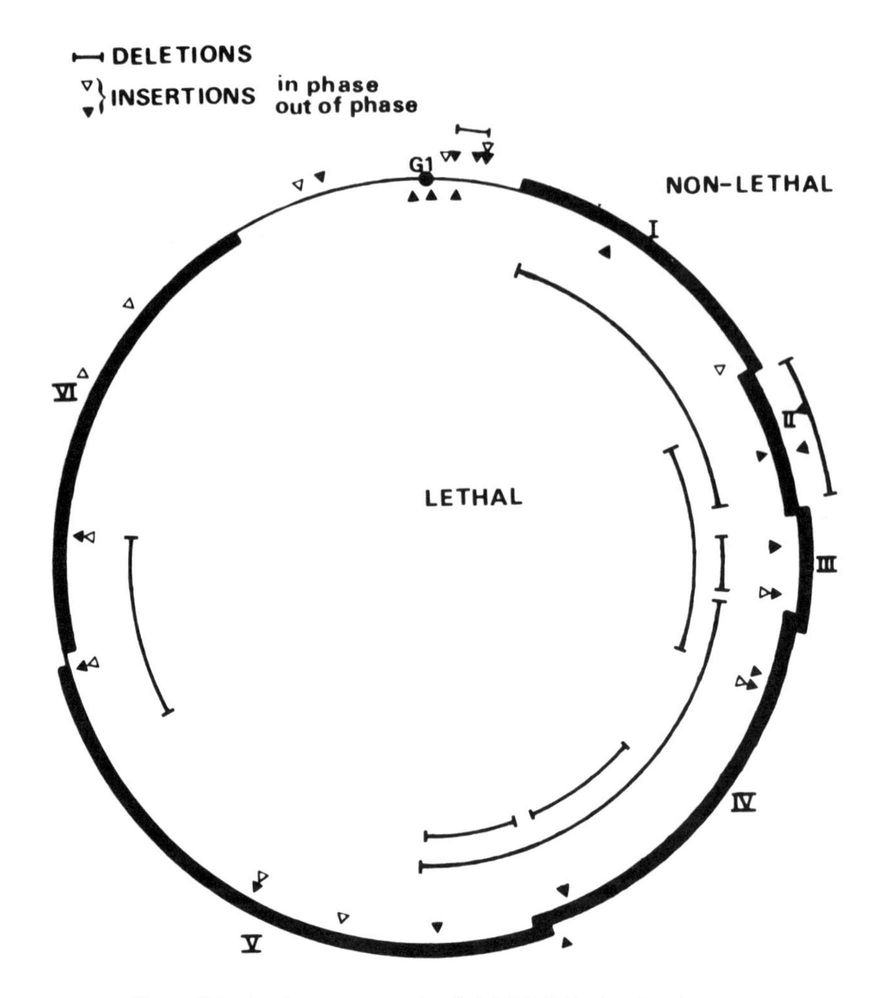

Figure 6.1. *In vitro* mutagenesis of CaMV DNA. On the circle representing the CaMV genome are indicated the six coding regions (thicker lines) and the position of gap 1. Deletions and insertions are as indicated on the diagram, those inside the circle resulting in loss of infectivity, those outside the circle retaining infectivity. Based on data from Hull (1983), updated with data from Daubert et al. (1983) and Dixon et al. (1983).

parts. Some insertions can be made in the large intergenic region. Coding region II does not appear to be essential for replication, and sizable parts of it can be removed without deleterious effects on virus multiplication (Armour et al. 1983; Howarth et al. 1981; Woolston et al. 1983). However, insertions (and presumably deletions) in this region that cause a reading-frame shift are characterized by delayed symptom appearance (Dixon et al. 1983; Sieg and Gronenborn 1982). Analysis of the progeny DNA showed that the reading frame had been restored.

Interactions between mutagenized DNAs

In view of the loss of infectivity caused by mutagenesis, several workers attempted to design complementation systems, as reviewed elsewhere (Hull and Davies 1983). However, it was found that pairs of defective molecules recombined to give wild-type DNA. Walden and Howell (1982) and Howell and Walden (1983) suggested that there are two mechanisms of recombination. In the first, the more common mechanism, there is a single crossover event that recombines out full-length unmodified DNA from end-to-end concatamers. The second, the rare mechanism, involves double crossover and probably occurs after some limited DNA replication. Lebeurier and associates (1982) and Walden and Howell (1983) showed that intragenomic recombination can also take place. They found that recombinant plasmids of tandem full-length CaMV dimers or single full-length CaMV molecules flanked by homologous arms of viral DNA can infect plants without prior excision of the viral DNA. There is intragenomic recombination that produces a single full-length viral DNA molecule and that is dependent on the presence of homologous arms.

Thus, the studies on mutagenesis and complementation of CaMV DNA have all shown that there is a high rate of recombination, presumably associated with viral replication. The reverse transcription mode of replication is characterized by a high recombination rate (Coffin 1979).

Single-stranded DNA viruses

As with double-stranded DNA viruses, there has been only one plant virus group, the geminiviruses, recognized as having single-stranded DNA. Geminiviruses are characterized by geminate particles; their host ranges include both dicotyledonous and monocotyledonous plants. The geminiviruses have been reviewed by Goodman (1981*a*, 1981*b*) and Bock (1982), and their potential as vectors has been discussed by Hull and Davies (1983).

Since this recent review, Stanley (1983) and Hamilton and associates (1983) have unequivocally demonstrated that both DNA molecules are needed for infection in the case of each of two geminiviruses, cassava latent virus (CLV) and tomato golden mosaic virus (TGMV). Thus, together with the more circumstantial evidence from bean golden mosaic virus (Ikegami et al. 1981), three whitefly-transmitted geminiviruses have bipartite genomes. On the other hand, there is an indication from restriction endonuclease mapping and sequencing that the leafhopper-transmitted geminivirus, chloris striate mosaic virus, has only one DNA species (Marriott and Symons 1983); this DNA is about the same size as each of the species from the whitefly-transmitted geminiviruses. Very recently, sequencing of the DNA of another leafhopper-transmitted geminivirus, maize streak virus, has shown that this also has only one DNA species (Howell 1984; Mullineaux et al. in press).

There is little or no further published information on the expression and replication of geminiviruses. However, particles and other virus-specific structures are found in nuclei of infected plants, suggesting that at least part of the replication cycle occurs there.

Single-stranded RNA viruses

As noted earlier, the majority of plant viruses have single-stranded RNA in the plus sense as their genomes. What is known about their potential as gene vectors was also recently reviewed by Hull and Davies (1983). Among the salient points are the following:

1. The many plant RNA viruses include ones with wide host ranges, others with narrow host ranges, ones that infect dicotyledonous plants, others that infect monocots, and yet others that will infect dicots and monocots.

2. The nucleic acids of these viruses are infectious to plants. There is no requirement for an encapsidated enzyme, although the genomic RNAs of alfalfa mosaic virus (AlMV) and ilarviruses require some molecules of coat protein or the cistron for coat protein for replication.

3. Many plant viruses have the genetic information needed for full infection split between two or three pieces of nucleic acid; these nucleic acid species are often packaged in separate particles. In several viruses, such as tobacco rattle virus (TRV), cowpea mosaic virus

(CPMV), brome mosaic virus (BMV), and AlMV, a form of infection can be initiated by just the part of the genome that carries the information for nucleic acid replication. In these viruses, replication of the nucleic acid does not need the cistron for coat protein (except for AlMV, which needs some coat protein molecules to be present). However, only in the case of TRV does the viral nucleic acid spread from the initially infected cells. It is thought that there is also a virus-coded product that is involved in cell-to-cell movement and that is not involved in the replication in the initially infected cells.

4. The particles of plant RNA viruses can be isometric, bacilliform, or rod-shaped. In the latter two cases the size of the particle is related to the length of nucleic acid being encapsidated. Thus, there is no apparent impediment to forming stable capsids of RNA to which has been added foreign nucleic acid (another gene).

5. In tobacco mosaic virus (TMV) RNA there is a nucleotide sequence responsible for initiating the spontaneous self-assembly of virions with the cognate coat protein, as reviewed by Butler (1984). This sequence is of potential use in constructing a chimeric molecule that could be encapsidated using TMV coat protein.

Expression and replication of RNA viruses

That plant viral RNAs are expressed and replicate very efficiently is exemplified by the rapid production of large numbers of virus particles following infection. There are more than 10^7 TMV particles per infected cell, and many viruses have between 10^6 and 10^7 particles per cell. Thus, there is on the order of 10–100 pg of viral coat protein per cell. The other virus-coded proteins are produced in much smaller amounts, and thus there must be expressional control.

Plant viral RNAs code for several gene products, as reviewed elsewhere (Davies and Hull 1982). Because of the restriction on translation of internal cistrons by eukaryotic ribosomes, two major strategies have been evolved for the expression of plant viral RNAs. In one strategy the RNAs of some viruses are divided, at least during some stage of the replication cycle, so that each cistron is 5′ terminally located. In several groups of plant viruses, the multicomponent viruses referred to earlier, most of the cistrons remain separated throughout the replication cycle and are packaged individually. Thus, these viruses have their genomes divided between two or three particle types. In the second strategy, the viral RNA is translated as a monocistron, and the large protein produced is then cleaved to give the functional proteins. Other viruses have a mixture of these two strategies.

Gene products or potential gene products have been identified for several plant viruses from studies of in vitro translation, in vivo translation, and RNA sequences. In all cases the virus codes for its own coat protein. Little is known about other products, but there are now some pointers to functions. The involvement of virus-coded products in RNA replication will be discussed later. Recent observations on TMV have implicated a virus gene product (the 30K protein) in cell-to-cell movement (Leonard and Zaitlin 1982; Zimmern and Hunter 1983). There is circumstantial evidence that a protein of similar size is involved in the cell-to-cell spread of other viruses (Kiberstis et al. 1981; Nassuth et al. 1981), and experiments by Taliansky and associates (1982*a*, 1982*b*) suggest that virus-specific transport functions of one virus will facilitate the movement of another virus in an apparent nonhost.

Relatively little is known about the detailed molecular biology of RNA replication in most plant viruses. The replication of RNA viruses with tripartite genomes has recently been reviewed by Hull and Maule (in press), and the facts given there pertain to most other plant RNA viruses. RNA viruses replicate via a complementary or minus strand, but it is not known if the replication is conservative or semiconservative. Sequence data suggest that terminal sequences, or at least those at the 3′ termini of the plus strand, are important in presumably the initiation of replication. There has been considerable controversy over which enzymes are involved in viral RNA replication. However, recent evidence indicates involvement of virus-coded product(s) in the cases of CPMV, AlMV, BMV, TRV, and tomato blackring viruses (Dorssers et al. 1984; Kiberstis et al. 1981; Lister 1968; Nassuth et al. 1981; Robinson et al. 1980). In the two viruses CPMV and BMV, where most is known about the enzymology of RNA replication, the virus-coded product(s) involved are large (more than 100K) (Dorssers et al. 1984; Hall et al. 1982). The work of Dorssers and associates (1984) suggests that the RNA of CPMV is replicated solely by its own enzyme and that no host enzyme is involved. Most other plant viruses for which the genome strategy is known have at least one gene product of about 100K or greater. Thus, it is likely that such products are involved in viral RNA replication.

Potential of plant viruses as vectors

Having described features of plant viruses associated with their possible development as vectors, I shall now discuss the advantages and disadvantages of the use of plant viral nucleic acids in this role.

Potential advantages

It was noted earlier that plant viruses replicate and express at a high rate in plant cells. The DNA viruses, at least the caulimoviruses, undergo more than one cycle of replication per cell; because significant numbers of geminivirus particles accumulate in infected cells, it is likely that they also have more than one round of replication per cell cycle. Thus, the replication systems of viruses would seem to lend themselves to the production of nucleic acid in high copy numbers. However, with caulimoviruses it is questionable whether or not it is possible to dissect the information needed for replication from undesirable features; this point will be discussed later. The replication systems of RNA viruses, and especially of the multicomponent viruses that have the replication information in one component, look more promising. Once the features that determine recognition of a nucleic acid by a replication system have been elucidated, it may be possible to attach these to a "foreign" gene.

The transcription of the major RNAs of CaMV indicate that the promoter sequences on this nucleic acid are very efficient. The regions containing these promoter sequences have been recognized, but the true structure of the promoters has yet to be determined. Once this has been done, it should be possible to construct promoter sequences of different efficiencies for different situations.

Many plant viruses have the ability to spread from cell to cell and to replicate in most of the cells of a plant; there are some that are restricted to certain tissues. This ability to spread systemically has been shown, for some viruses, to be virus-coded. Also, the work of Taliansky and associates (1982*a*, 1982*b*) suggests that, at least for the viruses they studied, the virus-coded effect is host-specific. If this phenomenon is widespread, it might be possible to use it to tailor vectors to specific hosts. Also, the ability to introduce nucleic acid into one part of a plant and have it spread to the rest of the plant would overcome the problems of growing plants from infected protoplasts or cell cultures.

No plant viral nucleic acids have yet been shown to integrate into the host genome. This might be thought to be a disadvantage, because it would not be possible to transmit in a Mendelian fashion characters introduced using viral vectors. However, as pointed out by Hull (1978, 1980*b*, 1981, 1983) and by Hull and Davies (1983), this is not necessarily a disadvantage in plants that can be vegetatively propagated. In fact, it is in many circumstances an advantage to have a character constitutively expressed and not under chromosomal control.

Potential disadvantages

Plant viruses cause disease, with infected plants showing symptoms and having reduced yields. Although symptom production has proved to be a useful selection marker for studying the mutagenesis of CaMV, it is obviously not a desirable character in a vector. Little is known about how viruses induce symptoms, and so it is not possible to determine at present if this feature can be removed from a potential vector without affecting advantageous characters. As noted by Grill (1983), some plant viruses, the cryptic viruses, which do not produce detectable symptoms in plants, might be of use in vector development. However, very little is known about the molecular biology of these viruses. Furthermore, the fact that they have isometric capsids might raise problems as to the amounts of nucleic acid that could be manipulated.

Perhaps the most serious problem in developing plant virus vectors lies in their modes of replication. Replication of caulimoviruses and of RNA viruses is by enzyme systems that do not have "proofreading." Thus, the error rate is high: 10^{-3}–10^{-4} misincorporations per base compared with 10^{-11} misincorporations in the proofreading DNA-dependent DNA polymerase systems. The error rates of different polymerase systems have been reviewed recently by Reanney (1984). Obviously, viruses are faced with this problem of errors in replication, and this leads to considerable variation within a virus population (Garcia-Arenal et al. 1984). However, there is selection pressure on viruses in that any errors that affect replication and expression are rapidly lost. Unless there is similar selection pressure on foreign nucleic acid introduced using a viral vector, errors will accumulate rapidly. It can be estimated that for an average-sized gene, at least one amino acid change will occur every five replication cycles.

Although little is known about the replication cycle of geminiviruses, it is quite possible that it involves a DNA-dependent DNA polymerase stage. If this is so, the problems of error rates will not apply to these viruses.

With CaMV, this inherent error rate presumably has led to the wide range of minor variations between isolates of this virus (Hull 1980*a*). On top of this is the high rate of recombination discussed earlier, which is thought to be a feature of replication by reverse transcription. This recombination has proved already to be a considerable obstacle to the introduction of foreign genes into CaMV DNA. Furthermore, as noted earlier, many of the gene functions of CaMV appear to be necessary for viral replication and systemic spread. The replication cycle also involves sequences in the intergenic region. Thus, it would prove difficult, if not impossible, to isolate a replicon.

This exemplifies one further feature of plant viruses: Their gene expression and functioning are likely to be highly controlled and integrated. It is unlikely that in such efficiently replicating and expressing molecules there is to be found much redundancy. Thus, attempts to use all or most of a viral genome as a vector are likely to be confounded, because perturbations of a highly integrated system would be lethal.

Conclusions

We are now reaching a stage of appraising the potential of plant viral nucleic acids as vectors in which it would appear unlikely that one could simply insert a foreign gene and have it expressed. It may be that by careful tailoring one could do so in exceptional cases, but widespread use of such plant viral vectors would seem to be improbable. However, there are features of plant viral nucleic acids that will be used in chimeric vectors. There are vectors already constructed in which the foreign genes are expressed from the CaMV promoter regions. Other characters that will be of use include the replicons of some RNA viruses, the sequences that make some viral RNAs efficient messengers, and the sequences involved in assembly of coat protein around TMV RNA. Thus, the future of plant viruses in vector development would seem to be mainly as sources of important "spare parts" for the construction of composite vectors.

Acknowledgments

I thank Drs. J. W. Davies and T. M. A. Wilson for helpful comments on this chapter.

References

Al Ani, R., Pfeiffer, P., Whitechurch, O., Lesot, A., Lebeurier, G., and Hirth, L. (1980). A virus specified protein produced upon infection of cauliflower mosaic virus. *Ann. Virol (Inst. Pasteur)* **131E,** 33–53.

Armour, S. L., Melcher, U., Pirone, T. P., Lyttle, D. J., and Essenberg, R. C. (1983). Helper component for aphid transmission encoded by region II of cauliflower mosaic virus DNA. *Virology* **129,** 25–30.

Balazs, E., Guilley, H., Jonard, G., and Richards, K. (1982). Nucleotide sequence of DNA from altered-virulence isolate D/H of the cauliflower mosaic virus. *Gene* **19,** 239–49.

Bock, K. R. (1982). Geminivirus diseases in tropical crops. *Plant Disease* **66,** 266–70.

Breathnach, R., and Chambon, P. (1981). Organization and expression of eukaryotic split genes coding for proteins. *Annu. Rev. Biochem.* **50,** 349–83.

Butler, P. J. G. (1984). The current picture of the structure and assembly of tobacco mosaic virus. *J. Gen. Virol.* **65,** 253–79.

Coffin, J. M. (1979). Structure, replication and recombination of retrovirus genomes; some unifying hypotheses. *J. Gen. Virol.* **42,** 1–26.

Condit, C., Hagan, T. J., McKnight, T. D., and Meagher, R. B. (1983). Characterization and preliminary mapping of cauliflower mosaic virus transcripts. *Gene* **25,** 101–8.

Condit, C., and Meagher, R. B. (1983). Multiple, discrete 35S transcripts of cauliflower mosaic virus. *J. Mol. Appl. Genet.* **2,** 301–14.

Covey, S. N., and Hull, R. (1981). Transcription of cauliflower mosaic virus DNA. Detection of transcripts, properties and location of gene encoding the virus inclusion body protein. *Virology* **111,** 463–74.

Covey, S. N., Lomonossoff, G. P., and Hull, R. (1981). Characterisation of cauliflower mosaic virus DNA sequences which encode major polyadenylated transcripts. *Nucl. Acids Res.* **9,** 6735–47.

Covey, S. N., Turner, D., and Mülder, G. (1983). A small DNA molecule containing covalently-linked ribonucleotides originates from the large intergenic region of the cauliflower mosaic virus genome. *Nucl. Acids Res.* **11,** 251–63.

Daubert, S., Richins, R., Shepherd, R. J., and Gardner, R. C. (1982). Mapping of the coat protein gene of cauliflower mosaic virus by its expression in a prokaryotic system. *Virology* **122,** 444–9.

Daubert, S., Shepherd, R. J., and Gardner, R. C. (1983). Insertional mutagenesis of the cauliflower mosaic virus genome. *Gene* **25,** 201–8.

Davies, J. W., and Hull, R. (1982). Genome expression of plant positive-strand RNA viruses. *J. Gen. Virol.* **61,** 1–14.

Dixon, L. K., Koenig, I., and Hohn, T. (1983). Mutagenesis of cauliflower mosaic virus. *Gene* **25,** 189–99.

Donson, J., and Hull, R. (1983). Physical mapping and molecular cloning of caulimovirus DNA. *J. Gen. Virol.* **64,** 2281–8.

Dorssers, L., van der Krol, S., van der Meer, J., van Kammen, A., and Zabel, P. (1984). The cowpea mosaic virus B-RNA-encoded 110K polypeptide is involved in viral RNA replication. *Proc. Natl. Acad. Sci. U.S.A.* **81,** 1951–5.

Dudley, R. K., Odell, J. T., and Howell, S. H. (1982). Structure and 5′-termini of the large and 19S RNA transcripts encoded by the cauliflower mosaic virus genome. *Virology* **117,** 19–28.

Franck, A., Guilley, H., Jonard, G., Richards, K., and Hirth, L. (1980). Nucleotide sequence of cauliflower mosaic virus DNA. *Cell* **21,** 285–94.

Garcia-Arenal, F., Palukaitis, P., and Zaitlin, M. (1984). Strains and mutants of tobacco mosaic virus are both found in virus derived from single-lesion-passaged inoculum. *Virology* **132,** 131–7.

Gardner, R. C. (1983). Plant viral vectors; CaMV as an experimental tool, in *Genetic Engineering of Plants,* ed. T. Kosuge, C. P. Meredith, and A. Hollaender, pp. 121–42. Plenum Press, New York.

Gardner, R. C., Howarth, A. J., Hahn, P., Brown-Luedi, M., Shepherd, R. J., and Messing, J. (1981). The complete nucleotide sequence of an infectious clone of cauliflower mosaic virus by M13mp7 shotgun sequencing. *Nucl. Acids Res.* **9,** 2871–88.

Goodman, R. M. (1981*a*). Geminiviruses, in *Handbook of Plant Virus Infections and Comparative Diagnosis,* ed. E. Kurstak, pp. 833–910. Elsevier/North-Holland, Amsterdam.

(1981*b*). Geminiviruses. *J. Gen. Virol.* **54,** 9–21.

Grill, L. K. (1983). Utilizing RNA viruses for plant improvement. *Plant Mol. Biol. Rep.* **1,** 17–20.

Guilfoyle, T. J. (1980). Transcription of the cauliflower mosaic virus genome in isolated nuclei from turnip *Brassica rapa* cultivar Just Right leaves. *Virology* **107,** 71–80.

Guilfoyle, T. J., and Olszewski, N. E. (1983). The structure and transcription of the cauliflower mosaic virus minochromosome, in *Plant Infectious Agents,* ed. H. D. Robertson, S. H. Howell, M. Zaitlin, and R. L. Malmberg, pp. 34–8. Cold Spring Harbor Laboratory, New York.

Guilley, H., Dudley, R. K., Jonard, G., Balazs, E., and Richards, K. E. (1982). Transcription of cauliflower mosaic virus DNA: detection of promoter sequences and characterization of transcripts. *Cell* **30,** 763–73.

Guilley, H., Jonard, G., and Richards, K. (1983*a*). Structure and expression of cauliflower mosaic virus DNA, in *Plant Infectious Agents*, ed. H. D. Robertson, S. H. Howell, M. Zaitlin, and R. L. Malmberg, pp. 17–22. Cold Spring Harbor Laboratory, New York.

Guilley, H., Richards, K. E., and Jonard, G. (1983*b*). Observations concerning the discontinuous DNAs of cauliflower mosaic virus. *EMBO Journal* **2,** 277–82.

Hahn, P., and Shepherd, R. J. (1982). Evidence for a 58-kilodalton polypeptide as precursor of the coat protein of cauliflower mosaic virus. *Virology* **116,** 480–8.

Hall, T. C., Miller, W. A., and Bujarski, J. J. (1982). Enzymes involved in the replication of plant viral RNAs. *Adv. Plant Pathol.* **1,** 179–211.

Hamilton, W. D. O., Bisaro, D. M., Coutts, R. H. A., and Buck, K. W. (1983). Demonstration of the bipartite nature of the genome of a single-stranded DNA plant virus by infection with the cloned DNA components. *Nucl. Acids Res.* **11,** 7387–96.

Hohn, T., Pietrzak, M., Dixon, L., Koenig, I., Penswick, J., Hohn, B., and Pfeiffer, P. (1983*a*). Involvement of reverse transcription in cauliflower mosaic virus replication, in *Plant Infectious Agents*, ed. H. D. Robertson, S. H. Howell, M. Zaitlin, and R. L. Malmberg, pp. 28–33. Cold Spring Harbor Laboratory, New York.

(1983*b*). Involvement of reverse transcription in cauliflower mosaic virus replication. *Plant Mol. Biol.* **1,** 147–55.

Hohn, T., Richards, K., and Lebeurier, G. (1981). Cauliflower mosaic virus on its way to becoming a useful plant vector. *Curr. Top. Microbiol. Immunol.* **96,** 193–236.

Howarth, A. J., Gardner, R. C., Messing, J., and Shepherd, R. J. (1981). Nucleotide sequence of naturally occurring deletion mutants of cauliflower mosaic virus. *Virology* **112,** 678–85.

Howell, S. H. (1982). Plant molecular vehicles; potential vehicles for introducing foreign DNA into plants. *Annu. Rev. Plant Physiol.* **33,** 609–50.

(1984). Physical structure and genome organisation of the genome of maize streak virus (Kenyan isolate). *Nucl. Acids Res.* **17,** 7359–75.

Howell, S. H., and Hull, R. (1978). Replication of cauliflower mosaic virus and transcription of its genome in turnip leaf protoplasts. *Virology* **86,** 468–81.

Howell, S. H., and Walden, R. M. (1983). Recombination of cloned CaMV genomes, in *Plant Infectious Agents,* ed. H. D. Robertson, S. H. Howell, M. Zaitlin, and R. L. Malmberg, pp. 39–43. Cold Spring Harbor Laboratory, New York.

Hull, R. (1978). The possible use of plant virus DNAs in genetic manipulation in plants. *Trends Biochem. Sci.* **3,** 254–6.

(1979). The DNA of plant DNA viruses, in *Nucleic Acids in Plants, Vol. 2,* ed. T. C. Hall and J. W. Davies, pp. 1–29, CRC Press, Boca Raton, Fl.

(1980*a*). Structure of the cauliflower mosaic virus genome. III. Restriction endonuclease mapping of thirty-three isolates. *Virology* **100,** 76–90.

(1980*b*). Genetic engineering in plants: the possible use of cauliflower mosaic virus DNA as a vector, in *Plant Cell Cultures: Results and Perspectives,* ed. F. Sala, B. Parisi, R. Cella, and O. Ciferri, pp. 219–24. Elsevier-North-Holland, Amsterdam.

(1981). Cauliflower mosaic virus DNA as a possible gene vector for higher plants, in *Genetic Engineering in the Plant Sciences,* ed. N. J. Panopoulous, pp. 99–109. Praeger, New York.

(1983). The current status of plant viruses as potential DNA/RNA vector systems, in *Plant Biotechnology,* ed. S. H. Mantell and H. Smith, pp. 229–312. Cambridge University Press.

(1984). *Caulimovirus Group. CMI/AAB Descriptions of Plant Viruses No. 295.*

Hull, R., and Covey, S. N. (1983*a*). Does cauliflower mosaic virus replicate by reverse transcription? *Trends Biochem. Sci.* **8,** 119–21.

(1983*b*). Characterization of cauliflower mosaic virus DNA form isolated from infected turnip leaves. *Nucl. Acids Res.* **11,** 1881–95.

(1983*c*). Replication of cauliflower mosaic virus DNA. *Sci Progr.* (*Oxford*) **68,** 403–22.

(1983*d*). Unencapsidated nucleic acids of cauliflower mosaic virus and their significance in virus replication, in *Plant Infectious Agents,* ed. H. D. Robertson, S. H. Howell, M. Zaitlin, and R. L. Malmberg, pp. 23–7. Cold Spring Harbor Laboratory, New York.

Hull, R., Covey, S. N., Stanley, J., and Davies, J. W. (1979). The polarity of the cauliflower mosaic virus genome. *Nucl. Acids Res.* **7,** 669–77.

Hull, R., and Davies, J. W. (1983). Genetic engineering with plant viruses, and their potential as vectors. *Adv. Virus Res.* **28,** 1–33.

Hull, R., and Donson, J. (1982). Physical mapping of the DNAs of carnation etched ring and figwort mosaic viruses. *J. Gen. Virol.* **60,** 125–34.

Hull, R., and Maule, A. J. (in press). Multiplication of viruses with tripartite genomes, in *Tripartite Viruses,* ed. R. I. B. Francki. Plenum, New York.:

Ikegami, M., Haber, S., and Goodman, R. M. (1981). Isolation and characterization of virus-specific double-stranded DNA from tissues infected with bean golden mosaic virus. *Proc. Natl. Acad. Sci. U.S.A.* **78,** 4102–6.

Kiberstis, P. A., Loesch-Fries, L. S., and Hall, T. C. (1981). Viral protein synthesis in barley inoculated with native and fractionated brome mosaic virus RNA. *Virology* **112,** 804–8.

Kleinhofs, A., and Behki, R. (1977). Prospects for plant genome modification by nonconventional methods. *Annu. Rev. Genet.* **11,** 79–101.

Lebeurier, G., Hirth, L., Hohn, B., and Hohn, T. (1982). *In vivo* recombination of cauliflower mosaic virus DNA. *Proc Natl. Acad. Sci. U.S.A.* **79,** 2932–6.

Leonard, D. A., and Zaitlin, M. (1982). A temperature sensitive strain of tobacco mosaic virus defective in cell-to-cell movement generates an altered viral-coded protein. *Virology* **117,** 416–24.

Lister, R. M. (1968). Functional relationship between virus-specific products of infection by viruses of the tobacco rattle type. *J. Gen. Virol.* **2,** 43–58.

Lurquin, P. F., and Kado, C. I. (1979). Recent advances in the insertion of DNA into higher plant cells. *Plant Cell Environment* **2,** 199–203.

Marco, Y., and Howell, S. H. (1984). Intracellular forms of viral DNA consistent with a model of reverse transcriptional replication of the cauliflower mosaic virus genome. *Nucl. Acids Res.* **12,** 1517–28.

Marriott, T. W., and Symons, R. H. (1983). Characterization of the DNA genomes of Australian geminiviruses, in *Plant Infectious Agents,* ed. H. D. Robertson, S. H. Howell, M. Zaitlin, and R. L. Malmberg, pp. 59–62. Cold Spring Harbor Laboratory, New York.

Menissier, J., Lebeurier, G., and Hirth, L. (1982). Free cauliflower mosaic virus supercoiled DNA in infected plants. *Virology* **117,** 322–8.

Menissier, J., Pfeiffer, P., Lebeurier, G., Guilley, H., Locroute, F., and Hirth, L. (1983). Cauliflower mosaic virus DNA in the elaboration of gene vectors in plants, in *Plant Infectious Agents,* ed. H. D. Robertson, S. H. Howell, M. Zaitlin, and R. L. Malmberg, pp. 44–8. Cold Spring Harbor Laboratory, New York.

Mullineaux, P. M., Donson, J., Morris-Krsinich, B. A. M., Boulton, M. I., and Davies, J. W. (in press). The nucleotide sequence of maize streak virus DNA. *EMBO Journal.*

Nassuth, A., Alblas, F., and Bol, J. F. (1981). Localization of genetic information involved in the replication of alfalfa mosaic virus. *J. Gen. Virol.* **53,** 207–14.

Odell, J. T., and Howell, S. H. (1980). The identification, mapping and characterisation of mRNA for P66, a cauliflower mosaic virus-coded protein. *Virology* **102,** 349–59.

Olszewski, N., Hagen, G., and Guilfoyle, T. J. (1982). A transcriptionally active, covalently closed minichromosome of cauliflower mosaic virus DNA isolated from infected turnip leaves. *Cell* **29,** 395–402.

Pfeiffer, P., and Hohn, T. (1983). Involvement of reverse transcription in the replication of cauliflower mosaic virus: a detailed model and test of some aspects. *Cell* **33,** 781–9.

Proudfoot, N. J. (1984). The end of the message and beyond. *Nature* (*London*) **307,** 412–13.

Proudfoot, N. J., and Brownlee, G. G. (1976). 3′ non-coding region in eukaryotic messenger RNA. *Nature* (*London*) **263,** 211–14.

Reanney, D. (1984). Genetic noise in evolution. *Nature* (*London*) **307,** 318–19.

Richards, K. E., Guilley, H., and Jonard, G. (1981). Further characterization of the discontinuities in cauliflower mosaic virus DNA. *FEBS Lett.* **134,** 67–70.

Richins, R. D., and Shepherd, R. J. (1983). Physical maps of the genomes of dahlia mosaic virus and mirabilis mosaic virus – two members of the caulimovirus group. *Virology* **124,** 208–14.

Robinson, D. J., Barker, H., Harrison, B. D., and Mayo, M. A. (1980). Replication of RNA-1 of tomato blackring virus independently of RNA-2. *J. Gen. Virol.* **51,** 317–26.

Shepherd, R. J. (1976). DNA viruses of higher plants. *Adv. Virus Res.* **20,** 305–39.

(1977). Cauliflower mosaic virus (DNA virus of higher plants), in *The Atlas of Insect and Plant Viruses,* ed. K. Maramorosch, pp. 159–66. Academic Press, New York.

(1979). DNA plant viruses. *Annu. Rev. Plant Physiol.* **30,** 405–23.

(1981). Caulimoviruses, in *Handbook of Plant Virus Infections and Comparative Diagnosis,* ed. E. Kurstak, pp. 847–78. Elsevier/North-Holland, Amsterdam.

Sieg, K., and Gronenborn, B. (1982). Introduction and propagation of foreign DNA in plants using cauliflower mosaic virus as a vector, in *Abstracts of N.A.T.O. Advanced Studies Institute/F.E.B.S. Advanced Course on Structure and Function of Plant Genomes,* Port Portese, Italy, p. 154.

Stanley, J. (1983). Infectivity of the cloned geminivirus genome requires sequences from both DNAs. *Nature (London)* **305,** 643–5.

Taliansky, M. E., Malyshenko, S. I., Pshennikova, E. S., and Atabekov, J. G. (1982*a*). Plant virus-specific transport function. II. A factor controlling virus host range. *Virology* **122,** 327–31.

Taliansky, M. E., Malyshenko, S. I., Pshennikova, E. S., Kaplan, I. B., Ulanova, E. F., and Atabekov, J. G. (1982*b*). Plant virus-specific transport function. I. Virus genetic control required for systemic spread. *Virology* **122,** 318–26.

Toh, H., Hayashida, H., and Mijota, T. (1983). Sequence homology between retroviral reverse transcriptase and putative polymerases of hepatitis B virus and cauliflower mosaic virus. *Nature (London)* **305,** 827–9.

Turner, D. S., and Covey, S. N. (1984). A putative primer for the replication of cauliflower mosaic virus by reverse transcription is virion-associated. *FEBS Lett.* **165,** 285–9.

van Vloten-Doting, L. (1983). Advantages of multipartite genomes of single-stranded RNA plant viruses in nature, for research and for genetic engineering. *Plant Mol. Biol. Rep.* **1,** 55–60.

Volovitch, M., Modjtakedi, N., Yot, P., and Brun, G. (1984). RNA-dependent DNA polymerase activity in cauliflower mosaic virus-infected plant leaves. *EMBO Journal* **3,** 309–14.

Walden, R. M., and Howell, S. H. (1982). Intergenomic recombination events among pairs of defective cauliflower mosaic virus genomes in plants. *J. Mol. Appl. Genet.* **1,** 447–56.

(1983). Uncut recombinant plasmids bearing nested cauliflower mosaic virus genomes infect plants by intragenomic recombination. *Plant Mol. Biol.* **2,** 27–31.

Woolston, C. J., Covey, S. N., Penswick, J. R., and Davies, J. W. (1983). Aphid transmission and a polypeptide are specified by a defined region of the cauliflower mosaic virus genome. *Gene* **23,** 15–23.

Xiong, C., Muller, S., Lebeurier, G., and Hirth, L. (1982). Identification by immunoprecipitation of cauliflower mosaic virus *in vitro* major translation product with a specific serum against viroplasm protein. *EMBO Journal* **1,** 971–6.

Zimmern, D., and Hunter, T. (1983). Point mutation in the 30-K open reading frame of TMV implicated in temperature-sensitive assembly and local lesion spreading of mutant Ni 2519. *EMBO Journal* **2,** 1893.

7 Some possibilities for genetic engineering of Rubisco

H. J. NEWBURY

For hundreds of years humans have endeavored to increase the productivity of crop plants. They met with great success initially by selecting high-yielding lines, and more recently by applying organized breeding programs to enlarge and improve the population from which to select. With the advent of "genetic engineering" it is only natural to find that crop productivity is proposed as a subject for the new technology, and it may be assumed in some quarters that results from such projects will soon be forthcoming. However, the experienced plant scientist will recognize that even allowing for the enormous advances that continue in molecular biology, rapid results from genetic improvement of crop productivity are not likely. The simplicity of the aim and the enormous potential benefits should not blind us to the complexity of such a research program. Neither should we assume that the explosion of new techniques pioneered using microorganisms can immediately be employed for improvement of crop plants. The tremendous progress achieved in understanding and manipulating microbial genes has been underpinned by the accumulated wisdom of decades of research into the physiology, biochemistry, and genetics of a relatively small number of bacteria and viruses. Such detailed knowledge is not available for plants, largely because they have not been exposed to such intensive investigation. Also, being eukaryotic, multicellular, and highly differentiated, they are clearly more complex in their molecular organization and, especially, in the regulation of their metabolism. Our understanding of plant molecular biology is patchy; one important reason for a discussion of the possibilities of improving crop productivity using genetic engineering is that the exercise does serve to highlight some areas of intense ignorance. This introduction may seem unnecessarily negative, especially because much of the remainder of the chapter is devoted to one particular possibility for improving crop productivity using recombinant DNA technology. The term *possibility* is used deliberately; such projects will depend not only on continued investigations into methods of plant gene isolation, manipulation, and transfer but also on more basic research into a number of aspects of the plant sciences about which we have suddenly discovered we know very little.

Before any process can be considered for genetic engineering, it is a great advantage if the gene(s) involved are identified, at least by ascribing to them a name and putative function. In a consideration of genetic improvement of crop productivity, the most important process would seem to be photosynthetic carbon fixation. The aim of our enterprise would be to increase the efficiency of fixation so that we achieve higher yields of harvested crops or, alternatively, the same yield in a shorter time. All of the plant materials that we harvest are composed of fixed carbon, although we use these materials in a wide variety of industries concerned with food or drink, timber and paper, fibers and textiles, rubber, drugs, and so forth. Although we would expect an increase in photosynthetic efficiency to result in increased yields of plant-derived commodities, we must recognize that other methods may also prove useful. In most cases the material valuable to humans may consist of only certain plant organs or tissues, the remainder of the plant being discarded. It may be possible to alter the balance of assimilate partitioning within such plants so that more photosynthate is directed toward the harvested product. We have some information about the role of hormones in directing nutrient transport in plants, and in some specific examples, such as grain filling in wheat, we possess valuable data concerning the movements of assimilates around the plant. However, a general ignorance of both the identities and products of the genes controlling this and other aspects of plant development severely inhibits a consideration of assimilate partitioning genes as subjects for genetic engineering at present.

We do know enough about photosynthesis, however, to make some initial suggestions as to how it may be improved at a molecular level. We are in no position to propose in any meaningful way improvements by genetic engineering that might be made in the efficiency of light adsorption or of leaf CO_2 uptake; these processes are controlled by large numbers of genes, and with few exceptions we know little about the genes involved. We are on safer ground when we consider enzymes of the dark reactions of photosynthesis. It is well known that in conditions of adequate illumination and moderate temperature and water supply, photosynthesis can often be limited by the rate of fixation of carbon dioxide. Under these circumstances an artificial increase in the CO_2 concentration of the environment will promote light-driven CO_2 fixation. This method is impractical (and uneconomic) in most commercial situations, and a clear alternative is to produce plants that use the low concentrations of carbon dioxide naturally available in a more efficient manner. The enzyme that recognizes CO_2 as one of its substrates and fixes it into a solid (dissolved) compound is ribulose bisphosphate carboxylase (E.C. 4.1.1.1.39), also known as RubPC and Rubisco. Rubisco, although confined to chloroplasts, can represent up to 50% of the soluble protein of a leaf and has therefore been proposed as the most abundant protein on earth. Its very abundance has suggested to some authorities that it is not a very efficient enzyme (i.e., it does not have a high specific activity); it is the hope that we can increase its efficiency, linked to the comparatively large amount of information we have on this particular plant protein, that has attracted biotechnological interest to this area.

One could argue that because plants have been under intensive selective pressure to increase their rates of carbon fixation for millions of years, we may have little chance of improving on the productivity rate of a complex and closely integrated photosynthetic system. However, this probably ascribes to the stumblings of the (Darwinian) evolutionary process an efficiency that it does not deserve and also ignores the tremendous strides humans have already taken in crop improvement using classic genetic methods. On the other hand, photosynthesis is such a fundamentally important process to the individual plant that it must have evolved in a very conservative manner; the implication that deviation from the norm has quickly been "weeded out" does not augur particularly well for the genetic engineer. However, humans do have some advantages over nature in this situation. For example, we may be able to induce specific changes at two or more genetic loci effectively simultaneously. Such multiple complementary modifications are unlikely in nature, even allowing for the enormous time span in which such events could have occurred. Also, although it seems likely in this case, we cannot automatically assume that a character that we consider to be improved would necessarily be advantageous in nature. We change characters for our own benefit, not to benefit the plants. We are able to partially relieve the pressures of natural selection from the plant in the agricultural situation, and to an even greater extent during the course of our genetic manipulations, so that intermediate forms that would be nonviable in nature may survive in the laboratory.

During photosynthesis, Rubisco catalyzes the chemical fusion of CO_2 and ribulose bisphosphate (RubP) to form phosphoglyceric acid (PGA). Because this is a three-carbon compound, those plants in which it represents the first product of photosynthesis are called C_3 plants (although in fact all plants utilize this enzyme in PGA production at some stage of photosynthesis). One approach to the improvement of net CO_2 fixation rates has the simple aim of increasing the rate of this carboxylation reaction. However, it is an alternative idea that will form the theme of the remainder of this chapter. Rubisco has been shown to catalyze either the carboxylation of RubP or its oxygenation (Bowes et al. 1971; Lorimer 1981). The reaction between RubP and molecular oxygen produces PGA and 2-phosphoglycollate; the latter molecule represents the first stage of a series of reactions involving chloroplasts, peroxisomes, and mitochondria that result in the loss of CO_2 (Ellis 1979; Jensen and Bahr 1977; Somerville and Ogren 1982). Because of its dependence on light and the fact that it involves a net uptake of O_2 and release of CO_2, this process has been dubbed photorespiration. Although it has been suggested that photorespiration may fulfill a useful function within the plant (Heber and Krause 1980; Woolhouse 1978), it is usually regarded as a wasteful process, because it results in the loss of fixed carbon and diverts Rubisco from its role in CO_2 fixation. This is not a trivial loss of efficiency, because net CO_2 fixation can be increased by 50% when photorespiration is suppressed (Ogren 1976). Some plants, such as C_4 plants, have overcome this problem. In these, the CO_2 is initially fixed into a four-carbon organic acid (hence C_4 plants) (Coombs et al. 1976). Energy is expended in pumping this into deeper-seated regions of the leaf, where CO_2 is released for refixation by Rubisco. The CO_2 concentration is increased in the deep-seated tissue, and CO_2 outcompetes O_2 for the attention of the enzyme. C_4 plants thus have an advantage over C_3 plants at the molecular level in their net CO_2 fixation rates; this is often apparent in field comparisons of crop productivities, although environmental factors have important effects on such results. The crop plants of many temperate parts of the world are exclusively of the C_3 type, so that photorespiratory loss is an important factor in their carbon balance.

It may be that genetic engineering techniques can be successfully employed in helping to solve this problem. If the oxygenase function of this enzyme could be reduced without adversely affecting its carboxylase activity, one might expect significant increases in crop productivity to result even though other constraints on photosynthesis would remain the same. The remainder of this chapter consists of a detailed appraisal of the possibilities of altering the oxygenase-carboxylase ratio of Rubisco. This is not meant to imply that this is the only possible project for improving the net CO_2 fixation rate of crop plants, or even that it is the most likely to succeed. It has been selected because it has a well-defined aim and because a step-by-step discussion of the suggested procedures seems to illustrate some of the achievements that have been made in molecular biology, as well as indicating certain areas where further knowledge is desperately required.

Ogren (1976) has reported that in soybeans at 25°C the ratio of O_2 to CO_2 fixed under standard atmospheric conditions is about 1:4. He has calculated that if this ratio were reduced to 1:8, then net photosynthesis would increase by about 20%; a complete loss of O_2 fixation would result in an increase of 43% in net photosynthesis. Experiments designed to discover whether or not oxygenase and carboxylase activities of Rubisco can be made to vary independent of one another have yielded mixed results. Variation was observed during a survey of Rubisco enzymes in photosynthetically active species (Jordan and Ogren 1981, 1983). However, Chollet and Anderson (1976) failed to find any differential regulation of the two functions of tobacco Rubisco in *in vitro* experiments using different temperature pretreatments and chloroplast metabolites. In other cases it has been reported that both temperature (Laing et al. 1974) and pH (Bahr and Jensen 1974) produce different responses for the carboxylation and oxygenation functions. Kung and Marsho (1976) and Nelson and Surzycki (1976) have reported nuclear mutations (inherited in a Mendelian fashion) that have led to alterations in the ratio of oxygenase/carboxylase activities in Rubisco. However, these latter results have been criticized on technical grounds by Akazawa (1979). Some workers have suggested that O_2 fixation is an inevitable consequence of the carboxylation reaction of this enzyme, so that photorespiration is unavoidable (Lorimer and Andrews 1973, 1978).

Taken together, these results do not provide a particularly clear picture, but it may be that variations in these two enzymic functions do occur in nature. Let us turn then to essentially practical considerations. Do we know enough about the molecular biology of Rubisco to enable us to specifically induce a heritable change in the oxygenase/carboxylase ratio in a crop plant without producing other adverse effects? Here we are fortunate. The abundance of the enzyme means that it is relatively easy to purify. In some situations the concentration of Rubisco is sufficiently high to allow enzyme crystallization (Kung et al. 1980). These practical advantages, along with the important role of Rubisco in photosynthesis, have led to a large literature on this protein. In this particular case, some of the underpinning molecular biology studies essential to effective crop improvement by genetic engineering have already been carried out. We know that in higher plants Rubisco is a very large enzyme (around 550,000 daltons) and consists of eight large subunits (LSUs; 52,000–60,000 daltons) and eight small subunits (SSUs; 12,000–18,000 daltons) (Ellis 1979) (Figure 7.1).

Where in this enzyme are the active sites for carboxylation and oxygenation of RubP located? The information on this subject has been admirably reviewed by Akazawa (1979). He lists four lines of evidence strongly implicating the LSU as the site of the carboxylase function, and these will now be briefly described. First the enzyme can be induced to dissociate into subunits and reassociate to form the original enzyme (Nishimura and Akazawa 1974*a*, 1974*b*). If an octameric form of LSU is produced, it retains partial carboxylase activity (specific activity about 20%) in the complete absence of SSUs. Second, it has been found that some bacterial enzymes contain only LSUs (two in the case of *Rhodospirillum rubrum* and eight for *Thiobacillus intermedius*), again demonstrating that activity is not dependent on the presence of SSUs. Third, various reagents have been used to demonstrate that certain -SH groups and lysyl residues that are essential for the function of the spinach and *Chromatium* enzymes are located on the LSU. This line of research assumes added significance later in the chapter. Fourth, antisera have been raised separately against purified LSUs and SSUs from spinach Rubisco. Addition of antiserum specific to the LSU inhibits the carobxylase activity of the whole enzyme, whereas addition of antiserum specific to the SSU is totally ineffective (Nishimura and Akazawa 1974*c*; Takebe and Akazawa 1975). Enzyme "immunotitration" frequently results in inhibition of activity, usually because the antibodies block the active site. The experiment with Rubisco antisera also indicates the lack of

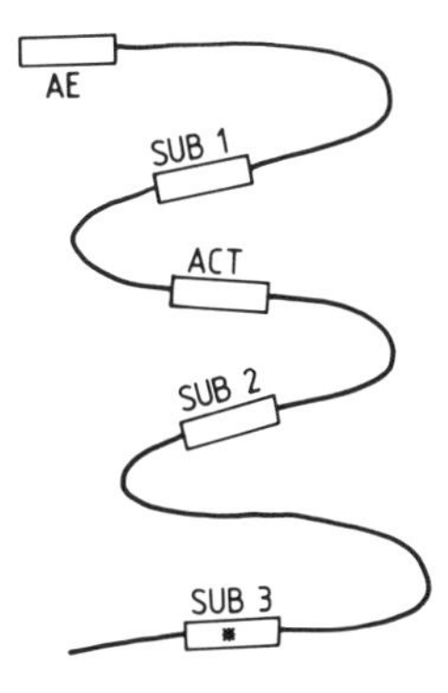

Figure 7.1. Diagrammatic representation of the structure of the LSU of Rubisco. AE is the amino-terminal extension or leader sequence consisting of 14 amino acids in tobacco. This peptide is not present in the mature LSU. SUB 1, 2, and 3 are short peptides that have been shown to be involved in the binding of RubP or substrate CO_2 and that contain essential thiol or amino groups. ACT is a short peptide involved in the binding of an activator CO_2 molecule. It is not yet possible to indicate the portions of the subunit that are involved with the binding of SSUs or other LSUs. The asterisk indicates the site in the subunit at which variation in amino acid sequences between species has been tentatively related to altered activity (see Figure 7.2).

homology between the two subunits. We can add a fifth and much less direct piece of evidence consistent with the information presented earlier. One might expect the subunit containing the active site to have evolved in a much more conservative manner than its partner, because of the extreme selective pressure on the carboxylase activity of Rubisco. Comparative studies of a large range of angiosperms indicate that the LSU is much less variable in the 40 amino acids of its *N*-terminal sequence than is the SSU (Martin and Jennings 1983).

Together, these data indicate that the LSU contains the active site for the carboxylase function. Kinetic and immunochemical studies have demonstrated that the oxygenase function is also located here (Nishimura and Akazawa 1974*c*; Takebe and Akazawa 1975). Perhaps, because both reactions involve the binding of RubP, we should expect the carboxylase and oxygenase functions to occur at the same site on the LSU; this is certainly what the evidence suggests (Akazawa 1979; Lorimer 1981). Kinetic analyses of competitive inhibition of the carboxylase reaction by O_2 and of the oxygenase reaction by CO_2 using purified soybean enzyme have demonstrated that K_m

(CO_2) (carboxylase) = K_i (CO_2) (oxygenase) and that K_m (O_2) (oxygenase) = K_i (O_2) (carboxylase) (Laing et al. 1974). CO_2 and O_2 thus act in a mutually competitive manner at what is generally assumed to be the same site.

Information is also available concerning the genes encoding the two types of Rubisco subunits. It has now been clearly demonstrated that the two subunit types are encoded in different compartments of the cell. The gene for the SSU forms part of the nuclear genome, and the LSU gene resides in the plastid. The mechanisms that lead to the production of active enzyme have proved of great interest to plant molecular biologists and are the subject of continuing research. The fact that the LSU gene is part of the plastid genome also has great significance in any project that aims to alter the structure of this protein. Evidence that the Rubisco LSU gene occurs in the plastid has been accumulated from investigations showing its maternal inheritance in breeding experiments (Chan and Wildman 1972), from demonstrations that the LSU is synthesized by chloroplasts in vitro (Blair and Ellis 1973), and from cell-free protein synthesis systems supplied with chloroplast RNA (Hartley et al. 1975; Sagher et al. 1976). Strictly speaking, the experiments showing that synthesis of the LSU occurs in chloroplasts do not prove that the gene is to be found in these organelles. There is always the possibility, albeit unlikely, that the messenger RNA is synthesized within the nucleus and subsequently moves into the plastid for translation. Direct evidence for the location of the LSU gene was obtained from some of the first experiments using recombinant DNA technology with plant genetic material. Two of the earliest publications on this subject showed that the LSU of Rubisco was encoded in the chloroplast DNA of *Chlamydomonas* (Gelvin et al. 1977) and maize (Coen et al. 1977). The techniques used in both studies for handling, cutting, and hybridizing DNA were pioneered using microorganisms and were exploited in what are now regarded as quite simple experiments. In both cases the basic aim of the experiment was achieved without recourse to gene cloning, but this method greatly enlarges the scope for experimentation. DNA cloning is merely the production of large numbers of copies of specific pieces of DNA (DNA sequences) that are all exactly identical. There may be certain constraints on the sizes of DNA sequences that can be cloned, but the first stage of a generalized protocol involves integration of the sequence of interest into a specially selected "carrier" piece of DNA known as a "cloning vehicle" or "vector." Such vectors are natural components of some microorganisms and have been pressed into service by genetic engineers. To be useful in this respect they must be maintained and replicated during bacterial growth and division. For practical reasons it is also necessary for the vectors to contain genes that produce a selectable phenotype. Cloning vehicles include a large range of plasmids and some viruses; in many cases the naturally occurring entity has been manipulated by researchers in order to produce a more convenient or effective vector. Following insertion of the foreign sequence, the cloning vehicle is introduced into an appropriate microorganism, and it is the rapid multiplication of cells that contain the foreign DNA that leads to its availability in large quantities. There are currently severe limitations on the range of microorganisms that can be used in such a procedure, and the fact that *Escherichia coli* is involved in the vast majority of cloning experiments reflects the accumulation of a wealth of information on the physiology and biochemistry of this organism. Other bacteria are, however, currently under investigation, with a view to exploitation as cloning organisms; yeasts provide further alternatives that may be especially useful because of their eukaryotic nature.

It was not coincidental that some of the first effective experiments on plants using the new DNA manipulation technology were carried out on chloroplast DNA (ct-DNA). There are great advantages to using organelle rather than nuclear DNA: ct-DNA exists in vivo in the form of double-stranded closed circles; there is some variation between species, but the size is usually around 1×10^8 daltons (Bedbrook and Kolodner 1979). Many identical copies of these circles exist in each plastid, and each circle is known as a "plastome" (Herrmann and Possingham 1980). Although care must be taken to avoid contamination of ct-DNA with nuclear DNA, such samples have a great advantage over the isolated nuclear genome because of the relative simplicity of the DNA sequences represented. Higher plant nuclear genomes are variable in size, but in addition to containing a very large number of genes, all contain a lot of nontranscribed DNA. Much of this consists of sequences repeated thousands or even millions of times. Identification and isolation of genes that may be represented only once per haploid nucleus in an enormous background population of other sequences present far greater technical problems than are encountered during investigations of the less abundant but simpler plastid genome.

Coen and associates (1977) isolated ct-DNA from maize leaves and wished to discover whether or not a gene for the LSU of Rubisco existed somewhere on the closed circle. They therefore employed a linked transcription-translation system that used *E. coli* RNA polymerase to synthesize mRNAs encoded by their DNA sample along with a rabbit reticulocyte lysate translation system (including radioactive amino acid) in order to produce the polypeptides encoded by the mRNAs. In this way they were able to show that maize ct-DNA included a sequence coding for a polypeptide product that possessed the same electrophoretic mobility as Rubisco LSU isolated from maize chloroplasts. However, the amount of this polypeptide produced in the linked transcription-translation system was low, presumably because there was a dilution effect caused by the presence of many other genes. It was at this point that cloning technology was employed. The object was to purify large amounts of a small portion of the ct-DNA that contained the LSU gene.

The restriction enzyme Bam-H1 (isolated from *Bacillus amyloliquefaciens,* strain H) was used to degrade the ct-DNA, and this led to the production of 19 restriction fragments of different sizes. These fragments were cloned into the plasmid RSF 1030 without any intermediate selection stage. RSF denotes "resistance specifying factor," and the plasmid does confer resistance to ampicillin. It has a copy number of 20–40 in *E. coli* and contains a single Bam-H1 restriction site. The plasmid was linearized using this endonuclease, and ct-DNA fragments, produced by the action of the same enzyme, were integrated using the "cohesive tail" strategy, followed by ligation with T4 DNA ligase. The DNA sample was then used to transform a culture of *E. coli* strain HB101. This strain is deficient in the normal restriction enzyme system; this means that foreign DNA entering such cells will not be degraded as rapidly as would otherwise be the case. Another feature of HB101 is the absence of plasmids. This avoids difficulties in differentiating between introduced and native nonchromosomal DNA and potential problems of plasmid incompatibility. Cell colonies transformed by the plasmid were selected using their ampicillin resistance. Those colonies containing recombinant plasmid were identified by plasmid sizing using agarose-gel electrophoresis; plasmids containing inserted DNA are obviously larger than nonrecombinant plasmids.

The ct-DNA fragments had been numbered, so that fragrant 9 was the ninth largest restriction product. One clone, pZmc37 (i.e., plasmid *Zea mays*, clone 37), was shown to contain Bam-H1 fragment 9 of maize ct-DNA. This was achieved by demonstrating that ^{32}P-labeled copy RNA (cRNA) transcribed form pZmc37 using the *E. coli* enzyme RNA polymerase would hybridize to this ct-DNA restriction fragment. Using the linked transcription-translation system described earlier, it was then shown that this fragment directed the production of a polypeptide that both comigrated with authentic LSU in polyacrylamide gels and was recognized by an antiserum raised against authentic Rubisco LSU. The real achievement of cloning in this case was to create a much more convenient and amenable source of the Rubisco LSU gene than is provided by leaf material. A subsample representing about 5% of the chloroplast genome was obtained in a pure form in very large quantities. In this situation, although extremely useful, DNA cloning was not essential, because evidence for the location of the LSU gene was obtained at first using DNA sequences extracted directly from the plant. However, when we consider the requirements for plant DNA sequencing, it is clear that cloning is a prerequisite. Large quantities of highly purified DNA ssquences are necessary. A heritable change in the oxygenase/ carboxylase ratio of the LSU of Rubisco can be induced in a controlled manner only if the structure (sequence) of the LSU gene is elucidated, if the regions of the coding sequence corresponding to the active site of the LSU are recognized, if these can be changed without adversely affecting the activity of the polypeptide, and if the altered gene can be reintroduced into the plant. As we shall see, some of the steps have already been performed, whereas the others vary in their apparent feasibility.

Following the initial work in which ct-DNA fragments containing the LSU gene were cloned, there have been a number of further developments. The chloroplast genome has been mapped for many more genes than the one of interest here, and interest in the plastome has increased (Bedbrook and Kolodner 1979; Bohnert et al. 1982). Also, the sequence of the LSU gene in some plants has been established. The first report of this kind concerned again the maize LSU gene (McIntosh et al. 1980). Once again a fragment of ct-DNA containing the gene was cloned in plasmid RSF 1030. It is not possible using present methods to obtain the complete sequence of a chain of more than a few hundred deoxyribonucleotides. The ct-DNA insert in this case comprised 2.7 kilobase pairs. The normal procedure in such a case is to obtain short overlapping fragments of the large sequence by first mapping the sites at which a variety of restriction enzymes cleave the double-stranded DNA and then use selected enzymes and agarose-gel electrophoresis to isolate specific fragments. A choice of methods exists for determining the sequence within these subfragments (Gaastra and Oudega 1983), and in this case the method of Maxam and Gilbert (1980) was employed. In this way the complete sequence of the Rubisco LSU gene for maize was obtained, and this information has proved very useful for a number of reasons. First, using the genetic code, the amino acid sequence of the LSU could be deduced, and it was found to consist of 475 amino acid residues. At the time, the complete maize amino acid sequence was useful in aligning amino acid sequences totaling 244 residues known for fragments of the barley and spinach LSU polypeptides (Hartman et al. 1978; Poulsen et al. 1979). Since that original article dealing with the maize Rubisco LSU gene sequence (McIntosh et al. 1980) there have been reports of this gene sequence from other higher plants. Comparison of the LSU gene sequences determined for tobacco (Shinozaki and Sugiura 1982) and spinach (Zurawski et al. 1981) shows the very conservative mode of evolution of this polypeptide; this will be reinforced when the sequences for the same gene from photosynthetic organisms other than higher plants are considered later in this chapter.

Other points of interest include the absence of introns from any of the four LSU genes thus far sequenced. Introns are very common in plant nuclear genes, but not in the chloroplast genome, which shares many features in common with prokaryotes. The sequencing studies have included DNA on either side of the LSU coding sequence and have allowed recognition of some of the short regions of DNA that are involved in the control of transcription and translation. Some of these are reviewed by Bohnert and associates (1982). Briefly, they consist of untranscribed promoter sequences that are recognized by RNA polymerase and to which this enzyme binds; also, untranscribed terminator sequences exist that halt transcription following the reading of the coding sequence. Within a short transcribed region just before the coding sequence lies a ribosome binding site whose function is self-explanatory. This short sequence is complementary to one end of the RNA within the SSU of plastid ribosomes; plastid and bacterial ribosomes are similar enough to share this characteristic. Without a ribosome binding site, a plastid messenger RNA would not be translated. An understanding of the various sequences that control the expression of the LSU gene is necessary so that during any gene manipulations these elements can be retained.

Examination of the coding sequences of the four Rubisco LSU genes has revealed several features relating to the function of the polypeptide. First, all four reports remarked on a difference in the *N*-terminal end of the polypeptide as defined by direct amino acid sequencing by Edman degradation (e.g., Martin and Jennings 1983), and as predicted by the nucleotide sequence of the gene. It has been suggested that the polypeptide is synthesized with a *N*-terminal leader sequence of about 14 amino acid residues that is removed during a processing step before the formation of the multimeric enzyme. It may be that the leader sequence of the LSU holds it in an appropriate conformation until it is assembled into the holoprotein (Langridge 1981).

These pieces of information are important in a general way to the biotechnologist because they provide an insight into the sensitivity of various parts of the polypeptide to induced change. However, the main features of interest relating to our original project concern those regions of the polypeptide (and coding sequence) that form the sites at which ribulose bisphosphate and carbon dioxide/oxygen bind. In order to define these portions of the complete sequences that have been obtained, it is necessary briefly to turn aside from studies involving DNA manipulation and consider experiments involving the localization of active sites on enzymes. At an earlier stage in this chapter, a review of the evidence that the active site of Rubisco resides on the LSU included the information that certain -SH groups and lysyl residues that are essential for the function of the enzyme are located on this polypeptide. An expansion of this line of inquiry at this stage proves to be extremely profitable. Initial experiments indicated that iodoacetamide, by binding to -SH groups, completely inhibited the activity of Rubisco. Use of ^{14}C-labeled iodoacetamide helped in the localization of some of these thiol groups on the LSU (Sugiyama and Akazawa 1970). Similarly, ^{14}C-labeled potassium cyanate was useful in demonstrating that essential lysyl residues (free -NH_2 groups) were present in the LSU (Chollet and Anderson 1978). In this kind of experiment it is useful to add ribulose bisphosphate to some of the treatments, because binding of the substrate protects reactive groups at the active site, whereas those elsewhere on the polypeptide are available for modification. Thiol and free

-NH_2 groups at the active site can then be distinguished by comparison of results obtained in the presence and absence of substrate. Other investigations have involved the use of a ribulose bisphosphate analogue (3-bromo-1,4-dihydroxy-2-butanone-1,4-bisphosphate) that binds irreversibly at the active site. Following the reaction between this compound and the enzyme, tryptic digestion was performed, and the peptides containing the bound analogue were isolated and subjected to amino acid sequencing using Edman degradation (Stringer and Hartman 1978). The two sequenced peptides both contained an internal lysyl residue and were 13 and 20 residues long; they were reported to be the first sequences determined for regions encompassing a portion of the active site of Rubisco. Further investigations involving the sequencing of selected tryptic peptides following chemical modification of reactive groups revealed one peptide that contained both essential -SH and -NH_2 groups (Schloss et al. 1978); furthermore, this was identical with one of the tryptic peptides previously sequenced following reaction of the enzyme with an analogue of RubP (Stringer and Hartman 1978). A second sequenced peptide containing an essential cysteine residue was quite different, so that the two reports together provide the sequences of three regions of the LSU, of 13, 13, and 20 residues, that are involved in the reaction between CO_2 and ribulose bisphosphate. There is also a fourth site that has been identified using similar technology, and this is concerned with the binding of CO_2, which acts not as a substrate but as an activator molecule. Both the carboxylase and oxygenase activities require activation by Mg^{2+} and CO_2 before binding of ribulose bisphosphate occurs (Badger and Lorimer 1976). The molecule of CO_2 involved in the activation reaction is distinct from that which becomes fixed during catalysis (Lorimer 1979; Miziorko 1979). Following reaction of the activator CO_2 with the free -NH_2 group of a lysyl residue, a carbamate is formed; this was chemically modified to form a stable product that again could be identified within a peptide produced after tryptic digestion. A 17-residue sequence was established for this peptide that is associated with the CO_2 activation site.

It has proved a very simple matter to recognize the sequences of all four short peptides defined earlier within the three complete amino acid sequences predicted following the elucidation of the higher plant LSU primary gene structures (Lorimer 1981). The three sites associated with the binding of ribulose bisphosphate and the substrate CO_2 molecule are quite widely separated, with the essential -SH and -NH_2 groups being associated with amino acid residues 173, 175, 339, and 459. Presumably the numbered amino acid residues are in close proximity when the subunit takes up its native conformation.

At the amino acid level the four short sequences hardly vary at all between the three higher plant examples thus far available. Recently, however, the LSU gene sequences for three cyanobacteria and a green alga have also been determined (Curtis and Haselkorn 1983; Dron et al. 1982; Shinozaki et al. 1983). There is amazing homology among all seven sequences, even though they are taxonomically the most widely separated examples of photosynthetic organisms that could be compared. The four short sequences known to be associated directly with enzyme activity are, as one might expect, among the most highly conserved regions of the polypeptide. Nevertheless, some variation does occur. Shinozaki and associates (1983) pointed out that a difference in the LSU gene of the cyanobacterium *Anacystis nidulans* may have some significance. The cysteine residue identified by Schloss and associates (1978) as being required for activity and which is alkylated in the absence but not the presence of ribulose bisphosphate is found at position 459 in the complete amino acid sequence for the three higher plants. However, in *Anacystis* the equivalent position (not the same number residue because of a slight difference in overall polypeptide lengths) is occupied by a leucine residue. These authors suggested that this change may correlate with the reports of Jordan and Ogren (1981, 1983) that the ratio of oxygenation to carboxylation catalyzed by Rubisco differs in cyanobacterial and higher plant enzymes. Using kinetic analyses of simultaneously performed carboxylase and oxygenase assays on purified Rubisco preparations, those workers reported (taking their two studies together) that the oxygenase/carboxylase ratio was much higher in two cyanobacteria (neither of which coincided with the species that have had their LSU genes sequenced) than it was in 16 higher plants (including spinach, tobacco, and maize).

This possible correlation between sequence and function obviously requires further supporting evidence before it can be accepted. The report of Curtis and Haselkorn (1983) describes the sequencing of the LSU gene of *Anabaena* (strain 7120) and also provides the gene sequence of another cyanobacterium, *Synechococcus*,

	451								459						466	
A	TRP	SER	PRO	GLU	LEU	ALA	ALA	ALA	CYS	GLU	VAL	TRP	LYS	GLU	ILE	VAL
B	TRP	SER	PRO	GLU	LEU	ALA	ALA	ALA	CYS	GLU	ILE	TRP	LYS	GLU	ILE	LYS
C	TRP	SER	PRO	GLU	LEU	ALA	ALA	ALA	CYS	GLU	VAL	TRP	LYS	GLU	ILE	LYS
D	TRP	SER	PRO	GLU	LEU	ALA	ALA	ALA	CYS	GLU	VAL	TRP	LYS	GLU	ILE	LYS
E	TRP	SER	PRO	GLU	LEU	ALA	ALA	ALA	LEU	ASP	LEU	TRP	LYS	GLU	ILE	LYS
F	TRP	SER	PRO	GLU	LEU	ALA	ALA	ALA	LEU	ASP	LEU	TRP	LYS	GLU	ILE	LYS
G	TRP	SER	PRO	GLU	LEU	ALA	VAL	ALA	CYS	GLU	LEU	TRP	LYS	GLU	ILE	LYS

Figure 7.2. The amino acid sequence for one of the three short peptides that are involved in the binding of substrates to the LSU of Rubisco is known to vary between certain photosynthetic organisms. A, tobacco; B, maize; C, spinach; D, *Chlamydomonas;* E, *Anacystis;* F, *Synechoccocus;* G, *Anabaena.* The alternative of cysteine or leucine at position 459 (numbering uses tobacco sequence) is particularly interesting, although there are also changes at other positions.

which was determined by Reichelt and Delaney. Figure 7.2 shows the amino acid sequences of the region of interest in the LSU polypeptides of the seven species thus far examined. The change from cysteine to leucine is also apparent in *Synechococcus* at the appropriate position and is coupled with an alteration of a neighboring residue from glutamic to aspartic acid in both cases. In the third cyanobacterium, *Anabaena,* however, these alterations do not occur, but the middle residue of a string of three alanines is replaced by valine. A comparison of the oxygenase/carboxylase ratios for Rubisco purified from these three species will be necessary to define more clearly the effect of these changes. Jordan and Ogren (1981) also reported that the oxygenase/carboxylase ratio of *Chlamydomonas* was higher than that of higher plants, although not as high as that of cyanobacteria. The reason for this is not apparent in Figure 7.2, because in the region shown this green alga has the same sequence as spinach. To summarize the conclusions, establishment of the primary sequences of the Rubisco LSU for a few species has provided what might be a molecular basis for differences in activity. However, we require more sequences and more kinetic analyses in order to determine what changes are correlated with what kinetic characteristics. These data do, however, lend some support to the idea that the oxygenase/carboxylase ratio can be altered by manipulating the LSU gene to produce changes in amino acid residues at specific positions.

Let us consider an experiment in which we intend to change the cysteine at position 459 of a higher plant to a leucine residue. This would be achieved by altering the TGT codon found in tobacco or spinach or the TGC codon of maize, all of which code for cysteine, to one of the six possible codons for leucine. The proposed experiment is only a model, because replacement of cysteine with leucine would increase the relative rate of photorespiration in our plant if our previous assumptions are correct. However, as an academic exercise it may serve to highlight as important point. It may not be a good idea to choose one of the six possible leucine codons at random, because it has been found that there is considerable variability in their usage in some chloroplast proteins (Bohnert et al. 1982). The codons TTA, TTG, CTT, and CTA are used with reasonable frequency to encode leucine, but the other two possibilities, CTC and CTG, are used very rarely. This probably relates to the availability of the six tRNA molecules that can accept leucine for use in the translation system; they may not all be present within the chloroplast in sufficient quantities. Naturally we would not wish to manipulate our gene in such a way that the rate of synthesis of its product was diminished.

How would we achieve an alteration of a base in order to change the encoded amino acid? Such site-specific mutagenesis can be achieved, for example, by the use of nucleotide analogues (Lathe et al. 1983). For this procedure one requires a relatively short fragment of single-stranded DNA containing the "target" base, which can be used as a template for the formation of a complementary strand using *E. coli* DNA polymerase. However, synthesis of this second strand would be allowed to occur only in short controlled steps; at each stage only a short additional complementary sequence would be allowed to form by omitting one or more of the four deoxyribonucleotides that are required for complete synthesis. The exact protocol is designed with the aid of the sequence of the strand to be copied, and the objective is to arrest synthesis at the position immediately prior to the target base. At this stage a nucleotide analogue would be incorporated opposite the target base, and then the remainder of the second strand would be produced following the addition of a full set of deoxyribonucleotides. The double-stranded DNA sequence could then be relegated into its appropriate position with the LSU gene. The technique does not actually alter the target base, but by introducing an analogue as its base pairing partner it can produce a change at this position during subsequent DNA replication. For example, N^4-hydroxy dCTP exists in two tautomeric forms; the amino form pairs with guanine, and the imino form pairs with adenine. Although the analogue may be incorporated into DNA as one form, it may act as its alternative tautomeric form during subsequent replication, so that a different nucleotide is incorporated in the target position.

More recently, methods of site-directed mutagenesis have been developed that involve chemical synthesis of a short single-stranded region of DNA (oligonucleotide) that spans the region of interest within a gene and contains the base alteration of interest. In spite of a slight mismatch, this can then be used to prime the synthesis of a second strand of DNA using the original gene sequence (or its noncoding complementary copy) as a template while it is present as an insert in the single-stranded vector M13. In this way the gene can be resynthesized containing a single base change.

The techniques of site-directed mutagenesis thus provide a means of altering the structure of the Rubisco LSU in one of the four regions involved directly in catalysis. However, it would be very useful to have a rapid method for assessing the effects of these alterations before attempts are made to reintroduce the manipulated gene back into a plant. It has been shown that the oxygenase/carboxylase ratio can be calculated using isolated Rubisco (Jordan and Ogren 1981, 1983), so that what is required is (1) a means of expressing the manipulated LSU gene, (2) a source of purified SSU, and (3) a method for the assembly of a mixture of the subunits to produce active holoenzyme. With regard to the first point, much progress has been made. It has been shown that the gene for the LSU of Rubisco can be not just cloned but expressed in *E. coli* (Gatenby et al. 1981). Fragments of DNA produced by restriction endonuclease digestion of the wheat and maize chloroplast genome were integrated into pBR322. This is a plasmid that is commonly used because it has been designed to offer the molecular biologist a number of advantages (Old and Primrose 1980). It contains genes conferring resistance to ampicillin and tetracycline; by cloning a foreign sequence into a restriction site within one of these genes, the latter will be inactivated (this is termed insertional inactivation). In this case the fragments were cloned (independently) into the single Bam-H1 site of pBR322 that lies within the tetracycline resistance gene. *E. coli* cells that had taken up the plasmid containing the foreign sequence were thus sensitive to tetracycline but resistant to ampicillin and could readily be selected. Although such transformed cells could be shown to contain the LSU gene, its expression would depend on the presence of sequences involved in the regulation of transcription and translation (i.e., a promoter and a ribosome binding site) on the fragment of ct-DNA transferred, and also the recognition of these controlling sequences by the *E. coli* protein synthesis machinery. If such genes are expressed at low levels, there can also be technical problems in the detection of gene products. To minimize this problem, a minicell-producing strain of *E. coli* was used (Dougan et al. 1979). Because these minicells contain no chromosomal DNA, it is much easier to detect proteins encoded by plasmids. On incubating minicells containing recombinant plasmids with a labeled amino acid, a polypeptide of 52,000 to 53,000 daltons was observed among the translation products. This species was not present in minicells containing nonrecombinant plasmid, and its identity as the LSU of Rubisco was confirmed by a demonstration that it reacted with Rubisco antiserum. These results emphasize the prokaryotic nature of the chloroplast protein-synthesizing system and support the long-term aim of producing large quantities of LSU in a bacterial culture. However, the yields of LSU were very low in this experiment, and further investigations revealed that greater expression could be achieved if the LSU gene could be linked to a promoter more powerful (i.e., more efficient in *E. coli*) than its own. Phage lambda represents an alternative vector to the plasmid systems that have been described thus far, and it was chosen here because its promoters (P_L and P_R) could be used to enhance expression of the LSU gene. Lambda genes are expressed in a regulated order during the lytic infection cycle in *E. coli,* and by integrating the LSU gene into a Bam-H1 site within the lambda DNA, it could be observed that LSU accumulated rapidly during the first 10 min following infection of the bacterium by the engineered virus. This was consistent with regulation by P_L, and the chloroplast protein accounted for 0.5–1% of total protein synthesis at this stage (Gatenby et al. 1981). However, after that time, LSU synthesis was decreased, because during natural lambda infection the production of protein regulated by P_L is later inhibited by one of the lambda gene products. However, this inhibition could be removed (Gatenby and Castleton 1982), and under conditions of unmoderated transcription the LSU of Rubisco accumulated to levels representing 2% of the total cell protein. In a later study it was reported that levels of up to 60,000 LSU molecules per cell can be attained (Gatenby 1984).

It seems possible, therefore, that if the Rubisco LSU gene were subjected to site-directed mutagenesis, the altered gene product could be produced in reasonable quantities in *E. coli.* For evaluation of the effect of the change in LSU structure we would also require a source of SSUs. Possibly these could be extracted directly from plant material and mixed with a LSU preparation isolated from *E. coli.* However, a neater solution would be to introduce into the bacterium the genes for both the LSU and SSU of Rubisco and hope that both synthesis and assembly would occur within these cells. Such a proposal introduces new technical difficulties. Because of the greater problems associated with isolation of specific unique sequences from nuclear DNA as compared with ct-DNA, it is only recently that the gene for the SSU of Rubisco has been cloned and sequenced (e.g., Barry-Lowe et al. 1982). Prior to this, investigations had been carried out using DNA sequences that were complementary to the mRNA for the SSU (e.g., Bedbrook et al. 1980; Broglie et al. 1981). Apart from the difficulty of obtaining full-length DNA copies of SSU mRNA, comparison of these copy DNA sequences with the gene sequence highlights an added complication. As is common in the plant nucleus, the SSU gene contains introns. The machinery for the removal of introns does not exist in prokaryotes, so that if an SSU "gene" were to be transferred to *E. coli,* it would need to be intronless. In this respect, a full-length DNA copy of the mature mRNA would be a more appropriate "gene" for use in *E. coli.* We have seen that the ct-DNA promoter was not very efficient in *E. coli* in driving LSU synthesis; it is much less likely that the nuclear promoter for the SSU would be effective, and there is certainly no binding site for bacterial ribosomes available. Consequently, manipulation of regulatory sequences would be required before expression could be obtained in the prokaryotic system.

These problems cannot be ignored, but one can envisage ways in which they might be overcome. Further careful consideration has to be given to the posttranslational changes that occur in the subunits within the plant but that are unlikely to occur within a bacterium. It is well known that the SSU is synthesized as a precursor polypeptide that is considerably larger than the mature subunit, and it has been shown that a large leader sequence is removed before assembly of the holoenzyme in the chloroplast (Chua and Schmidt 1979). It appears that the leader sequence is important for recognition of the plastid membrane (Dobberstein et al. 1977; Highfield and Ellis 1978). The leader sequence is removed following cleavage by a metallo-endoprotease found in the chloroplast stroma (Ellis and Gatenby 1984). Naturally this would not occur within a bacterium, and it would seem wise to remove the coding sequence for the leader portion of the SSU from the "gene" introduced into *E. coli.* Because the leader sequence occurs at the amino terminus of the polypeptide (the end that is synthesized first), it should be possible to achieve this by removal of only the AUG codon that initiates translation. Translation could then be initiated at the beginning of the sequence encoding the mature SSU. Conveniently, the coding sequence for the mature SSU commences with a methionine codon in soybean (Berry-Lowe et al. 1982), pea (Takruri et al. 1981), spinach (Martin 1979), and *Chlamydomonas* (Schmidt et al. 1979); presumably this is not chance but an indication of the way in which the leader sequence has been added during evolution. If we assume that it will be possible to produce populations of both LSU and SSU of Rubisco in *E. coli,* we must next consider the possibilities of holoenzyme assembly within the bacterium. Although this process is not fully understood, there is some information concerning the assembly mechanism within chloroplasts (Ellis 1981; Ellis and Gatenby 1984). It has not been possible to reproducibly dissociate purified Rubisco into subunits and then reassociate them to produce active enzyme in vitro (in spite of the report described earlier that gave evidence for the location of the active site on the LSU; Nishimura and Akazawa, 1974*a*, 1974*b*). One problem here is that isolated LSUs are insoluble in aqueous media, although they are soluble immediately after synthesis in isolated intact chloroplasts (Blair and Ellis 1973). It seems that in the chloroplast the newly synthesized LSU is bound to a "large subunit binding protein" (Barraclough and Ellis 1980). There is some evidence that the binding protein is synthesized on cytoplasmic ribosomes and presumably therefore is nuclear-encoded (R. Ellis, unpublished data). Until the process of Rubisco assembly in chloroplasts is elucidated, clearly it will not be possible to re-create the mechanism in *E. coli.* To avoid some of these problems, some workers have preferred to work with Rubisco from *Rhodospirillum rubrum.* Here the enzyme consists of only two LSUs, so that there are no problems with SSU synthesis, and difficulties of assembly may not be so prominent. Expression of the *R. rubrum* Rubisco gene in *E. coli* has been achieved using the lac operon promoter. Taken overall, it is apparent that some important progress has been made in Rubisco gene expression in *E. coli,* and we may expect further, possibly rapid, developments in this field (Gatenby 1983); however, there are still processes, like the assembly of Rubisco subunits, about which we are partly ignorant. We must await the results of further basic research in this area before this section of the overall Rubisco project can be brought to a satisfactory conclusion.

Whether or not we are able to evaluate changes in LSU structure following enzyme production in *E. coli,* a proposal with a stated aim of crop improvement eventually requires the introduction of an "improved" gene into a crop plant. Significant advances have been made in this field, and the use of *Agrobacterium tumefaciens* as a vector is explained elsewhere in this book. Unfortunately, the impressive results obtained with this bacterium are of little use in our system, because the former integrates foreign genes only into the nuclear genome, and we are concerned here with a piece of ct-DNA. This is a very serious obstacle to the development of our proposal. If we produce an altered sequence for the LSU of Rubisco, we currently have no way of delivering it into chloroplasts within plant cells.

It is not easy to predict the type of technology that will eventually be developed to allow chloroplast gene transfer. By analogy to the situation in the nucleus, it would be extremely useful if we could use a bacterial or viral pathogen that introduces DNA into the organelle and, better still, integrates it into ct-DNA. Such vectors are not currently available. Possibly, techniques of microinjection may advance to a stage where foreign DNA can be injected directly into large chloroplasts within cells. Plant microinjection techniques are in their infancy, although there is likely to be rapid progress in this field (Flavell and Mathias 1984). However, higher plants normally contain relatively large numbers of plastids, and each contains a large number of copies of the ct-DNA circle; thus, each cell is very highly polyploid for the chloroplast genome. If single cells or protoplasts were used as the receiver material, it is conceivable that one could inject a portion of the chloroplast population with foreign sequences, but it is not possible to predict what would happen to these either in the injected plastids or during subsequent rounds of plastid replication. Other investigations, using protoplasts, have demonstrated that under certain conditions they can take up chloroplasts from the surrounding medium. However, this does not appear to be very efficient, and we are then faced with the problem of introduction of foreign DNA into isolated chloroplasts.

What, then, is our final conclusion concerning this line of research? Is it worthwhile to proceed with a project (or parts of a project) aimed at reducing photorespiration by altering the relative rates of carboxylation and oxygenation by Rubisco using recombinant DNA technology? We may well receive different answers from merchant bankers and molecular biologists. In expensive biotechnological projects it is inevitable that money raised by nonscientists is used to fund such targeted research. For obvious reasons such financial backers prefer shorter-term projects that contain fewer technical problems. A multidisciplined assault in this area with an indefinite time scale and the definite possibility of failure will probably not appeal to such speculators, even though enormous profits could be made in the event of success. However, breakthroughs in a few key areas could quickly change this scenario. A molecular biologist would be much more likely to participate in such a research program because of the questions it could answer concerning basic biological processes. The present biotechnology boom is based on decades of research into pure science; it is interesting to note that, at least with respect to recombinant DNA methodology, the major product of biotechnology projects has been the wealth of further basic information for the biologist. It is to be hoped that over the next decade the output will satisfy both the sponsor and research worker.

References

Akazawa, T. (1979). Ribulose-1,5-bisphosphate carboxylase. *Enc. Plant Physiol.* **6,** 208.

Badger, M. R., and Lorimer, G. H. (1976). Activation of ribulose-1,5-bisphosphate oxygenase. The role of Mg^{2+}, CO_2 and pH. *Arch. Biochem. Biophys.* **175,** 723.

Bahr, J. T., and Jensen, R. G. (1974). Ribulose bisphosphate oxygenase activity from freshly ruptured spinach chloroplasts. *Arch. Biochem. Biophys.* **164,** 408.

Barraclough, R., and Ellis, R. J. (1980). Protein synthesis in chloroplasts. IX. Assembly of newly-synthesised large subunits into ribulose bisphosphate carboxylase in isolated intact pea chloroplasts. *Biochim. Biophys. Acta* **608,** 19.

Bedbrook, J. R., and Kolodner, R. (1979). The structure of chloroplast DNA. *Annu. Rev. Plant Physiol.* **30,** 593.

Bedbrook, J. R., Smith, S. M., and Ellis, R. J. (1980). Molecular cloning and sequencing of CDNA encoding the precursor to the small subunit of the chloroplast enzyme ribulose-1,5-bisphosphate carboxylase. *Nature* **287,** 692.

Berry-Lowe, S. L., McKnight, T. D., Shah, D. M., and Meagher, R. B. (1982). The nucleotide sequence, expression and evolution of one member of a multigene family encoding the small subunit of ribulose-1,5-bisphosphate carboxylase in soybean. *J. Mol. Appl. Genet.* **1,** 483.

Blair, G. E., and Ellis, R. J. (1973). Protein synthesis in chloroplasts. I. Light-driven synthesis of the large subunit of fraction 1 protein by isolated pea chloroplasts. *Biochim. Biophys. Acta* **319,** 223.

Bohnert, H. J., Crouse, E. J., and Schmitt, J. M. (1982). Organisation and expression of plastid genomes. *Enc. Plant Physiol.* **14B,** 475.

Bowes, G., Ogren, W. L., and Hageman, R. H. (1971). Phosphoglycollate production catalysed by ribulose diphosphate carboxylase. *Biochem. Biophys. Res. Commun.* **45,** 716.

Broglie, R., Bellemaire, G., Bartlett, S. G., Chua, N.-H., and Cashmore, A. R. (1981). Cloned DNA sequences complementary to mRNAS encoding precursors to the small subunit of ribulose-1,5-bisphosphate carboxylase and a chlorophyll a/b binding polypeptide. *Proc. Natl. Acad. Sci. U.S.A.* **78,** 7304.

Chan, P., and Wildman, S. G. (1972). Chloroplast DNA codes for the primary structure of the large subunit of fraction 1 protein. *Biochim. Biophys. Acta* **277,** 677.

Chollet, R., and Anderson, L. L. (1976). Regulation of ribulose-1,5-bisphosphate carboxylase-oxygenase activities by temperature pretreatment and chloroplast metabolites. *Arch. Biochem. Biophys.* **176,** 344.

(1978). Cyanate modification of essential lysyl residues in the catalytic subunit of tobacco ribulose-bisphosphate carboxylase. *Biochim. Biophys. Acta* **525,** 455.

Chua, N.-H., and Schmidt, G. W. (1979). Transport of proteins into mitochondria and chloroplasts. *J. Cell Biol.* **81,** 461.

Coen, D. M., Bedbrook, J. R., Bogorad, L., and Rich, A. (1977). Maize chloroplast DNA fragment encoding the large subunit of ribulose bisphosphate carboxylase. *Proc. Natl. Acad. Sci. U.S.A.* **74,** 5487.

Coombs, J., Baldry, C.W., and Bucke, C. (1976). C4 photosynthesis, in *Perspectives in Experimental Biology, Vol. 2,* ed. N. Sunderland, pp. 117–88. Pergamon Press, New York.

Curtis, S. E., and Haselkorn, R. (1983). Isolation and sequence of the gene for the large subunit of ribulose 1,5-bisphosphate carboxylase from the cyanobacterium *Anabaena* 7120. *Proc. Natl. Acad. Sci. U.S.A.* **80,** 1835.

Dobberstein, B., Blobel, G., and Chua, N.-H. (1977). In vitro synthesis and processing of a putative precursor for the small subunit of ribulose-1,5-bisphosphate carboxylase of *Chlamydomonas reinhardtii. Proc. Natl. Acad. Sci. U.S.A.* **74,** 1082.

Dougan, G., Saul, M., Twigg, A., Gill, R., and Sherratt, D. (1979). Polypeptides expressed in *Escherichia coli* K-12 mini cells by transposition elements Tn1 and Tn3. *J. Bact.* **138,** 48.

Dron, M., Rahire, M., and Rochaix, J.-D. (1982). Sequence of the chloroplast DNA region of *Chlamydomonas reinhardtii* containing the gene of

the large subunit of ribulose bisphosphate carboxylase and part of its flanking genes. *J. Mol. Biol.* **162,** 775.
Ellis, R. J. (1979). Fraction 1 protein. *Trends Biochem. Sci.* **4,** 241.
(1981). Chloroplast proteins: synthesis, transport and assembly. *Annu. Rev. Plant Physiol.* **32,** 111.
Ellis, R. J., and Gatenby, A. A. (1984). Ribulose bisphosphate carboxylase: properties and synthesis, in *Genetic Manipulation of Plants and Its Application to Agriculture,* ed. P. Lea and G. R. Stewart. Oxford University Press, London.
Flavell, R., and Mathias, R. (1984). Prospects for transforming monocot crop plants. *Nature* **307,** 108.
Gaastra, W., and Oudega, B. (1983). The determination of DNA sequences, in *Techniques in Molecular Biology,* ed. J. M. Walker and W. Gaastra. Croom-Helm, London.
Gatenby, A. A. (1983). The expression of eukaryotic genes in bacteria and its application to plant genes, in *Plant biotechnology,* ed. H. Smith and S. Mantell, pp. 269–97. Cambridge University Press.
(1984). The properties of the large subunit of maize ribulose bisphosphate carboxylase/oxygenase synthesised in *Escherichia coli. Eur. J. Biochem.* **144,** 361.
Gatenby, A. A., and Castleton, J. A. (1982). Amplification of maize ribulose bisphosphate carboxylase large subunit synthesis in *E. coli* by transcriptional fusion with the Lambda N Operon. *Mol. Gen. Genet.* **185,** 424.
Gatenby, A. A., Castleton, J. A., and Saul, M. W. (1981). Expression in *E. coli* of maize and wheat chloroplast genes for large subunit of ribulose bisphosphate carboxylase. *Nature* **291,** 117.
Gelvin, S., Heizmann, P., and Howell, S. H. (1977). Identification and cloning of the chloroplast gene coding for the large subunit of ribulose bisphosphate carboxylase from *Chlamydomonas reinhardtii. Proc. Natl. Acad. Sci. U.S.A.* **74,** 3193.
Hartley, M. R., Wheeler, A., and Ellis, R. J. (1975). Protein synthesis in chloroplasts. V. Translation of messenger RNA for the large subunit of fraction 1 protein in a heterologous cell-free system. *J. Mol. Biol.* **91,** 67.
Hartman, F. C., Norton, I. L., Stringer, C. D., and Schloss, J. V. (1978). Attempts to apply affinity labeling techniques to ribulose bisphosphate carboxylase/oxygenase, in *Photosynthetic Carbon Assimilation,* ed. H. W. Siegelman and G. Hind, pp. 245–69. Plenum Press, New York.
Heber, U., and Krause, G. H. (1980). What is the physiological role of respiration? *Trends Biochem. Sci.* **5,** 32.
Herrmann, R. G., and Possingham, J. V. (1980). Plastid DNA – the plastome, in *Chloroplasts,* ed. J. Reinert, pp. 45–96. Springer-Verlag, New York.
Highfield, P. E., and Ellis, R. J. (1978). Synthesis and transport of the small subunit of chloroplast ribulose bisphosphate carboxylase. *Nature* **271,** 420.

Jensen, R. G., and Bahr, J. T. (1977). Ribulose-1,5-bisphosphate carboxylase-oxygenase. *Annu. Rev. Plant Physiol.* **28,** 379.

Jordan, D. B., and Ogren, W. L. (1981). Species variation in the specificity of ribulose bisphosphate carboxylase/oxygenase. *Nature* **291,** 513.

(1983). Species variation in kinetic properties of ribulose 1,5-bisphosphate carboxylase/oxygenase. *Arch. Biochem. Biophys.* **227,** 425.

Kung, S. D., Chollet, R., and Marsho, T. V. (1980). Crystallisation and assay procedures of tobacco ribulose-1,5-bisphosphate carboxylase oxygenase. *Methods Enzymol.* **69,** 326.

Kung, S. D., and Marsho, T. V. (1976). Regulation of RuDP carboxylase-oxygenase activity and its relationship to plant photorespiration. *Nature* **259,** 325.

Laing, W. A., Ogren, W. L., and Hageman, R. H. (1974). Regulation of soybean net photosynthetic CO_2 fixation by the interaction of CO_2, O_2, and ribulose 1,5-diphosphate carboxylase. *Plant Physiol.* **54,** 678.

Langridge, P. (1981). Synthesis of the large subunit of spinach ribulose bisphosphate carboxylase may involve a precursor polypeptide. *FEBS Lett.* **123,** 85.

Lathe, R. F., Lecocq, J. P., and Everett, R. (1983). DNA engineering: the use of enzymes, chemicals and oligonucleotides to restructure DNA sequences in vitro, in *Genetic Engineering, Vol. 4* pp. 1–56. Academic Press, New York.

Lorimer, G. H. (1979). Evidence for the existence of discrete activator and substrate sites for CO_2 and ribulose-1,5-bisphosphate carboxylase. *J. Biol. Chem.* **254,** 5599.

(1981). The carboxylation and oxygenation of ribulose-1,5-bisphosphate; the primary events in photosynthesis and photorespiration. *Annu. Rev. Plant Physiol.* **32,** 349.

Lorimer, G. H., and Andrews, T. J. (1973). Plant photorespiration – an inevitable consequence of the existence of atmospheric oxygen. *Nature* **243,** 359.

(1978). Photorespiration – still unavoidable? *FEBS Lett.* **90,** 1.

McIntosh, L., Poulsen, C., and Bogorad, L. (1980). Chloroplast gene sequence for the large subunit of ribulose bisphosphate carboxylase of maize. *Nature* **288,** 556.

Martin, P. G. (1979). Amino acid sequence of the small subunit of ribulose-1,5-bisphosphate carboxylase from spinach. *Aust. J. Plant Physiol.* **6,** 401.

Martin, P. G., and Jennings, A. C. (1983). The study of plant phylogeny using amino acid sequences of ribulose-1,5-bisphosphate carboxylase. I. Biochemical methods and the patterns of variability. *Aust. J. Bot.* **31,** 395.

Maxam, A. M., and Gilbert, W. (1980). Sequencing end-labelled DNA with base-specific chemical cleavages. *Methods Enzymol.* **65,** 499.

Miziorko, H. M. (1979). Ribulose-1,5-bisphosphate carboxylase. Evidence in support of the existence of distinct CO_2 activator and CO_2 substrate sites. *J. Biol. Chem.* **254,** 270.

Nelson, P. E., and Surzycki, S. J. (1976). Characterization of the oxygenase activity in a mutant of *Chlamydomonas reinhardtii* exhibiting altered ribulosebisphosphate carboxylase. *Eur. J. Biochem.* **61,** 475.

Nishimura, M., and Akazawa, T. (1974*a*). Structure and function of chloroplast proteins. XXII. Dissociation and reconstitution of spinach leaf ribulose-1,5-diphosphate carboxylase. *J. Biochem.* **76,** 169.

(1974*b*). Reconstitution of spinach ribulose-1,5-diphosphate carboxylase from separated subunits. *Biochem. Biophys, Res. Commun.* **59,** 584.

(1974*c*). Studies on spinach leaf ribulose-bisphosphate carboxylase. Carboxylase and oxygenase reaction examined by immunochemical methods. *Biochemistry* **13,** 2277.

Ogren, W. L. (1976). Search for higher plants with modifications of the reductive pentose phosphate pathway of CO_2 assimilation, in *CO_2 Metabolism and Plant Productivity,* ed. R. H. Burris and C. Black, pp. 19–29. University Park Press, Baltimore.

Old, R. W., and Primrose, S. B. (1980). *Principles of Gene Manipulation. An Introduction to Genetic Engineering. Studies in Microbiology, Vol. 2.* Blackwell Scientific, London.

Poulsen, C., Martin, B., and Svendsen, I. (1979). Partial amino acid sequence of the large subunit of ribulosebisphosphate (EC 4.1.1.39) carboxylase from barley. *Carlsberg Res. Commun.* **44,** 191.

Sagher, D., Grosfeld, H., and Edelman, M. (1976). Large subunit ribulose bisphosphate carboxylase mRNA from *Euglena* chloroplasts. *Proc. Natl. Acad. Sci. U.S.A.* **73,** 722.

Schloss, J. V., Stringer, C. D., and Hartman, F. C. (1978). Identification of essential lysyl and cysteinyl residues in spinach ribulosebisphosphate carboxylase/oxygenase modified by the affinity label *N*-bromoacetylethanolamine phosphate. *J. Biol. Chem.* **253,** 5707.

Schmidt, G. W., Devillers-Thiery, A., Desruisseaux, H., Blobel, G., and Chua, N.-H. (1979). NH_2-terminal amino acid sequences of precursor and mature forms of the ribulose-1,5-bisphosphate carboxylase small subunit from *Chlamydomonas reinhardtii. J. Cell Biol.* **83,** 615.

Shinozaki, K., and Sugiura, M. (1982). The nucleotide sequence of the tobacco chloroplast gene for the large subunit of ribulose-1,5-bisphosphate carboxylase/oxygenase. *Gene* **20,** 91.

Shinozaki, K., Yamada, D., Takahata, N., and Sugiura, M. (1983). Molecular cloning and sequence analysis of the cyanobacterial gene for the large subunit of ribulose-1,5-bisphosphate carboxylase/oxygenase. *Proc. Natl. Acad. Sci. U.S.A.* **80,** 4050.

Somerville, C. R., and Ogren, W. L. (1982). Genetic modification of photorespiration. *Trends Biochem. Sci.* **7,** 171.

Somerville, C. R., and Somerville, S. C. (1984). Cloning and expression of the *Rhodospirillum rubrum* ribulosebisphosphate carboxylase gene in *E. coli. Mol. Gen. Genet.* **193,** 214.

Stringer, C. D., and Hartman, F. C. (1978). Sequences of two active site peptides from spinach ribulosebisphosphate carboxylase/oxygenase. *Biochem. Biophys. Res. Commun.* **80,** 1043.

Sugiyama, T., and Akazawa, T. (1970). Subunit structure of spinach leaf ribulose 1,5-diphosphate carboxylase. *Biochemistry* **9,** 4499.

Takebe, T., and Akazawa, T. (1975). Further studies on the subunit structure of *Chromatium* ribulose-1,5-bisphosphate carboxylase. *Biochemistry* **14,** 46.

Takruri, I. A. H., Boulter, D., and Ellis, R. J. (1981). Amino acid sequence of the small subunit of ribulose-1,5-bisphosphate carboxylase of *Pisum sativum. Phytochemistry* **20,** 413.

Woolhouse, H. W. (1978). Light-gathering and carbon assimilation processes in photosynthesis; their adaptive modifications and significance for agriculture. *Endeavour* **2,** 35.

Zurawski, G., Perrot, B., Bottomley, W., and Whitfield, P. R. (1981). The structure of the gene for the large subunit of ribulose 1,5-bisphosphate carboxylase from spinach chloroplast DNA. *Nucl. Acids Res.* **9,** 3251.

8 Genetic engineering of seed proteins: current and potential applications

R. R. D. CROY AND J. A. GATEHOUSE

Introduction

This review is divided into four sections. In this first section the reasons why seed proteins and their encoding genes have been a major subject of research are outlined, and the application of genetic engineering to the biotechnology of seed proteins is introduced. The methods employed in the study of seed proteins and their genes are described in the second section, and the third section summarizes present knowledge on seed proteïns and their genes in a number of major food crops. Finally, the fourth section describes some ways in which genetic engineering can be applied to the manipulation of specific seed protein genes in order to improve nutritional and other properties of seeds, on the basis of projects currently under study and on more speculative suggestions.

Seed proteins, and notably the storage proteins, have been the subject of many relevant general reviews and conference proceedings, including those by Derbyshire and associates (1976), Thomson and Doll (1979), Nelson (1980), Miflin and Shewry (1981), Larkins (1981), Boulter (1981), Payne and Rhodes (1982), Miège (1982), Mossé and Pernollet (1983), Pernollet and Mossé (1983), Thompson and Casey (1983), Gatehouse and associates (1984*b*), and Higgins (1984). Several recent reviews have referred to the use of recombinant DNA techniques for the study of seed protein structure and biosynthesis and in the elucidation of the structure and evolution of the genes (Brown et al. 1982; Chlan and Dure 1983; Gatehouse et al. 1984*b*; Higgins 1984; Sorenson 1984); one review, by Larkins (1983), deals with some applications of genetic engineering specifically in studies of maize and soybean seed proteins.

The last decade has seen a dramatic increase in research into many aspects of seed proteins that has been a result of both scientific and economic interests. Unfortunately (or perhaps fortunately for researchers in the field) the former interests are not always completely independent from the latter.

The economic aspects of seed protein research are related to the fact that most of the protein component of human food is derived directly or indirectly from the proteins stored in the seeds of cereals, legumes, or other crop plants, with other plant storage tissues such as roots and tubers contributing lesser amounts; see Figure 2 in Payne (1983). Whereas in developing countries seed proteins (e.g., those from beans and rice) provide virtually the only protein component of the diet (Payne 1983), in developed countries such as in Western Europe and North America, a large proportion of human food is derived from animals reared on diets enriched either directly with cereal and legume seeds or with protein concentrates or additives containing large amounts of seed proteins. Relatively few countries produce sufficient food crops to be self-sufficient in protein and calories, and fewer still are able to produce enough for export. Western Europe, for example, has a substantial deficit between seed protein and requirement and production; the United Kingdom alone has to import over 70% of its crude protein, primarily for addition to animal feeds (Payne and Rhodes 1982). With a continuously increasing world population there is an increasing demand for high-protein food for both human and animal consumption, and this is commercially exploited by those countries able to export surplus production. Thus, research is, on the one hand, geared to improve crop yields and quality in varieties that can be grown by producers in the developed countries for increased profit and, on the other hand, to produce better crop varieties for developing countries that will provide a more appropriate dietary protein in combination with other desirable characteristics. Plant breeders have sought to increase the efficiency of crop harvests by developing higher-yielding, higher-quality varieties that also carry the appropriate resistances to adverse climatic or pathogenic conditions likely to be encountered in the field or, after harvesting, in storage.

Higher or better quality, with reference to seed proteins, can have many different meanings, depending on the crop and its utilization. Most often, protein quality refers to the amino acid composition of the relevant seed proteins or seed meal/flour, and how this compares with the balanced amino acid composition reckoned to be optimal for human or animal diets in question. Cereals as protein sources are generally deficient in lysine and tryptophan, whereas the quality of legume seed protein is generally poor because it tends to be deficient in methionine and cystine; see Table 3 in Burr (1975) and Table 2 in Payne (1983). Generally, seeds as protein sources fall far below the standard of nutritionally excellent proteins, such as casein and ovalbumin, not only because of poor amino acid composition (Eggum and Beames 1983) but also because of poor digestibility and the presence of toxic or antinutritional factors, many of which are themselves proteins. This latter group of seed proteins includes the lectins and protease inhibitors, some of which, although implicated in the poor nutritional value of seeds, may be useful in conferring resistance to insect pests and pathogens, as described in a later section. The properties and nutritional aspects of these proteins have been extensively reviewed (Gatehouse 1984; Pusztai et al. 1983).

Apart from a nutritional role, seed proteins are playing an ever-increasing role as food additives in providing functional properties such as foam and emulsion stability, and texture. As such they are being intensively investigated as cheap substitutes for the relatively expensive animal proteins (casein) in current use. Bread-making quality is another important property attributed to seed proteins, almost exclusively those from wheat and rye, and current research in this field is aimed at elucidation of the physicochemical roles played by certain proteins in good bread-making varieties (Miflin et al. 1983*a*; Payne 1983.

Apart from their agricultural importance, seeds have long been an attractive source of biological material for scientific research, for many reasons. First, seed proteins are available in large amounts; storage proteins variously represent between 5% and 50% of the seed dry weight, with individual types accounting for as much as 10% to 20%. Moreover, most plant species that have been studied are crop plants and therefore tend to grow well in the field or under artifical conditions and reproducibly produce good yields of protein. Secondly, many researchers have proposed the developing seed as an excellent model system for studying the control and expression of plant genes. In particular, the storage protein genes represent a group of developmentally regulated genes "switched on" and expressed only during a precise period in the plant life cycle. They are in many respects analogous to the developmentally regulated animal genes such as those encoding the hemoglobin and ovalbumin proteins. As will be discussed later, in order to study gene control mechanisms it is essential to have an endogenous situation or external stimulus that directly influences the activity of one gene or a set of genes. Developmental or temporal regulation is one of the most studied systems, being the "physiological" or "inherent" control, although other effects on seed protein genes based on hormonal, tissue-specific (spatial), osmotic, and genetic controls have been described. Mutant and variant genotypes selected or produced by plant breeding programs and showing aberrant storage protein gene expression should prove of value in studying these controls.

For the purposes of this review we interpret genetic engineering as the recently developed recombinant DNA technology, which has as a major element in vitro manipulation of DNA sequences to produce new "unnatural" combinations. We specifically exclude manipulations involving chemical or radiological mutagenesis of whole plant geneomes, chromosome engineering (Law 1983), and "legitimate" or "illegitimate" fusion of genomes by sexual or artificial means, which have been defined by some authors as "genetic engineering"; see definitions by Phillips (1983) and Pomeranz (1980). The new methods offer the possibility of producing very specific and precisely engineered changes in the plant genome, and although all the elements of a viable technology are not quite yet perfected, genetic engineering of seed proteins is under active consideration in a number of laboratories. Broadly, the aims of such projects can be summed up as follows: to introduce new and potentially valuable gene products, to alter existing gene products to improve their nutritional, functional, or other properties, or to alter the expression of existing genes so that their products are removed if undesirable or increased if desirable.

Methodology

Identification and characterization of storage proteins

Proteins extracted from seeds and grains have been divided into four classes on the basis of their solubility properties (Osborne

Table 8.1. *Solubility classification of seed proteins*

Class	Definition	Notes
Albumin	"Water-soluble"	Now often taken to mean soluble in low-ionic-strength buffer ($I<0.05$) at pH 4.8 (e.g., Croy et al. 1984*a*), because water is an ill-defined solvent.
Globulin	"Soluble in dilute salt solutions at neutral pH"	Phosphate- or borate-buffered saline (0.15-M NaCl) at pH 7.5 is often used as solvent. Usually regarded as water-insoluble. Globulins are precipitated by dialysis against low-ionic-strength acid buffers (Pusztai and Stewart 1980).
Prolamin	"Alcohol-soluble"	60–70% ethanol (Osborne 1924) originally used; 50% propanol-1-ol and 55% propanol-2-ol (Shewry et al. 1984) now preferred. Some storage proteins previously classified as glutelins have been shown to be soluble as prolamins after reduction of disulfide bonds and are now classified as such (Shewry et al. 1984).
Glutelin	"Alkali-soluble"	Ill-defined, because alkali denatures most proteins. Generally used for proteins soluble only in strongly denaturing solvent. Proteins of this solubility class are probably structural and metabolic proteins (Wilson et al. 1981).

Source: Osborne (1924).

1924). These solubility classes are defined in Table 8.1. A common procedure has been to extract seed sequentially with successively more denaturing solvents and to label abundant components as "storage proteins." However, strictly speaking, material should be designated as storage protein only after it has been demonstrated that its components are degraded rapidly after seed germination, to fulfill the necessary role of a storage protein (i.e., to supply nutrient to the developing seedling) (Boulter 1982). The solubility fractionation schemes do not generally produce fractions containing single protein species (e.g., pea globulin fraction contains at least three distinct protein types) (Gatehouse et al. 1984*b*); consequently, further protein purification is usually necessary if detailed studies of homogenous proteins are to be carried out. All the common techniques of protein purification, including salt precipitation, column chromatography, electrophoresis, isoelectric focusing, and so forth, have been employed (Derbyshire et al. 1976)

Identification of storage proteins

Considerable problems exist in attempting to define the precise nature of a storage protein. This is a result of the heterogeneity of the polypeptides that are assembled into storage protein molecules – a direct consequence of multigene families, as will be described later. The practical result is that the polypeptide composition of a storage protein, as determined by fully dissociating techniques such as SDS polyacrylamide-gel electrophoresis (SDS-PAGE), is difficult to define and of a complexity that obscures a simpler structure at the molecular level. For example, vicilin-type globulin proteins are composed of molecules each containing three polypeptides of approximately 50,000 M^r, but because of a heterogeneous selection of these polypeptides being produced of closely similar but not identical amino acid sequences, and because of optional posttranslational glycosylation and proteolysis, many bands may be observed on SDS-PAGE for what is essentially the same protein (Gatehouse et al. 1981). In the case of proteins insoluble except in denaturing solvents it is even more difficult to identify distinct protein species, as it is impossible to determine which polypeptides associate with each other in vivo to form molecular assemblies; it is apparent that some cereal proteins are deposited as macromolecular arrays cross-linked by disulfide bonds (Payne et al. 1984). The only clear definition of a protein species must therefore be based on sequence homology of purified component polypeptides, although in the case of proteins soluble as molecules, assembly of polypeptides into molecules can be used as the basis of an operational definition (i.e., if two molecular species contain no common polypeptides, they are not the same) (Croy et al. 1980*c*).

Characterization of storage proteins

Common analytical techniques of protein chemistry have been employed successfully in studies of storage proteins, although study of molecular properties (molecular weight, etc.) is difficult in the case of many water-insoluble cereal storage proteins. One- and two-dimensional gel electrophoresis systems have become the analytical techniques of choice largely because of convenience, reproducibility, and high resolution. Most systems are based on the high-resolution discontinuous Laemmli SDS-PAGE method (Laemmli 1970); SDS in the presence of reducing agent is the preferred dissociating medium because it will solubilize virtually all seed proteins and eliminate most, if not all, interactions between seed protein polypeptides. Other electrophoretic methods that have been used either alone or in combination with SDS-PAGE include nondissociating gel electrophoresis at pH 8.3 for soluble proteins (Davis 1964; Ornstein 1964) and electrophoresis in gels containing acetic acid and urea (Bonner et al. 1980) or at pH 4.0 (Gabriel 1971) or in gels containing aluminium lactate (Mechan et al. 1978). Isoelectric focusing (IEF) and nonequilibrium pH-gradient electrophoresis (NEPHGE), usually in urea-containing gels, are further highly resolving techniques often used alone or in combination with SDS-PAGE (O'Farrell 1975; Payne et al. 1984; Rhighetti and Drysdale 1974). Examples of these are shown in Figure 8.2. Numerous reviews of these techniques are available (Smith and Nicholas 1983).

Complexity of storage proteins

When analyzed by the foregoing techniques, storage proteins normally show heterogeneity in terms of polypeptide size, charge, and composition. By using two-dimensional IEF/SDS-PAGE analyses, comparatively simple storage proteins such as pea legumin or vicilin can be shown to contain up to 30 separable polypeptides (Matta et al. 1981), whereas a complex fraction such as wheat storage protein contains over 60 (Payne et al. 1984) (Figure 8.2). Genetically determined differences between band or spot patterns of storage proteins by gel analysis are well documented for many species (e.g., Casey 1979*a*; Hynes 1968; Shepherd 1968) and have been used for genetic analysis (e.g., Payne and Lawrence 1983) and for mapping genetic loci (e.g., Matta and Gatehouse 1982; Wrigley and Shepherd 1973).

This complexity of storage proteins implies a large number of encoding genes, although it has been suggested that some of the observed heterogeneity of polypeptides may be introduced after translation (Gatehouse et al. 1984*b*). The complexity of storage protein polypeptides has made sequence data difficult to obtain in many cases, because it is very difficult to isolate individual polypeptides, although limited sequence data have been obtained for several legume proteins by conventional techniques (Hirano et al. 1982); see also reviews by Ramshaw (1982) and Pernollet and Mossé (1983).

Physical properties of storage proteins

Storage proteins that are soluble in nondenaturing solvents are amenable to techniques that examine their physical properties, such as optical rotatory dispersion and circular dichroism to determine α-helix content (Chen et al. 1974), ultracentrifugation studies to determine molecular shape and size heterogeneity (e.g., Pusztai and Stewart 1980), low-angle X-ray scattering to determine molecular symmetry (Damaschum et al. 1979; Plietz et al. 1984), and quasi-elastic light scattering to determine diffusion coefficients (Beone and Pecora 1976). Crystallization of storage proteins is generally not easy to achieve, and thus X-ray diffraction patterns from crystals are not widely available. A tertiary structure for a legume vicilin (7S) type of protein (canavalin, from jack bean) at 3.0 Å resolution has been published (McPherson 1980).

Other seed proteins

Besides proteins that have well-defined roles as storage proteins, other proteins are accumulated by seeds, although usually to a lesser extent; whereas a typical storage protein can constitute up to 50% of total seed protein, other proteins may account for less than 5%. Some of these proteins are enzymes (e.g., urease in jack beans or β-amylase in cereals) or enzyme inhibitors such as protease inhibitors, whereas others do not yet have defined functions (Croy et al. 1984*a*). Common types of proteins to be accumulated by seeds are the lectins (phytohemagglutinins), which are carbohydrate-binding proteins; although these proteins are thought to be involved in cell-cell interactions in other tissues, their role in seeds is more likely to be that of a protective antimetabolite against pathogens. It seems likely that proteins accumulated in seeds generally do not have roles in the physiology of the developing seed and, if they do not function as storage reserves, have functions concerned with seed protection, viability, or germination; this conclusion is supported by failure to demonstrate any involvement of those enzymes accumulated by seeds in seed metabolism. Furthermore, the amounts of those proteins that are accumulated in seeds vary considerably from species to species, and even from line to line in the same species, without affecting the development of viable seeds; so, clearly, any role these proteins play is in secondary, rather than primary, metabolism.

The minor seed proteins often have amino acid compositions that are more balanced than those of the storage proteins; they are also often freely soluble and are thus isolated as the albumin protein fraction from seed tissues. The nonessential nature of these minor seed proteins, their variability, and their secondary metabolite functions, where established, make them of potential value in plant genetic engineering.

Seed protein nomenclature

Plant storage protein nomenclature is very confusing and nonsystematic. To avoid confusing the subject further, we have tried to follow established usage in this review. However, a few remarks may prove helpful to the reader. Ideally, the name given to a storage protein should convey two pieces of information: the species from which the protein has been extracted and its relatedness to other storage proteins. Most (but not all) trivial names give the first but not the second piece of information, because they are derived from the generic species name, such as hordein from barley (*Hordeum vulgare*). A major exception is wheat, in which the storage proteins are called gliadins and glutenins, although "triticins" would appear to be more appropriate and systematic. Few trivial names, however, give any idea of the relations of storage proteins to each other, and often such relationships as are implied are misleading, such as "conglycinins" or "concanavalin" [which are different proteins, completely unrelated to the major storage proteins of *Glycine max* (soybean) and *Canavalia ensiformis* (jack bean): glycinin and canavilin, respectively]. By way of illustration of the problem, we have listed in Table 8.2 some examples of seed storage proteins and other proteins that have been studied, along with any designations and identities established. In cases such as the legumes, where generic types of proteins have been established, the most sensible system would be to give the species name and the generic type, such as pea (or *Pisum sativum*) legumin, and jack bean (or *C. ensiformis*) vicilin. Those trivial names that do not obscure this system could be retained (e.g., phaseolin for *Phaseolus vulgaris* vicilin, glycinin for *Glycine max* legumin), but those that merely confuse (e.g., conglycinin for *Glycine max* vicilin) should now fall out of use. The additional information now available on

Table 8.2. *Some examples of characterized seed protein types*

Species (common name)	Protein class	Trivial name	Type	References
Legumes[a]				
Arachis hypogaea (peanut)	Globulin	Arachin (13S)	Legumin	Neucere and Ory (1970)
	Globulin	Conarachin (7S)	Vicilin	
Canavalia ensiformis (jack bean)	Globulin	Canavalin	Vicilin	Sammour et al. (1984)
	Albumin	Concanavalin A	Lectin	Wang et al. (1975)
Glycine max (soybean)	Globulin	Glycinin (11S)*	Legumin	Hill and Breidenbach (1974)
	Globulin	β conglycinin (7S)*	Vicilin	Thanh and Shibasaki (1978)
	Globulin	γ conglycinin (7S)	Vicilin	Fukushima and Koshiyama (1976)
	Globulin	α conglycinin (2S)	–	
	Albumin	Soybean lectin*	Lectin	
	Albumin	Kunitz trypsin inhibitors*	Trypsin inhibitors	
Lens culinaris (lentil)	Albumin	Lentil lectin	Lectin	Fliegerová et al. (1974)
Lupinus albus (lupin)	Globulin	Conglutin α	Legumin	Blagrove et al. (1976)
		Conglutin β	Vicilin	
		Conglutin γ	–	Elleman (1977)
Phaseolus vulgaris (French bean)	Globulin	Gl, phaseolin (7S),* glycoprotein II	Vicilin	Pusztai and Watt (1970)
	Globulin/albumin	PHA, G2*, Phaseolus lectin	Lectin	Pusztai and Watt (1974)
	Albumin	–	Trypsin inhibitors	Pusztsi (1966)
	Albumin	Glycoprotein I	–	Pusztai and Duncan (1971)
Pisum sativum (pea)	Globulin	Pea legumin (11S)*	Legumin	Casey (1979*b*)
	Globulin	Pea vicilin (7S)*	Vicilin	Gatehouse et al. (1981)
	Globulin	Pea convicilin*	Vicilin	Croy et al. (1980*c*)

	Albumin	Pea lectin*	Lectin	Trowbridge (1974)
	Albumin	PMA*	–	Croy et al. (1984*a*)
	Albumin	I_A*	–	Gatehouse et al. (1984*a*)
Psophocarpus tetragonolobus (winged bean)	Globulin	Psophocarpin A	–	Gillespie and Blagrove (1978)
		Psophocarpin B	–	
		Psophocarpin C	–	
Vicia faba (broad bean)	Globulin	Broad bean legumin	Legumin*	Bailey and Boulter (1970)
		Broad bean vicilin	Vicilin*	Bailey and Boulter (1972)
		Broad bean convicilin	Vicilin	Croy et al. (1980*c*)
	Albumin	Favin, broad bean, lectin	Lectin	Cunningham et al. (1979)
Vicia narbonensis	Globulin	Narbonin	–	Schlesier et al. (1978)
Vigna unguiculata (cowpea)	Globulin	–	Vicilin	Khan et al. (1980)
Cereals and others				
Avena sativa (oats)	Globulin	–*	Legumin	Burgess et al. (1983)
	Prolamin	Avenin	Prolamin	Kim et al. (1978)
Brassica napus (rapeseed)	Globulin	12S	Legumin	Goding et al. (1970)
		Napins	–	Londerdal and Janson (1972)
Cannabis sativus (hempseed)	Globulin	Edestin (13S)	Legumin	St. Angelo et al. (1968)
Cocus nucifera (coconut)	Globulin	Cocosin (11S)	Legumin	Wallace and Dieckert (1976)
		Concocosin (7S)	Vicilin	
Cucurbitaceae sp.	Globulin	Cucurbitins	Legumin	Blagrove and Lilley (1980)
Eleusine coracana (millet)	Prolamin	Pennisetin	Prolamin	Pernollet and Mossé (1983)
Gossypium hirsutum (cottonseed)	Globulin	α globulin (52 kD)*	–	Dure and Chlan (1981)
	Globulin	β globulin (48 kd)*	–	
Helianthus annuus (sunflower)	Globulin	Helianthin (11S)	Legumin	Schwenke et al. (1974)
Hordeum vulgare (barley)	Prolamin	Hordein* B, C, D	Prolamin	Shewry et al. (1984)
	Glutelin	–	–	Wilson et al. (1981)

Table 8.2. (*cont.*)

Species (common name)	Protein class	Trivial name	Type	References
Oryzae sativa (rice)	Prolamin	Oryzin	Prolamin	–
	Glutelin	–	Legumin	Zhao et al. (1983)
Secale cereale (rye)	Prolamin	Secalins* γ and ω	Prolamin	Shewry et al. (1982)
	Glutelin	–	–	
Sorghum vulgare (sorghum)	Prolamin	Kafirin	Prolamin	Payne and Rhodes (1982)
Triticum aestivum (wheat)	Prolamin	Gliadin* α, β, γ, and ω	Prolamin	Kasarda (1980)
	Glutelin	Glutenin	Prolamin	Payne et al. (1984)
Zea mays (maize)	Prolamin	Zeins*	Prolamin	Rhighetti et al. (1977)
	Globulin	α and β Globulins	–	Dierks-Ventling (1981)
	Albumin	–	–	Dierks-Ventling (1981)

Note: Asterisk indicates studies performed using genetic engineering techniques; dash indicates information not available.
[a]See Derbyshire et al. (1976) for a more exhaustive list of legumin- and vicilin-type proteins.

cereal proteins might permit generic terms to be devised for them also. The situation is further complicated by the multiple subunits and polypeptides that are now being described, isolated, and in some cases sequenced. As will be seen later, it is apparent that the seed storage proteins are encoded by multigene families, and with the rapid increase in gene cloning and characterization, a systematic scheme of nomenclature is urgently required to avoid further series of independent and confusing designations; see Tables 8.2, 8.4, and 8.6. We would therefore suggest that researchers working with plant seed proteins and their genes should consider possible ways of improving the nomenclature system currently in use. Suggestions for such systems would be welcomed by the authors.

Production of cDNAs encoding storage proteins

Purification of mRNA from seed tissues

Plant tissues in general tend to be poor sources of nucleic acids, because the cells are difficult to break open and contain high levels of nucleases; many plant cells also contain secondary metabolites (phenols, carbohydrates) that interact with nucleic acids. Thus, methods for purifying RNA from seeds generally rely on strongly denaturing and dissociating conditions to prevent such interactions and to inactivate RNAses as rapidly as possible. Some representative methods are listed in Table 8.3. It is usually necessary to ensure that the RNA preparation actually does encode the desired polypeptides before transcription into cDNA. Examination of translation products of RNA preparations is commonly carried out by in vitro translation followed by immunoprecipitation and analysis by SDS-PAGE; relevant methods have been reviewed by Grierson and Spiers (1983). It may be advisable to carry out in vivo labeling studies or other assays to determine at what stage of seed development RNA should be isolated, because some components are synthesized only at certain stages. Similarly, size fractionation of RNA preparations coupled with in vitro translation may indicate discrete size fractions encoding the desired polypeptides (Evans et al. 1979, 1980). Careful choice of developmental stage or RNA size can optimize the cloning of a desired cDNA clone, as will be described later. Alternatively, the cDNA to be cloned can be enriched for the desired sequences using two mRNA preparations that are similar but for the presence of the

Table 8.3. *Methods for isolation of RNA from seed tissue*

Method	References	Comments
Total RNA		
Isolation of polysomes	Evans et al. (1979), Vodkin (1980)	Product is active in translation systems, but poly(A)$^+$ RNA isolated from polysomes often shows degradation. Specific mRNA sequences may be enriched by immunoprecipitation of polysomes in the presence of heparin as an RNase inhibitor (Shapiro and Young 1981)
Isolation by extraction in EGTA-borate, followed by SDS treatment and proteinase K digestion	Hall et al. (1978)	A very good method, producing pure nondegraded RNA, even after oligo(dT) cellulose chromatography; rather lengthy procedure.
Isolation by extraction in guanidinium thiocyanate, followed by certrifugation in CsC1	Chirguin et al. (1979)	Good quick method, but RNA obtained contains carbohydrate and does not translate well. It is, however, nondegraded.
Isolation by extraction in buffer containing iodoacetic acid	Langridge et al. (1982)	Very quick small-scale method described for maize endosperm, but satisfactory for pea cotyledons also.

desired sequence (Alt et al. 1977; Gorecki and Rozenblatt 1980). The mRNA population lacking the desired sequence is used to hybridize out all cDNAs common to both mRNA preparations, leaving cDNA enriched for the required sequences, which are then cloned. Mutants lacking the desired seed protein are ideal for this approach, but mRNA preparations at different stages of seed development or from different tissues may also be usable.

Preparation of cDNA clone banks

Methods involved in the preparation of cDNA clone banks have been adequately reviewed by Williams (1981), Maniatis and associates (1982), and Forde (1983), among others, and need little further comment here. The construction of a cDNA clone bank is advantageous for isolation and study of gene sequences encoding seed proteins. Because many seed proteins are translated from abundant mRNAs, cDNA clones for these proteins will make up a relatively large proportion of the cloned sequences and are therefore easily isolated. Such clones are essential as probes for gene bank screening, besides many other uses.

Poly(A) RNA, purified from total RNA preparations by chromatography on oligo(dT) cellulose or poly(U) Sepharose, is commonly used as a template for reverse transcriptase in preparing the first strand in complementary DNA synthesis, although satisfactory results can also be obtained from total RNA preparations (King et al. 1979) provided the subsequent screening method takes into account ribosomal-RNA-derived clones that are produced even in the presence of oligo(dT) as a primer (Williams and Lloyd 1979). If RNA degradation is a serious problem, it is often preferable to avoid preparation of poly$(A)^+$ RNA and to routinely use RNase inhibitors such as human placental inhibitor (RNasin) in the first-strand synthesis (Martynoff et al. 1980). Second-strand synthesis is normally carried out with either DNA polymerase or reverse transcriptase after hydrolyzing away the RNA template. The "hairpin loop" formed at the end of the first DNA strand can be used as a primer, or alternatively the DNA-RNA hybrid can be given an oligo(dC) "tail" using terminal transferase, prior to RNA removal, and second-strand synthesis can then be primed with oligo(dG). The double-stranded cDNA is most often made "blunt-ended" with S1 nuclease and DNA polymerase I (or a similar enzyme) before cohesive termini are added either by restriction enzyme oligonucleotide linkers (via blunt-end ligation with T_4 DNA ligase followed by restriction) or homopolymer tails (via terminal transferase). The cDNA is then ligated into a suitable restricted cloning vector. Alternative schemes and variations have been described, some of which allow one-step synthesis of the second cDNA strand and incorporation into the cloning vectors (Okayama and Berg 1982). Transformation of a suitable bacterial strain with the circularized vector produces the cDNA bank. A comprehensive selection of methods used to prepare and isolate cloned cDNAs encoding seed proteins is given in Table 8.4.

Screening cDNA clone banks

The cDNA bank is plated out under antibiotic selection and screened initially by the Grunstein-Hogness colony hybridization (GHC) procedure (Grunstein and Hogness 1975) or its modification (Hanahan and Meselson 1980). Labeled mRNA is a good probe for abundant sequences, but less abundant sequences may require enrichment of the mRNA probe, as described previously. cDNA probes are also widely used in this procedure and can be labeled to higher specific activities. Enriched or differential probes are particularly useful for identifying certain components: poly(A) RNA (or transcribed cDNAs) derived from different plant tissues (Bartels and

Table 8.4. *Examples of cloning and isolation of cDNAs encoding seed proteins*

mRNA source[a]	cDNA synthesis	Cloning method	Vector/site	Host
Maize endosperm– poly(A) polysomal RNA	i. Reverse transcriptase ii. DNA polymerase I iii. S1 nuclease	Homopolymer, GC tailing	[1][d] pBR322/Pst-I	c600$^-$
Maize endosperm protein bodies – poly(A) RNA	i. Reverse transcriptase ii. DNA polymerase I iii. S1 nuclease	Homopolymer, AT tailing	[2] pMB9/Eco-RI	HB101
Maize endosperm – total membrane-bound polysomal RNA	i. Reverse transcriptase ii. DNA polymerase I iii. S1 nuclease iv. Size fractionation	Homopolymer, GC tailing	pBR322/Pst-I	HB101
Maize endosperm – poly(A) RNA from total and polysomal zein RNA	N.S.[c]	Oligonucleotide, Eco-RI linkers	[3] λ641/Eco-RI	[4] POP101
Maize endosperm protein bodies – mRNA	i. Oligo (dT)-tailed pUC9 primer ii. Reverse transcriptase iii. Terminal transferase iv. Size fractionation v. Annealed with øligo(dC)- tailed pUC9 vi. DNA polymerase I	Homopolymer, GC tailing, combination method	[5] pUC9/Pst-I	[5] JM83
Barley endosperm–poly(A) RNA from membrane- bound polysomes – size fractionated	i. Reverse transcriptase ii. DNA polymerase I iii. S1 nuclease	Homopolymer, AT tailing	pBR322/Hind-III	HB101
Barley endosperm – poly- (A) RNA from membrane- bound polysomes	N.S.	Homopolymer, GC tailing	pBR322/Pst-I	HB101
Barley endosperm – poly- (A) RNA from membrane- bound polysomes	i. Reverse transcriptase ii. DNA polymerase I iii. S1 nuclease	Homopolymer, GC tailing, AT tailing	pBR322/Pst-I [6] pPH207/Hind-III	HB101 HB101
Wheat endosperm – poly (A) RNA	i. Reverse transcriptase ii. DNA polymerase I iii. S1 nuclease iv. Size fractionation	Homopolymer, GC tailing	pBR322/Pst-I	HB101
Soybean seeds – poly(A) RNA – size-fractionated	i. Reverse transcriptase ii. DNA polymerase I iii. S1 nuclease	Oligonucleotide, Hind-III and Eco-RI linkers Homopolymer, AT tailing	[7] pTR262/Hind-III [8] pBR325/Eco-RI pBR322/Hind-III	LE392 LE392 HB101
Soybean embryos – poly(A) polysomal RNA	i. Reverse transcriptase ii. DNA polymerase I iii. S1 nuclease	Homopolymer, AT tailing	pBR322/Pst-I	c600
French bean – poly(A) RNA – size-fractionated	i. Kinniburgh et al. (1978)	Oligonucleotide, Eco-RI linkers	[9] λ Charon 16A/Eco-RI	DP50 SupF
French bean – poly(A) RNA – total and size- fractionated	Different methods assessed for improved yields of full-length cDNAs with and without size fractionation	Homopolymer, GC tailing	pBR322/Pst-I	HB101
French bean – poly(A) RNA– size-fractionated	Non-S1-nuclease method of Land et al. (1981), as in Murray et al. (1983)	Homopolymer, GC tailing	pBR322/Pst-I	HB101
Pea cotyledon – poly(A) RNA	i. Reverse transcriptase ii. DNA polymerase I iii. S1 nuclease iv. Size fractionation	Oligonucleotide, Bam-H1 linkers	pBR322/Bam-H1	RecBC$^-$ (91[illegible] RecA$^-$ (GL8[illegible]
Pea cotyledon – poly(A) RNA from membrane- bound polysomes – size- fractionation	i. Reverse transcriptase ii. DNA polymerase I (Klenow) iii. S1 nuclease iv. Size fractionation	Homopolymer, GC tailing	[11] pAT153/Pst-I	HB101

Screening methods[b]	Seed protein(s) encoded	cDNA size range	Clone designation	References
Tet^R, Amp^S, cDNA/mRNA-GCH	Zeins	200–800 bp	pFW13, pFW19	Weinand et al. (1979)
Tet^R, Amp^S, cDNA/mRNA-GCH, HRT	Zeins	90–1,450 bp	A20, A30, B59, B14	Burr and Burr (1980) Burr et al. (1982, 1978)
Tet^R, cDNA-GCH, HRT	Zeins	500–1,000 bp	p2.15, pZ19.1, pZ22	Pedersen et al. (1982) Marks and Larkins (1982)
Clear plaques, cDNA-PLH	Zeins	200–1,000 bp	pcM1, pcM4 (not zein cDNA), pcM6, pcMG1	Viotti et al. (1982)
GCH with zein genomic clones or cDNAs	Zeins	60% > 400 bp 8% > 800 bp 100–1,200 bp	ZG 31A ZG 124 ZG 19 ZG 7	Hu et al. (1982) Heidecker and Messing (1983)
Amp^R, Tet^S, cDNA-GCH, HART	Hordeins	Av. 850 bp	N.S.	Brandt (1979)
N.S.	Hordeins	153–720 bp	pcHor-2-1, pcHor-2-2 pcHor-2-3, pcHor-2-4	Rasmussen et al. (1983)
Tet^R, mRNA-GCH, HRT, Cmp^R, Tet^S	Hordeins	230–900 bp	pc901, pc1-73, pc101-300	Forde et al. (1981)
Tet^R, Amp^S, mRNA-GCH, HART, HRT	Gliadins	350–1,500 bp	pTag24, pTag38, pTag544	Bartels and Thompson (1983*a*)
Tet^R, HRT	β conglycinin	> 450 bp	pGmc236 (Gmcα^{a1}236) pGmc232	Beachy et al. (1981)
Tet^R, Amp^R, Cmp^S, cDNA-GCH, HRT,	Glycinin	125 bp, 650 bp	pGmc40, pGmc73	Barton et al. (1982)
Amp^R, Tet^S, DNA-GCH	β conglycinin	N.S.	GMCα^{a1}1,Gmcα^{a1}2,α^{a1}2 Gmcp^{60}53.58	Schuler et al. (1982*a*, 1982*b*)
Tet^R, Amp^S, RNA-GCH, HRT	β conglycinin Glycinin Kunitz trypsin inhibitor	650 bp N.S. N.S.	A-16, A-28 A36, A-37 (A = abundant)	Goldberg et al. (1981*a*)
	Soybean lectin	370 bp	L-9	Goldberg et al. (1983*b*)
cDNA-PLH, HART	Phaseolin	1,200 bp	Ch16A.G1.9-5	Hall et al. (1980)
Tet^R, Amp^S, phaseolin cDNA-GCH	Phaseolin	Up to 1,850 bp 2,000bp	– cDNA31	Murray et al. (1983) Slightom et al. (1983)
Tet^R, Amp^S, lectin cDNAs-GCH, pea lectin cDNA-GCH	French bean Lectin	190–940 bp	pPVL134	Hoffman et al. (1982)
RNA-GCH = $Tet^S Amp^R$, cDNA-GCH, HRT	Pea legumin Pea vicilin	638 pb, 788 bp 300 bp, 900 bp	pRC2.2.4, pRC2.11.7 pRC2.2.10, pRC2.2.1 pDUB1–pDUB12	Croy et al. (1982) Gatehouse et al. (1982*a*) Lycett et al. (1983*a*,*b*) Lycett et al. (1984*b*)
Tet^R, Amp^S, RNA-GCH, HRT	Pea convicilin Pea legumin Pea vicilin	650 bp – 1,200 bp	pCD59 pCD40 pCD4, pCD48	Domoney and Casey (1983)
	Pea legumin	1,300 bp, 1,000 bp	pCD32, pCD40	Domoney and Casey (1984)

Table 8.4 (*cont.*)

mRNA source[a]	cDNA synthesis	Cloning method	Vector/site	Host
Pea cotyledon – poly(A) polysomal RNA	i. Reverse transcriptase ii. Reverse transcriptase iii. S1 nuclease iv. Size fractionation	Homopolymer, GC tailing	pBR322/Pst-I	RR1
Rapeseed embryos in culture – poly(A) RNA treated with methyl-mercury hydroxide	i. Reverse transcriptase ii. DNA polymerase I (Klenow) iii. Reverse transcriptase iv. S1 nuclease	Homopolymer, GC tailing	pBR322/Pst-I	HB101

[a]All tissues isolated from developing (immature) seeds.
[b]PLH = plaque-lift hybridization (Benton and Davies 1977). GCH = Grunstein colony hybridization (Grunstein and Hogness 1975). HRT = hybrid-release translation. HART = hybrid-arrested translation. Cmp = chloramphenicol. Amp = ampicillin. Tet = tetracycline. mRNA = poly(A) RNA or purified RNA fraction.

Thompson 1983*a*), from different stages of seed development (Dure et al. 1983), or from abscisic-acid-induced (ABA-induced) or noninduced cultured seed embryos (Crouch et al. 1983).

A secondary screen is carried out on selected clones, after purification of plasmids from these clones, by hybrid-release translation (HRT) or hybrid-arrested translation (HART) of the mRNAs that hybridize to the cDNA. Single-stranded cDNA plasmid chemically attached to diazobenzyloxylmethylcellulose paper is the most convenient and reliable means of carrying out this operation (Alwine et al. 1977, 1979; Smith et al. 1979). The translation product of the hybridized mRNA produced in an in vitro translation system is identified by SDS-PAGE either with or without immunoprecipitation. Alternatively, if appropriate expression vectors are used for cloning, it may be possible to screen for bacteria producing the protein of interest by using radiolabeled antibodies (Ehrlich et al. 1979). Final characterization of the cDNA clones is carried out by nucleic acid sequencing after restriction mapping, as will be described later.

Production of genomic clones containing seed protein genes

Purification of genomic DNA

It is not advantageous to purify DNA from developing seeds in order to produce clones containing storage protein genes, because there is no evidence of storage protein gene amplification in these

Screening methods[b]	Seed protein(s) encoded	cDNA size range	Clone designation	References
Tet^R, Amp^S, cDNA-GCH, HRT [10], HART	Pea legumin Pea vicilin Pea lectin	1,680 cp N.S. ~800 bp	pPS15-75 pPS15-84 pPS15-50, pPS15-104	Chandler et al. (1983) Spencer et al. (1984) Higgins et al. (1983*a*)
Tet^R, cDNA-GCH, HRT	Napins	583 739	pN1 pN2	Crouch et al. (1983)

[c] N.S. = not specified
[d] [1] Bolivar et al. (1977*b*). [2] Bolivar et al. (1977*a*). [3] Murray et al. (1977). [4] Lathe and Lecocq (1977). [5] Vieira and Messing (1982). [6] Bishop and Davies (1980). [7] Roberts et al. (1980). [8] Bolivar (1978). [9] Williams and Blattner (1980). [10] Chandler (1982). [11] Twigg and Sherratt (1980).

tissues (Croy et al. 1982). As a consequence, plant DNA is often purified from early leaves of young seedlings, because these tissues are readily obtained and easy to process, although DNA has been successfully purified from other tissues (Table 8.5). In the authors' experience, most species give good DNA preparations from leaves, but some (e.g., *Vicia faba*) do not, and in these cases other tissues may be preferable.

For the purposes of gene bank production it is desirable to isolate DNA at least 100 kb in length, and methods of DNA isolation must therefore avoid degradation of nucleic acids, either by nucleases or by physical shearing. This frequently is not a trivial problem. In addition, the isolated DNA must be of sufficient purity to be readily and reproducibly digested by restriction endonucleases and must not inhibit the enzymes involved in the cloning processes. Whether or not a particular DNA sample meets these criteria must be established on a trial-and-error basis. Representative purification methods for plant DNA are given in Table 8.5.

Choice of vector

Besides vectors based directly on λ bacteriophage, plasmids containing the cohesive termini of λ, which are packaged and transfected by components of the λ transfection system ("cosmids"), can also be employed; λ-based vectors have been reviewed by Williams and Blattner (1980), Brammar (1982), and Murray (1983), and cos-

Table 8.5. *Methods for isolation of plant genomic DNA*

Method	Starting tissue	References	Comments
Extraction in guanidine hydrochloride, followed by banding on CsC1 gradients	Dry embryos	Sung and Slightom (1981), Slightom et al. (1983), Sun et al. (1981)	Methods for jack bean, soybean, and French bean embryos given
Extraction in SDS/ perchlorate/phenol chloroform, followed by banding on CsC1 gradients	Leaves	Graham (1978)	Works for seed cotyledons also; a successful method
Extraction in the presence of ethidium bromide	Leaves	Bendich et al. (1980), Kislev and Rubenstein (1980)	Gives long DNA fragments, but yield usually poor
DNA precipitation from cell lysates by cetyltrimethyl-ammonium bromide	Cultured plant cells	Darby et al. (1970)	Not successful for ordinary tissues; tissue culture may be difficult to set up

mids by van Embden (1983); comprehensive reviews of both techniques are given in the manual of Maniatis and associates (1982).

To a large extent, the choice of vector in cloning genomic DNA will depend on the purpose envisaged. If a relatively small segment of repeated-sequence DNA is required, one of the normal cloning plasmids may be satisfactory. If specific genomic restriction fragments of known size are to be cloned, it is advantageous to enrich the DNA sample to be cloned by size fractionation after restriction, and then use a λ vector suitable for fragments of this size range (e.g., λgtWES λB, Table 8.6). On the other hand, construction of a full gene bank will require random digestion of genomic DNA to an appropriate size range with a frequently cutting restriction enzyme (e.g., MboI or Sau 3A) followed by ligation into a λ vector that will accept fragments as large as possible (up to 20–22 kb, such as λL47, λ1059, or λEMBL3, Table 8.6) to minimize the number of clones necessary for complete representation of the genome. Cosmids may seem advantageous for this purpose, because they will accept inserted DNA fragments up to 40 kb; however, production of 40-kb random fragments of plant genomic DNA with intact ends for cloning is not easy, and the cosmid transfection system does not give high numbers of plants genomic clones reliably (although it is not clear why this is so). λ vectors have therefore been employed in most plant gene banks for seed protein gene isolation. A representative survey of the plant gene banks thus far constructed for this purpose, the techniques and vectors used, and the seed protein genes thus far isolated is given in Table 8.6.

Production and screening of a gene bank

This topic has been reviewed by Murray (1983). For the average plant species (genome size $\sim 10^9$ bp), a gene bank in bacteriophage λ that gives a 99% probability of obtaining a chosen sequence will contain of the order of 10^6 recombinants (Clarke and Carbon 1976). Even though λ vector systems have been designed so that only recombinant phages are viable (λL47 and λ1059 and their derivatives λ2001 and λEMBL 3 and 4) (Frischauf et al. 1983; Karn et al. 1983), there is still a considerable screening problem. Phages are plated out as plaques on a lawn of bacteria and are screened by hybridization with specific DNA probes (usually a specific cDNA species) after transfer to nitrocellulose filters (Benton and Davies 1977); a phage amplification step on the filters may be advantageous

Table 8.6. *Examples of cloning and isolation of seed protein genes*

DNA source	Genomic DNA treatment	Vector site	Host	Selection/screening
Maize 8-day	Eco-RI limit seedlings	[1][a] λgtWES λB/Eco-RI digest	N.S.[b] arms	Zein cDNA-PLH[c]
Maize tissue culture	Eco-RI partial and limit digests	[2] Charon 4A/Eco-RI	N.S. N.S.	Zein cDNA-PLH Zein cDNA-PLH
Maize leaf nuclei	Size-fractionated Eco-RI partial and limited digests	[3] Charon 4/Eco-RI	DP50supF	Zein cDNA-PLH
Maize unfertilized ears	Size-fractionated Bam-H1 partial digests	[4] λL47.1/Bam-H1	N.S.	P_2 lysogen (Spi-) Zein cDNA-PLH
Soybean leaf nuclei	Eco-RI partial digests	[3] λCharon 4/Eco-RI	N.S.	cDNA-PLH
French bean embryo axes from dry seeds	Size-fractionated Eco-RI partial digests	[5] λCharon 24A/Eco-RI	N.S.	Phaseolin cDNA-PLH
	Mbo-I partial digests	[6] λ1059/Bam-H1	N.S.	N.S.
Pea leaf tissue	Size-fractionated Eco-RI partial digests	[1] λgtWES λB/Eco-RI	LE392	Legumin cDNA-PLH
	Sau-3A partial digests	[4] λL47.1/Bam-H1	5k	Spi-; legumin cDNA-PLH

[a][1] Leder et al. (1977). [2]Blattner et al. (1977). [3] Williams and Blattner (1979, 1980). [4] Loenen and Brammar (1980). [5] Blattner et al. (1978). [6] Karn et al. (1980).
[b]N.S. = not specified.
[c]PLH = plaque-lift hybridization (Benton and Davies 1977).

Seed proteins encoded	Fragment size range	Clone designation	References
Zein (M_r 20,000)	4.4 kb	λMF7	Weinand et al. (1981)
Zein (M_r 20,000)	7.7kb	λZG99	Pedersen et al. (1982)
Zein (M_r 20,000 and 22,000)	4.3 kb 7.6 kb 7.5 kb	λ(W22)Z1 λ(W22)Z4 λ(W22)Z7	Lewis et al. (1981)
Zein (M_r 22,000)	N.S.	ZA1	Spena et al. (1982)
Zein (M_r 20,000)	N.S.	ZE	Spena et al. (1983)
Glycinin	Av. 16 kb	λA-28-5,λA28-16,	Fischer and Goldberg (1983)
15 kd protein		λA36-2,λA36-5	
Soybean lectin		(Le^+) λL9-4, λL9-9 (Le^-) λS-5, λS-3	Goldberg et al. (1983*b*)
β conglycinin	N.S.	N.S.	Goldberg et al. (1983*a*)
Kunitz trypsin inhibitor	N.S.	N.S.	
Phaseolin	11.1 kb	Ch24A-177.4 Ch24A-176.2	Sun et al. (1981)
Phaseolin	N.S.	Clones 1.9, 2.2, 4.10, etc.	Hall et al. (1983*a*, 1983*b*)
Pea legumin	13.5 kb, 11 kb	λLeg 1, λLeg 2	Croy et al. (1984*c*) Lycett et al. (1984*a*)
Pea legumin	14kb	λLeg 3	Shirsat (1984), Croy et al. (1984*c*)

(Woo 1979). It is often necessary to screen duplicate filters to eliminate false-positive hybridization signals, and it is always desirable to replate and rescreen putative specific recombinants. The reviews cited earlier give further details of these procedures.

A further factor in successful production of specific genomic clones via gene banks is careful selection of the strain of *Escherichia coli* used as the transfection host for recombinant bacteriophages. Practical experience has shown that bacterial strains that are only slightly different can show 10-fold differences in the number of recombinants produced per microgram of genomic DNA, and the reasons for this variation, which is dependent on the origin of genomic DNA, are not understood (Federoff 1983; R. Flavell, personal communication). The specific identified bacteriophage clones are purified by standard protocols, and after restriction mapping the desired fragments are normally subcloned into more convenient plasmid vectors for further analysis.

Use of cloned DNA sequences in characterizing seed storage proteins

Cloned DNA sequences have been used to demonstrate sequence homology between storage protein polypeptides by cross-hybridization of the nucleic acid sequences and to predict storage protein sequences, posttranslational modifications, and degrees of homology by direct sequencing of DNA and comparison of predicted with actual protein sequences. Because a particular DNA clone can often be identified with a particular polypeptide by hybrid-release translation or sequence comparison, this has enabled sequences of, and relationships between, seed protein polypeptides to be examined.

Hybridization techniques

Intraspecific hybridization. This technique is commonly used to classify DNA sequences in a cDNA clone bank by assessing cross-hybridization of a number of isolated cDNAs to other clones within the clone bank; the results are often used to construct a Venn diagram consisting of families of related clone sequences (Forde et al. 1981). Nick-translated cDNA sequences purified from the vector DNA are hybridized to clones (bacterial colonies in a colony hybridization) or DNA ("plasmid minipreps" in a "dot-blot" hybridization (Kafatos et al. 1979) immobilized on nitrocellulose filters under varying conditions of stringency. Hybridization temperatures from T_m − 10°C to T_m − 50°C { T_m = melting temperature calculated from the formula $T_m = (\%G + C)/2 + 81.6 + 16.6\log[\text{salt}] - 0.6(\%\text{ formamide})$} (Sharp et al. 1980) have been used (Marks and Larkins 1982). Thus, a complete picture of weak and strong cross-hybridizations can be built up indicating strong and less strong homologies between gene coding sequences and the encoded polypeptides. Similar studies can also be carried out by hybrid-release translations and Southern-blot analysis at different stringencies.

Interspecific hybridization. Cross-hybridization of cDNA sequences from different species can, in theory, be used to detect homologous polypeptides where other methods, such as immunologic cross-reactions, are difficult. Such cross-hybridizations are highly dependent on stringency, as discussed previously, and are therefore of limited use. However, related seed protein sequences have been detected in oats and peas using a pea seed storage protein cDNA (I. Altosaar, personal communication). Cross-hybridization between cDNAs and seed mRNAs from different species has also been employed, with some success, to detect homologies between cereal gene coding sequences and polypeptides (Bartels and Thompson 1983*b*).

Sequencing of cloned DNA species

Sequencing techniques. Both major techniques for DNA sequencing – the Maxam-Gilbert controlled chemical-degradation technique (Maxam and Gilbert 1977) and the variations of the Sanger dideoxynucleotide-chain-termination technique (Sanger and Coulson 1974; Sanger et al. 1977) – have been used to sequence plant DNA, and sequencing methods and strategies are identical with standard protocols, as reviewed by Gaastra and Oudega (1983). Three sequencing methods have been successfully used in the authors' laboratory: the Maxam-Gilbert method, the method of Seif and associates (1980) (a dideoxynucleotide method using DNAse and DNA polymerase to generate the sequence "ladder"), and the bacteriophage M13/dideoxynucleotide method of Messing and associates (1981) using M13 mp8 and 9 or later versions. Each method has advantages, but the lower radioactivity involved and consumed in the M13 method may make this the favored routine method, especially as good results are obtained with ^{35}S nucleotides (Biggin et al. 1983).

Comparison of nucleic acid and protein sequences. Although a cloned DNA sequence can be identified by hybrid-release translation, rigorous identification through comparison of amino acid sequences predicted by the DNA with protein sequences, yields much extra information, including the correct reading frame. However, there is only very limited amino acid sequence information, or none, available for most storage proteins, and in these cases predicted amino acid sequences may be compared to protein sequences from a homologous protein from another species. Information concerning posttranslational modification is then difficult to derive. In some extreme cases, in which no protein sequence is available, a coding sequence may have to be inferred by amino acid composition and distribution of "stop" codons in the three reading frames (Sun et al. 1981). Once established, the availability of predicted coding sequences enables many properties of storage proteins to be explored.

Structure prediction. Secondary structure may be predicted from an amino acid coding sequence by using residue conformation probability tables such as those of Chou and Fassman (1974) or Garnier and associates (1978). These methods rely on computer handling of the large number of parameters involved, but they are not conceptually complex. Correctness of prediction is not likely to be higher than 50–55%, although the availability of physical data that allow the α-helix content of the protein to be predicted will improve the accuracy of prediction. Although limited, a useful impression of the structure of the storage protein can be built up, as shown in Figure 8.3 for pea legumin and vicilin, and different domains of protein structure can be readily identified. Improved programs, currently under development, will both improve the accuracy and extend the usefulness of these techniques.

Another useful method is that of Hopp and Wood (1981) for plotting the hydrophilicity indices of different regions of a predicted protein sequence and for indicating possible antigenic determinants. This technique allows regions of the polypeptide chain to be assigned to the "inside" or "outside" of the molecule (see Figure 8.3).

Heterogeneity. Comparison of sequences of cloned DNAs immediately shows the degree of heterogeneity in the predicted protein sequences and may indicate the number of genes encoding the family of polypeptides. Comparisons of long lengths of nucleotide or amino acid sequences are best dealt with by computer programs, and the results are conveniently illustrated in the form of a matrix of dots corresponding to perfect or partial sequence matches. Programs such as that of Staden (1982) have been used to compare several storage protein gene and cDNA nucleotide sequences and their predicted protein sequences (Croy et al. 1984*b*, in press; Schuler et al. 1983).

Posttranslational modification. If a partial or total amino acid sequence of a storage protein is compared to the predicted sequence, several differences are usually apparent. First, the predicted sequence usually contains an extra *N*-terminal region corresponding to the "leader" or "signal" peptide sequence required for storage protein cotranslational secretion into the endoplasmic reticulum during its synthesis (Blobel and Dobberstein 1975; Spencer 1984), and because *N*-terminal sequences of proteins are the regions determined most often, it is often possible to define the leader sequence exactly. Some storage proteins show evidence for removal of a short polypeptide at the *C*-terminal end (Edens et al. 1982). Second, the predicted sequence may show where posttranslational proteolysis takes place at other regions in the polypeptide; the mature protein then generally contains two or more polypeptides corresponding to different sections of the same cDNA predicted sequence (i.e., the precursor) (Croy et al. 1982; Gatehouse et al. 1982). Third, the predicted sequence will show potential sites for glycosylation (Asn-*x*-Ser or Thr) (Hoffman et al. 1982; Struck et al. 1978).

By careful sequence comparisons, the processes involved in storage protein processing can be defined exactly at the sequence level; examples are given in later sections.

Protein evolution. Predicted sequences are of obvious value in comparing homologous seed proteins between species in order to attempt to gauge the trends in evolution of the functional properties and "domains" of storage proteins (as opposed to genome evolution). This work is at an early stage, but based on such comparisons it is possible to speculate how particular species have come to possess storage proteins of particular sequences and properties and how the various types of heterogeneity and sequence diversity may have arisen (Croy et al. 1984*b*).

Use of cloned DNA sequences to probe gene structure

Estimation of gene copy numbers

Hybridization of labeled DNA (usually cDNA) sequences to total genomic DNA has been used to estimate the copy number of sequences complementary to the probe. Estimates can be obtained from the kinetics of hybridization in solution (e.g., Viotti et al. 1979) but are more commonly carried out with the genomic DNA immobilized on nitrocellulose. Genomic DNA is digested to completion with one or more restriction enzymes, and the different digests are size-fractionated by agarose-gel electrophoresis. "Reconstruction" or "titration" experiments consisting of amounts of the linear cloned cDNA equivalent to various gene copy numbers in the genomic DNA are run on the same gel. The gel is "Southern blotted" to transfer the DNA to nitrocellulose, and highly labeled cDNA is then used as a hybridization probe to detect both the complementary sequences in genomic DNA and the standard amounts of its own sequence in the reconstruction experiment; intensities of bands on the autoradiograph may then be compared and the copy numbers of the hybridizing sequences in genomic DNA estimated (Young et al. 1981). Results obtained by using this method depend on the degree of stringency employed in the hybridization and subsequent washes, and thus care must be employed in interpretation. Partially homologous sequences give weaker hybridization, resulting in fainter bands in the final autoradiograph, and thus are not quantitatively estimated. The size of plant genomes necessitates that highly radioactive probes (specific activities $> 10^7$ cpm/μg DNA) be used for these experiments if single-copy genes are to be detected in a reasonable time of autoradiography. The probe must also be of reasonable length (>200 nucleotides) to obtain sufficient radioactivity in hybrids, and results may be dependent on probe length in that the larger a probe, the more partially homologous sequences will be detected (which may be a problem when using genomic clones as probes if the clone also hybridizes to repeated sequences). The number of different genomic clones obtained from a gene bank may also give an approximate estimate of gene copy numbers, provided the bank is complete.

Detection of coding and noncoding gene sequences

The presence of introns in plant genes (Sun et al. 1981), like those of other eukaryotes, gives rise to a requirement for careful comparison of protein, cDNA, and genomic sequences to identify their precise locations. Plant introns obey the Breathnach-Chambron rule for the intron boundaries (Breathnach et al. 1978; Messing et al. 1983), and the requirement for maintaining the reading frame may be used to deduce intron boundaries even if the cDNA sequence is not available.

Alternatively, introns may be located by the R- or D-looping technique in which genomic clones are hybridized to complementary mRNA or cDNA and the hybrids are visualized by electron microscopy (Kaback et al. 1979). The presence of "loops" where the mRNA sequence hybridizes across the ends of the intron is a clear indication of the presence of noncoding intervening sequences. However, this technique is not capable of detecting introns smaller than about 150 bp, and because plant gene introns are on the whole smaller and fewer than those in animal genes (Fischer and Goldberg 1983), they may not be detected. Restriction-map comparisons of gene and cDNA, or direct nucleotide sequencing, will reveal smaller intervening sequences.

The transcription start in a genomic clone is determined by S1 nuclease mapping techniques (Berk and Sharp 1977; Weaver and Weissmann 1979). The location of polyadenylation sites (-AATAAA-) in the 3′ noncoding regions is determined from cDNA or gene sequences; plant genes often have multiple and in some cases overlapping (AATAAATAAA) polyadenylation "signals" (Croy et al. 1982; Lycett et al. 1983*a*). The use of alternative signals can be inferred from different positions of poly(A) tails on otherwise homologous cDNAs, as will be described later; 5′ promoter and control sequences are usually inferred from their sequence positions and from existing consensus sequences (e.g., "TATA" and "CAAT" boxes) (Messing et al. 1983).

Genome structure and evolution

As representative regions of transcriptionally active DNA, storage protein genes may prove to be an ideal model system with which to explore plant genome structure in terms of organization of genes, arrangement with respect to repeated sequences, and gene evolution. The "genome-walking techniques" that employ overlapping genomic clones to study linked sequences progressively farther away from a starting point will be of advantage in this respect. Little work in this area has been carried out with seed protein genes (Goldberg et al. 1983*a*). Detailed comparisons of nucleotide sequences of different genes from the same and different species encoding single protein types can be used to examine genome evolution, as will be described later. The availability of variant lines and mutants can also be exploited to study the origins and nature of the phenotypic variations.

Use of cloned DNA sequences to measure gene expression

mRNA levels in vivo

Characterized cDNA or genomic DNA clones encoding parts or all of storage protein polypeptides have been used by many workers to measure the levels of specific mRNA species in RNA preparations from plant tissues. Although solution hybridization assays have been used for this purpose (Goldberg et al. 1981*a*, 1981*b*), most recent workers have used techniques based on hybridization of labeled cDNA probes to RNA immobilized on nitrocellulose or DBM paper (Alwine et al. 1977) in the form of "dot blots" (Kafatos et al. 1979) or conventional northern blots (Thomas 1980) of RNA fractionated by electrophoresis on agarose gels after denaturation (Efstratiadis et al. 1980; Lehrach et al. 1977; McMaster and Carmichael 1977; Schwinghamer and Shepherd 1980). Levels of a specific mRNA species can then be estimated by comparing the relative intensities of bands or dots after hybridizing with labeled DNA probes.

Use has been made of these techniques to demonstrate seed-tissue-specific accumulation of storage protein mRNA species in soybean, french bean, and pea. Isolation of RNA preparations from developing seeds at different stages of development has shown the presence of storage protein mRNA species predominantly during the cell expansion phase of seed development.

Measurement of relative levels of mRNA species can give only an estimate of the balance between mRNA synthesis and degradation, and because many eukaryotic mRNA species have been shown to be relatively long-lived, and posttranscriptional processing may operate to rapidly degrade certain RNA sequences (Bathurst et al. 1980), the steady-state mRNA level may not give a true reflection of the transcriptional activity of specific genes. However, the increases and decreases in the levels of storage protein mRNA species during seed development are commonly considered to indicate the onset and cessation of transcription of the corresponding genes (Goldberg et al. 1981*b*).

mRNA transcription in isolated nuclei

A more direct estimate of the transcriptional activities of specific genes in vivo can be made by isolation of nuclei (also chloroplasts or mitochondria) followed by incubation with radioactive uridine triphosphate (UTP). The highly labeled newly synthesized RNA can then be assayed for specific sequences using the hybridization techniques outlined earlier, but with the labeled RNA as a probe for unlabeled (and usually immobilized) cloned DNA species. This approach has been used by Gallagher and Ellis (1982) to show transcription of nuclear-encoded chloroplast genes in pea leaves and by Evans and associates (1984) to show specific transcription of storage protein genes in developing pea seeds. Because transcription of specific genes, even those whose protein products are very abundant, forms a very low ($< 10^{-4}$) proportion of the total nuclear transcription, quantitation of the expression of particular genes is difficult in plant systems (Gallagher and Ellis 1982). A high background level of hybridization of the labeled RNA transcribed by pea seed nuclei prevented quantitative assessment of storage protein gene expression (Evans et al. 1984).

Expression of cloned plant genomic DNA in vitro

Assay of endogenous promoters on cloned genomic DNA sequences has been attempted by measuring expression of those sequences in in vitro transcription systems. The systems employed include (1) transcription systems prepared from cell lysates, usually human HeLa cells, with or without the addition of extra protein components (Manley et al. 1980; Weil et al. 1979), (2) microinjection of genes into *Xenopus* oocytes (Metz and Gurdon 1977) (both methods have been reviewed by Wickens and Laskey 1982), and (3) transformation of animal cells by circular viral DNAs (e.g., SV40) incorporating the genomic sequences, as reviewed by Rigby (1982). These systems are potentially useful for investigation of the short-range controls of gene expression, the TATA and CAT sequences found in most seed protein genes, as will be discussed later.

A cloned storage protein gene from maize has been shown to be transcribed by RNA polymerase II in both a HeLa transcription system and when microinjected in *Xenopus* oocytes (Langridge and Feix 1983); the different transcription systems appeared to recognize different promoter sequences. The oocyte system has the potential advantage of being able to translate storage protein mRNA species efficiently and will also carry out some posttranslational modifications (proteolysis and glycosylation) on the polypeptides produced (Bassüner et al. 1983*a*; Hurkman et al. 1981; Matthews et al. 1981; Tumer et al. 1982).

Engineering of cloned DNA sequences

Alterations of cloned DNA sequences will be discussed in a later section; these include codon alteration by single base-pair changes to alter the amino acids coded, deletion of DNA sequences, addition of new DNA sequences, and substitutions (deletion and addition). Although we are not aware of any application of these methods to the manipulation of seed protein genes, the technology (DNA engineering)) is well established and has been demonstrated to work for other gene systems. The methodology and enzymology of DNA engineering have been extensively reviewed (Lathe et al. 1983; Maniatis et al. 1982; Mooi and Gaastra 1983; Singer 1979; Weissman et al. 1979).

Genetic engineering techniques for studying the structure and synthesis of seed proteins from major crop species

In this section we describe a number of examples of seed storage protein systems that have been studied using recombinant DNA technology and the associated methods discussed in the preceding section. The descriptions involve an account of seed protein structure and synthesis, followed by additional confirmatory information derived from work with cloned cDNAs and genes.

Maize (*Zea mays* L.)

The properties of maize seed proteins have been extensively reviewed by Wall and Paulis (1978). The major storage proteins are prolamins called zeins, which constitute up to 60% of the total grain protein (Table 8.2) (Soave and Salamini 1984). Glutamine, proline, leucine, and alanine make up more than half of the total amino acids in zein and account for its insolubility in aqueous buffers (Mossé 1966; Wall and Paulis 1978). Dierks-Ventling and Cozens (1982) have established immunological relatedness between zein polypeptides and certain wheat and barley prolamin polypeptides. Zeins are exclusively located in the endosperm cells (Dierks-Ventling and Ventling 1982) within membrane-bound protein bodies (Burr and Burr 1976; Khoo and Wolf 1970). Maize protein bodies are unusual in having polysomes attached to the outer surfaces of their membranes, which actively synthesize zeins and deposit them within the lumen of the protein body (Burr and Burr 1976; Burr et al. 1978; Viotti et al. 1978). Zein proteins are made up from a large heterogeneous family of related polypeptides that have been classified into two major types (M_r 22,000 and M_r 20,000), both of which exhibit some size and considerable charge heterogeneity, and at least two other minor types (M_r 14,000 and M_r 10,000), which show much less size and charge heterogeneity (Burr and Burr 1981; Lee et al. 1976; Park et al. 1980; Rhighetti et al. 1977; Soave and Salamini 1984). As many as 28 zein polypeptides have been revealed by two-dimensional electrophoresis (Hagen and Rubenstein 1980). Such heterogeneity is also shown in in vitro synthesized zein polypeptides, reflecting heterogeneity in the zein mRNAs and genes (Burr et al. 1978; Larkins and Hurkman 1978; Larkins et al. 1979).

Several groups have cloned and isolated cDNAs specifically encoding different zein polypeptides (Table 8.4). Such cDNAs have been used to study the complexity and relatedness of zein mRNAs by using restriction mapping (Burr et al. 1982; Feix et al. 1981; Marks and Larkins 1982; Weinand et al. 1979), hybrid-release translation (Marks and Larkins 1982; Park et al. 1980; Pedersen et al. 1982; Viotti et al. 1982), dot-blot hybridizations (Burr et al. 1982; Marks and Larkins 1982), northern-blot hybridizations (Viotti et al. 1982), and directly by nucleotide sequencing (Geraghty et al. 1981, 1982; Marks and Larkins 1982; Pedersen et al 1982). Generally these studies have established at least five or six different classes of zein sequences based on the different levels of cDNA-cDNA or cDNA-mRNA cross-hybridizations at various stringencies or by sequence homologies (Geraghty et al. 1982; Messing et al. 1983; Rubenstein 1982). The zein family tree devised by Messing and associates (1983) summarizes this classification and collates much of the diverse data published by differeng groups. Subfamilies B49 and B36 encode mainly M_r 22,000 zein polypeptides; A20, A30, and B59 encode mainly M_r 20,000 zein polypeptides. Homology is highest between members of each subfamily, and then within the subfamilies encoding the same size class, although sequences homologous to one size class may encode a polypeptide of another size class (Messing et al. 1983; Soave and Salamini 1984; Viotti et al. 1982). The cDNAs encoding the M_r 14,000 polypeptides may form a sixth subfamily possibly related to one of the M_r 20,000 subfamilies (Marks and Larkins 1982). Sequencing studies have shown a broad overall homology between all zein cDNAs and genomic clones thus far studied. Thus, there was more than 95% homology between DNA sequences within the A30 (M_r 20,000) zein subfamily (Hu et al. 1982), 85% homology between two members of the A20 (M_r 20,000) and A30 subfamilies, but only 65% homology between these and a B49 (M_r 22,000) cDNA (Geraghty et al. 1982; Marks and Larkins 1982).

Additional studies using characterized zein cDNAs have included quantitative assessment of zein-specific mRNAs and primary transcripts. Thus, Larkins and associates (1983) analyzed the levels of different mRNAs during maize seed development using northern-dot-blot hybridizations with cDNAs specific for the M_r 14,000, M_r 20,000, and M_r 22,000 zein polypeptides. In a normal maize line, all the mRNAs appeared 10–12 days after pollination, reached a maximum abundance at 18–22 days, and declined slowly after 28–35 days. In contrast, an opaque-2 mutant line showed a much slower accumulation of all zein mRNAs starting at 16 days, much lower mRNA levels, especially those for the M_r 22,000 zeins, and a very rapid decline after 22 days. These results confirm and extend previous observations on the opaque-2 mutant line (Burr and Burr 1982; Jones et al. 1976; Langridge et al. 1982; Pedersen et al. 1980; Tsai 1979). Furthermore, it is apparent that the different zein gene families are coordinately expressed and therefore presumably under the same developmental control mechanism (Di Fonzo et al. 1980; Larkins et al. 1983; Soave and Salamini 1984).

Northern-blot hybridizations of electrophoretically size-fractionated endosperm RNA, using zein cDNAs, have revealed the presence of multiple discrete zein precursor RNAs approximately 1,800, 2,800, and 3,800 nucleotides long, as well as lesser amounts of the mature zein mRNA about 900 bp in length (Langridge et al. 1982). The 1,800-base transcript has been shown to arise from a second active promoter (P_1) 900 bp upstream from the promoter (P_2) from which the "normal" zein mRNA is transcribed (Langridge and Feix 1983).

The complexity and organization of the zein genes have also been studied using zein cDNAs as probes in Southern-blot hybridizations, for in situ hybridization and for the isolation of genomic clones from gene libraries. Estimates of the number and sequence complexity of the different zein genes in the maize genome by restriction (e.g., Eco-RI, Bam-H1, and Hind-III) and Southern-blot analyses have been carried out by several groups (Burr et al. 1982; Pedersen et al. 1982; Weinand and Feix 1980). Because of the varying degrees of homology between zein sequences, conditions must be selected whereby cross-hybridization is minimized, and appropriate gene copy equivalents of standard sequences (reconstruction or "titration" analysis) must be included in such analyses for specific zein genes to be enumerated (Burr et al. 1982; Hagen and Rubenstein 1981; Larkins et al. 1983). Generally, specific zein cDNAs hybridized to a large number (10–20) of different sizes of genomic DNA restriction fragments, indicating that the related zein sequences are found within a wide range of different DNA sequences. Hybridization patterns with different cDNA probes were largely unique, although fragments apparently hybridizing to more than one cDNA were evident, indicating cross-homology or that different zein genes may be clustered together (Burr et al. 1982; Viotti et al. 1982; Weinand and Feix 1980). Whereas several restriction fragments contained only single gene copies, many others contained multiple gene copies or reiterated sequences (Burr et al. 1982; Hagen and Rubenstein 1981; Viotti et al. 1982). Various estimates of the total numbers of zein genes using these methods have been reported: 40–50 (Larkins 1983), 90 (Hagen and Rubenstein 1981), and 50 (Burr et al. 1982) copies per haploid genome, although it is by no means certain if all these genes are functional (Burr et al. 1982; Hagen and Rubenstein 1981). Such estimates are lower than those of the cDNA-mRNA and cDNA-genomic-DNA hybridization kinetic studies of Viotti and associates (1979) and Soave and associates (1980), who reported 100–160 gene copies and 120 gene copies, respectively, but they are much higher than the 6–15 gene copies proposed by Pedersen and associates (1980). Interestingly, whereas the M_r 22,000 and M_r 20,000 zein genes are present in many copies (26–30 copies), the M_r 14,000 genes are present in only two or three copies (Larkins et al. 1983).

Viotti and associates (1982) have used zein cDNA probes for in situ hybridization to microsporocyte squashes to specifically locate different zein genes on maize chromosomes. Thus, certain genes for the M_2 22,000 zeins were located on the distal end of the long arm of chromosome 4, and genes for M_r 20,000 zeins were located on the proximal part of the long arm of chromosome 4, on the long arm of chromosome 10, and on the short arm of chromosome 7, partly confirming and extending conclusions from classic genetic analyses (Soave et al. 1981; Soave and Salamini 1984).

Several zein genomic clones have been isolated using zein cDNA probes (Tables 8.4 and 8.6) and sequenced (Hu et al. 1982; Pedersen et al. 1982; Spena et al. 1982, 1983). Various structural features have emerged from the cloned genes and cDNAs. None of the genes thus far examined has any introns, as judged by size comparisons of gene and cDNA restriction fragments (Lewis et al. 1981), by R- and D-looping studies (Pintor-Toro et al. 1982; Weinand et al. 1981), or by direct nucleotide sequencing (Hu et al. 1982; Pedersen et al. 1982; Spena et al. 1982, 1983). The gene base compositions are unusual in having a GC content possibly as low as 30% (Weinand et al. 1981), and the codon usage is nonrandom, showing marked preferences for certain amino acid codons (Geraghty et al. 1981, 1982).

Calculations of the true M_r values of the zein polypeptides from the amino acids encoded by gene or cDNA sequences give values much higher than those estimated from SDS gels, presumably because of anomalous behavior of zeins due to their unusual amino acid compositions (Geraghty et al. 1982; Hu et al. 1982; Marks and Larkins 1982; Spena et al. 1982).

Full-length cDNAs and genes encode leader sequences of 21 amino acids (Geraghty et al 1982; Hu et al. 1982; Marks and Larkins 1982; Pedersen et al. 1982; Spena et al. 1982), confirming that zein polypeptides are synthesized as preproteins in vivo, and in vitro from mRNA and polysomes. Such leader sequences are cotranslationally removed by plant membrane systems or in zein-mRNA-injected oocyte systems (Burr and Burr 1982; Burr et al. 1978; Hurkman et al. 1981; Larkins and Hurkman 1978; Larkins et al.

1979; Park et al. 1980). Possibly the most interesting feature of the zein gene coding sequences is the presence of tandem repeats of nucleotide sequences that encode similarly repeated blocks of amino acid sequences about 20 residues long (Argos et al. 1982; Geraghty et al. 1981, 1982; Marks and Larkins 1982; Pedersen et al. 1982; Spena et al. 1982). Similar tandem repeats have been indicated by direct protein sequencing (Esen et al. 1982). Whereas the 5′ and 3′ coding regions of the cDNAs are unique within a particular sequence, the central regions encode up to nine of these amino acid sequence repeats that are largely conserved between and within different zein polypeptides (Marks and Larkins 1982). These features are clearly illustrated by dot-matrix comparisons of nucleotide and amino acid sequence data (Marks and Larkins 1982; Pedersen et al. 1982). It is apparent from the large number of these amino acid repeats and their conservation in different zein polypeptides that they probably form the major structural elements of the proteins. Based on this notion and on additional physical data, Argos and associates (1982) have proposed the three-dimensional structural model for zein proteins illustrated in Figure 8.1. Such a model assumes the formation of antiparallel α helices from the 20-amino-acid tandem repeats, linked by glutamine "spacers" at the ends of the repeats. Polar residues in adjacent helices are then in juxtaposition to form hydrogen bonds, and the nine helices are thus assembled and held together in a cylindrical rod-shaped molecule. Although the repeats are highly conserved, variations in the sequences such as deletions/additions and substitutions are apparent, but would not be expected to severely disrupt this structure (Marks and Larkins 1982). Similar suggestions regarding the zein structure have been advanced by Spena and associates (1982). Within the same region and superimposed on these repeats, a major tandem duplication of a 96-base-pair sequence encoding 32 amino acids has been described in one of the zein genes (Z4) and in one homologous cDNA (ZG7), but it was present in one copy only in other homologous cDNAs (A30, ZG31A) (Heidecker and Messing 1983; Hu et al. 1982). Pedersen and associates (1982) have noted a similar duplication in one of their zein cDNAs (pZ19.1) that is present in one copy in the homologous genomic clone (λZG99). The presence of this duplication represents a difference of about M_r 3,000 in the encoded polypeptides that corresponds to the size difference between the actual molecular weights of the major zein polypeptides. This may therefore be one of the

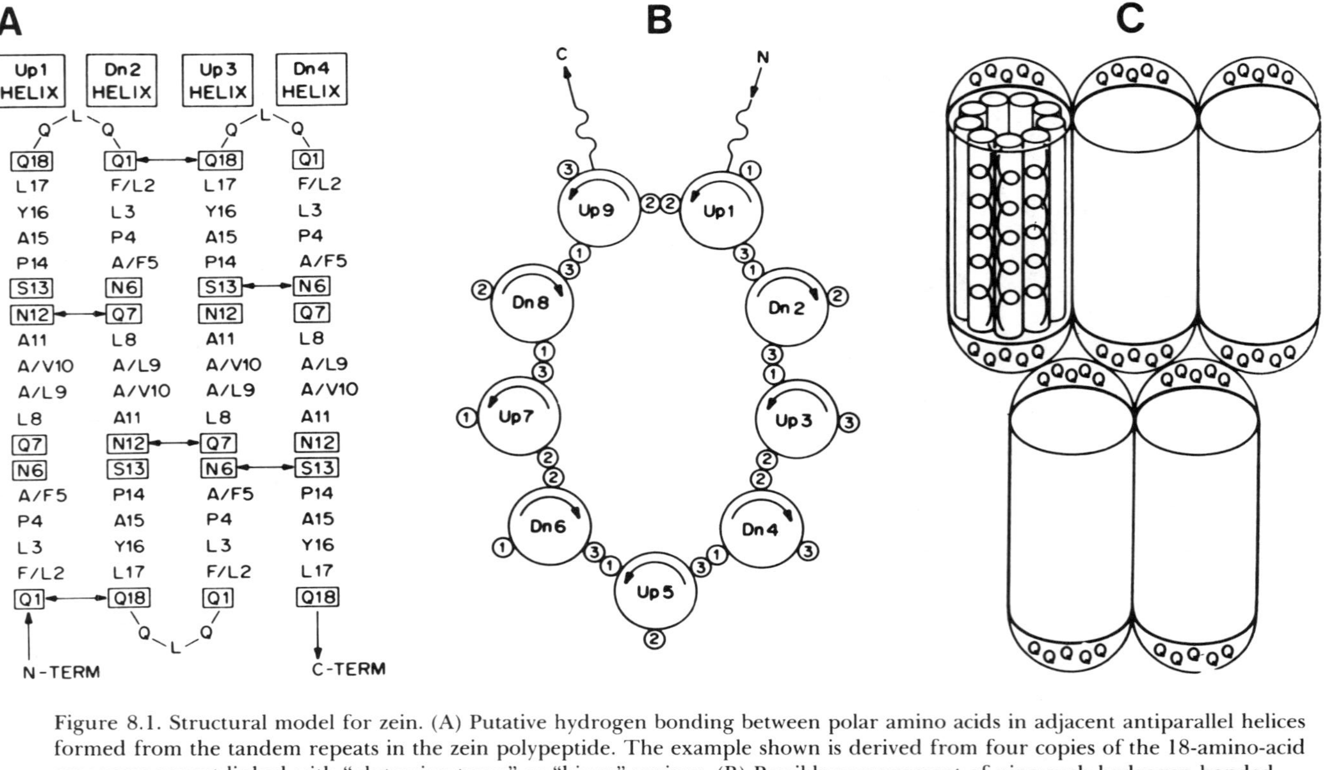

Figure 8.1. Structural model for zein. (A) Putative hydrogen bonding between polar amino acids in adjacent antiparallel helices formed from the tandem repeats in the zein polypeptide. The example shown is derived from four copies of the 18-amino-acid consensus repeat linked with "glutamine turns" or "hinge" regions. (B) Possible arrangement of nine such hydrogen-bonded antiparallel helices in the zein molecule. The additional two repeats mentioned by Messing et al. (1983) presumably would add an additional pair of helices. (C) Suggested packing arrangement of zein molecules allowing stabilization through hydrogren bonding involving the glutamines in the hinge regions. Reproduced from Argos et al. (1982), with permission.

underlying reasons for certain genes and cDNAs encoding the M_r 22,000 size class polypeptide cross-hybridizing to contain cDNAs or mRNAs encoding M_r 20,000 polypeptides (Geraghty et al. 1982; Hu et al. 1982; Park et al. 1980; Rubenstein 1982). Messing and associates (1983) have suggested that this duplication introduces an additional two helices in the foregoing protein structural model (Figure 8.1), giving it a total of 11 rather than 9 helices. On the other hand, Spena and associates (1982) have reported a single base change in the sequence of an M_r 22,000 zein gene (ZA1) that introduces a premature termination codon, so that the gene consequently may translate a polypeptide of only M_r 20,000. Similar single-base changes have been shown in two closely adjacent, possibly defective, zein (pseudo) genes (ZE19, ZE25), one producing a termination codon after only 39 amino acids, the other changing the initiation codon (ATG) to a leucine codon (CTG) (Spena et al. 1983).

The various putative consensus control sequences in the 5′and 3′ flanking regions of many eukaryotic genes (Messing et al. 1983) are also represented in one form or another in the zein genes. These include at least one TATA box and one AGGA/CAT box (Hu et al. 1982; Messing et al. 1983; Pedersen et al. 1982; Spena et al. 1982), though Langridge and Feix (1983) have found duplicates of these in the vicinity of the active P_1 promoter; polyadenylation sites have been found in the 3′ flanking regions, with variations in sequence, position, and copy number (Messing et al. 1983), and there is some evidence that suggests the use of alternative polyadenylation sites (Geraghty et al. 1982; Heidecker and Messing 1983; Marks and Larkins 1982; Messing et al. 1983). The structural features of the zein gene system, such as the internal tandem repeats, the close homology of sequences between and within different gene subfamilies, and the "clustering" of the genes, are all consistent with the idea that the zein gene family has evolved from a common short ancestral sequence (Soave and Salamini 1984). Such a process may have involved internal tandem duplications within the archetypal zein gene, followed by other evolutionary events such as gene duplication (amplification) and divergence by mechanisms such as accumulation of point mutations, insertion and deletion of sequences, and translocation of DNA sequences to other chromosomal sites (Geraghty et al. 1982; Hu et al. 1982; Spena et al. 1982). The possible involvement of transposon-like mechanisms in these rearrangements has been suggested (Spena et al. 1982) and may have been involved in the evolution of the complex, highly efficient double-promoter system proposed for at least one zein gene (Spena et al. 1983). The possible involvement of intrachromosomal duplication has been suggested by the patterns of zein genes on chromosomes 4 and 7 (Soave et al. 1981, 1982). The heterogeneity in the zein coding sequences and the high gene copy number indicate (1) that the zein protein structure is flexible enough to tolerate such sequence alterations without significantly interfering with its functional properties as a nitrogen store and (2) that the maize genome is also flexible (plastic), which, in conjunction with indication 1, and possibly low selection pressure, has led to the accumulation and maintenance of large numbers of the genes.

Barley (*Hordeum vulgare* L.)

Various aspects of barley seed proteins are dealt with in reviews by Cameron-Mills and associates (1980), Nelson (1980), Doll (1983), Miflin and associates (1983*a*), and Shewry and Miflin (1983). The major storage proteins of barley are prolamins (Table 8.2) called hordeins, which account for about 40–45% of the total grain protein and typically contain high levels of proline and glutamine and low levels of lysine (Doll 1983, 1984; Shewry and Miflin 1983). The hordeins, like the zeins, are endosperm-specific and are synthesized on membrane-bound polysomes as preproteins from which the leader sequences are cotranslationally removed; the proteins are subsequently deposited in protein bodies (Brandt and Ingversen 1978; Cameron-Mills et al. 1978; Matthews and Miflin 1980). Unlike the situation in maize, barley protein body membranes appear to lose their integrity during development (Miflin and Burgess 1982).

The hordeins form a large heterogeneous family of related polypeptides with a size heterogeneity more extensive than that of the zeins. Hordein classification is largely based on polypeptide size and consists of two main groups, B hordeins (M_r 30,000–51,000) and C hordeins (M_r 57,000–86,000), which are further divided into polypeptide subgroups and numbered in order of increasing molecular size (e.g., B_1, B_2, B_3 and C_1, C_2) (Holder and Ingversen 1978; Rahman et al. 1983). The B hordeins are sulfur-rich prolamins containing 2.5% cysteine and 0.6% methionine and thus have been further classified according to their CNBr cleavage patterns (dependent on the numbers and positions of methionine residues) to give classes I, II, and III (Faulks et al. 1981; Shewry et al. 1984). The two major B_1 polypeptides, for example, are of the class I type (Faulks et al. 1981). The C hordeins are almost devoid of sulfur amino acids and are classed as sulfur-poor prolamins (Shewry et al. 1978). The B and C hordeins together account for 95% of the total hordein fraction, and a third group, the D hordeins or high-molecular-mass (or -weight) prolamins ($M_r > 100{,}000$), occupies less than 5%. The "A hordein" class differs considerably from the B and C classes in solubility, amino acid composition, lack of genetic variability, and absence from protein bodies, and it is no longer considered a hordein, but nevertheless is of interest because of its high contents of lysine, arginine, methionine, and cysteine (Salcedo et al. 1980; Shewry and Miflin 1983).

The major hordein groups exhibit considerable size and charge heterogeneity; two-dimensional gels resolve 20–30 polypeptide components, depending on the cultivar (Faulks et al. 1981; Shewry et al. 1984). Such heterogeneity has also been reflected by the polypeptides from in vitro translation of barley endosperm poly(A)$^+$ RNA, confirming the extensive hordein polymorphism (Kreis et al. 1983*a*). The molecular weights of the hordein polypeptides as determined by SDS electrophoresis are likely to be erroneous because of their high proline content (Shewry and Miflin 1983). Partial homology between B and C hordeins has been reported (Holder and Ingversen 1978; Schmitt and Svendsen 1980), although their amino acid compositions differ appreciably, the C hordeins having little or no cysteine and methionine (Miflin et al. 1983*b*; Shewry and Miflin 1983), and B hordein cDNAs did not select any C hordein mRNA in hybrid-release translation experiments (Forde et al. 1981).

Results from protein and nucleic acid sequencing to be described later, indicate a high degree of structural homology between polypeptides within each of these groups (Shewry and Miflin 1983). There are also strong similarities between the same hordein groups in different barley cultivars, and these homologies extend to similar polypeptide classes in other cereals, such that Miflin and associates (1983*a*) have proposed a much broader classification scheme relating the prolamins of barley, wheat, and rye.

Classical genetic studies have established that gene loci encoding the B and C hordeins, designated *Hor-2* and *Hor-1*, respectively, are closely linked and lie on the short arm of chromosome 5. *Hor-3*, the D hordein locus, is on the long arm of chromosome 5 (Shewry et al. 1984).

cDNAs synthesized from barley endosperm poly(A) RNA have been cloned, characterized, and used for a number of studies (Table 8.4). Cross-hybridization experiments with these cDNAs have demonstrated the complexity and relatedness within the hordein mRNA population. Thus, a Venn classification of a library of cDNA clones by colony hybridization using a random sample of cDNA probes showed many clones hybridizing with two or more cDNAs (Forde et al. 1981). Moreover, several cross-hybridizing cDNAs hybrid-selected mRNAs encoding different groups of B hordeins, often of widely differing molecular sizes (Forde et al. 1981), confirming a similar result of Brandt (1979). Absolute confirmation of the identity of such clones was obtained by nucleotide sequencing, which showed exact matches of predicted amino acid sequences with the determined sequences of two isolated peptides and a C-terminal peptide of B_1 hordein (Forde et al. 1981). Detailed analysis of the types of hordein polypeptides encoded by mRNAs hybrid-selected by certain B hordein cDNAs has led Kreis and associates (1983*a*) to confirm that the *Hor-2* (B hordein) locus is very complex, containing at least two distinct, though related, multigene families.

Further confirmation of the multigenic nature of the *Hor-2* locus has been provided by Southern-blot analysis of the hordein genomic restriction fragments hybridized with cDNA probes. The results showed clearly that at least 13 copies of B-hordein-related genes are present per haploid genome and that there is no amplification of these genes during endosperm development (Kreis et al. 1983*b*). Comparison of these results with an identical analysis of DNA from the mutant derivative Risø 56 showed the loss of 11 of the 13 B hordein genes (as judged by the absence of restriction fragments hybridizing to cDNA), corresponding to a structural deletion of at least 80–96 kb and accounting for the absence of the main B hordein polypeptides from this line (Kreis et al. 1983*a*, 1983*b*). Such figures, although only minimum estimates, indicate that, like the zein gene families, the B hordein genes may be highly clustered – 11 genes per 80 kb of DNA.

Some of the structural features of the B_1 hordein genes have been revealed by the cDNA nucleotide sequences (Rasmussen et al 1983). Data for four non-full-length cDNAs have been presented. At least two of the cDNAs, although exhibiting a high degree of homology, encoded different B_1 hordeins, as judged by changes in coding sequence. Comparison of the predicted amino acid sequences with that determined for B_1 hordein shows close homology, with 61 of 74 amino acids identical. Like the zeins, the hordein nucleotide sequence encodes a number of amino acid sequence repeats, which consist of eight heptapeptide sequences in two groups of four homologous sequences. These do not all appear to be tandem repeats, although three pairs out of the eight sequences might represent such features, because none of the sequences are dispersed at regular positions. Similar octapeptide repeats have been reported in C hordein by means of protein sequencing (Shewry et al. 1984). Therefore, although the hordeins do not appear to have the regular primary structural features of the zeins, the repeat sequences almost certainly have a structural significance, and the prominence of glutamine residues in these repeats is probably also of importance (see the zein structural model in Figure 8.1). The high cysteine content and its inclusion in some of the repeats may indicate the involvement of disulfide bridges within and between hordein proteins, possibly in the final deposition and packaging of the protein in the protein bodies (Rasmussen et al. 1983).

The hordein mRNA is unusual in having three contiguous stop codons and a nonrandom codon usage (Forde et al. 1981; Rasmussen et al. 1983). Another feature shown in the 3′ noncoding regions is the occurrence of three putative polyadenylation signals, and evidence from two cDNAs shows that alternative signals 1 and 3 may be used in processing transcripts from the same gene (Rasmussen et al. 1983).

The various features of the hordein gene structure and the composition of its family are consistent with an origin and process of evolution similar to that described for zein.

Under conditions of limiting sulfur nutrition in barley, synthesis of the sulfur-rich B hordeins is markedly depressed, and the sulfur-poor C hordeins are increased (Shewry et al. 1983*b*). To elucidate the nature of the controls operating in this phenomenon, Rahman and associates (1983) used cDNAs encoding B_1, B_3, and C_1 hordeins to quantify the corresponding mRNAs in developing barley endosperm in sulfur-stressed and unstressed plants using northern-dot-blot hybridizations. The results indicate a two-tier mechanism operating under conditions of sulfur stress: One is a direct transcriptional effect whereby the synthesis of B_1 and B_3 mRNAs is markedly reduced in proportion to the sulfur amino acid content of the proteins, while C hordein mRNA is unaffected; the other is a translational effect whereby the B_1 and B_3 hordein mRNAs are translated less efficiently, presumably because of limiting sulfur amino acids, whereas the C hordein mRNA is more efficiently translated because of reduced competition by other mRNAs (Miflin et al. 1983*b*; Rahman et al. 1983).

Wheat (*Triticum aestivum* L.)

A brief description of the wheat storage proteins is difficult because of the large number of components, the high degrees of size and charge heterogeneity, and difficulties in defining solubility classes experimentally. However, two-dimensional electrophoresis systems such as nonequilibrium pH-gradient electrophoresis (NEPHGE), or isoelectric focusing (IEF), or low-pH aluminum lactate electrophoresis, coupled with SDS electrophoresis, are capable of resolving the main groups of polypeptides (Jackson et al. 1983; Payne 1983; Payne et al. 1984; Shewry et al. 1984). Two such systems are shown in Figure 8.2. Like maize and barley, the prolamin proteins in wheat form the major fraction (> 90% total grain protein) and consist of "gliadins" and "glutenins" (Tables 8.1 and 8.2). The proteins in the latter group are insoluble in the usual prolamin solvents (70% ethanol, 55% isopropanol), but they are now regarded as highly aggregated and disulfide-linked prolamin-type polypeptides, distinct from other insoluble proteins (the "true glutelin fraction") that comprise structural and metabolic proteins and are not deposited in protein bodies (Table 8.1) (Payne et al. 1984; Shewry et al. 1984). The prolamins and other wheat proteins have been extensively reviewed by Kasarda and associates (1976), Kasarda (1980), Miflin and Shewry (1981), Payne and Rhodes (1982), Porceddu and associates (1983), Payne and associates (1984), and Shewry and associates (1984).

The gliadins and glutenins are both endosperm-specific proteins and are deposited within protein bodies that, like those of barley, lose their integrity during development (Greene 1981; Miflin and Burgess 1982). Both protein groups are very rich in glutamine (approximately 1 in every 3 residues) and proline (approximately 1 in every 6 residues) and low in basic and acidic residues. The gliadins have undergone intensive study because of their ease of extraction and the importance of their nutritional and functional properties in bread making (Payne and Rhodes 1982). In contrast to the other cereal prolamins, the gliadins have been classified on the basis of

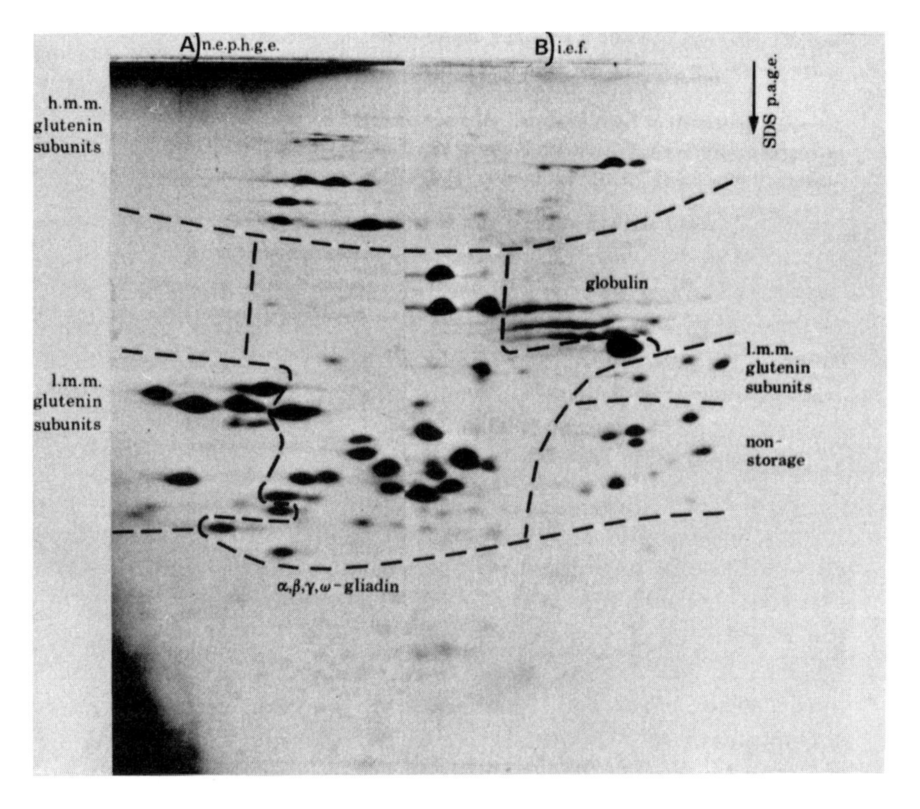

Figure 8.2. Fractionation of wheat total endosperm proteins by two different two-dimensional electrophoresis systems: (A) nonequilibrium pH-gradient gel electrophoresis (n.e.p.h.g.e.), followed by SDS polyacrylamide gel electrophoresis (SDS p.a.g.e.), and (B) isoelectric focusing (i.e.f.) followed by SDS p.a.g.e. Reproduced from Payne et al. (1984), with permission.

their mobilities in aluminum lactate electrophoresis into α (or A), β, γ, and ω gliadins, in decreasing order of mobility. The α, β, and γ gliadin polypeptides are sulfur-rich prolamins, and all have molecular weights in the range of M_r 30,000–45,000, whereas the ω gliadins are sulfur-poor prolamins devoid of sulfur amino acids and are larger, with M_r of 65,000–80,000 (Shewry et al. 1984). The polypeptides all appear to exist as single-chain monomeric molecules, with (α, β, and γ gliadins), or without (ω gliadins) internal disulfide bonds (Caldwell 1983). In hexaploid wheat the gliadins can be resolved into about 45 polypeptide components by two-dimensional systems (Payne and Rhodes 1982; Shewry et al. 1984). α, β, and γ gliadins are all related, but can be further classified into two groups (α, β, and γ_1 gliadins and γ_2 and γ_3 gliadins) based on homologies and differences in their *N*-terminal sequences (Payne and Rhodes 1982).

The glutenin polypeptides, solubilized with the aid of reducing agents and therefore present in wheat as intermolecular disulfide-linked aggregates, consist essentially of two groups: the high-molecular-weight (HMW) or high-molecular-mass (HMM) glutenins (M_r 95,000–145,000) (Payne et al. 1980*b*) and the low-molecular-weight (LMW) or low-molecular-mass (LMM) glutenins (M_r 36,000–44,000), also known as aggregated gliadins or HMW gliadins because of their similarities to the α, β, and γ gliadins (Payne et al. 1984; Shewry et al. 1983). Overall, the storage proteins of wheat endosperm consist of 50% gliadin, 10% HMM glutenins, and 40% LMM glutenins (Payne et al. 1984).

Payne and associates (1984) have summarized the results of classic genetic analyses to establish the chromosomal location of the genes encoding the various wheat storage protein groups. Thus, the genes occur at nine different loci: *Glu-A1, Glu-B1,* and *Glu-D1,* coding for the HMM glutenins, are found on the long arms of chromosomes 1A, 1B, and 1D; *Gli-A1, Gli-B1,* and *Gli-D1,* coding for the ω gliadins, γ gliadins, and LMM glutenins, are found on the short arms of chromosomes 1A, 1B, and 1D; *Gli-A2, Gli-B2,* and *Gli-D2,* encoding the α and β gliadins, are found on the short arm of chromosome 6. The loci are all complex each containing multiple genes clustered together (Payne et al. 1984).

α, β, and γ gliadin polypeptides are all synthesized on membrane-bound polysomes and appear to have leader sequences that are presumed to be cotranslationally removed before deposition (Bartels and Thompson 1983*a*; Donovan et al. 1982; Greene 1981; Okita and Greene 1982). Interestingly, gliadin mRNAs can be enriched either on poly(U) agarose because of their poly(A) tails or on poly (G) agarose because of the preponderance of cytosine bases – a reflection of the high proline content of gliadins (Okita and Greene 1982). This feature has also proved useful in identifying wheat and barley prolamins synthesized in vitro by differential incorporation of ^{3}H-proline (10–30% in prolamins) and ^{3}H-lysine (0.1–1.3% in prolamins) (Bartels and Thompson 1983*a*; Forde et al. 1981; Shewry et al. 1984). Greene (1983) has proposed that the gliadin mRNAs are long-lived species and that this factor and transcription are the major factors controlling the synthesis of wheat storage proteins.

Bartels and Thompson (1983*a*) have produced cDNA banks from wheat endosperm poly(A) RNA (Table 8.4). Selected wheat cDNA clones were used in colony and Southern-blot filter hybridizations with endosperm poly(A) RNAs from different, though related, cereals, to assess the homologies between the abundant mRNAs transcribed during seed development. The results obtained with 100 wheat cDNAs showed 50% unique to wheat, 12% cross-hybridizing to barley only, 12% to rye only, 25% to both rye and barley, and none to oats (Bartels and Thompson 1983*a*, 1983*b*). Additionally, one group of wheat cDNAs strongly cross-hybridized to a barley B hordein cDNA (Bartels and Thompson 1983*b*; Forde et al. 1981). Subsequent characterization of these clones by hybrid-release translation showed that several different cDNAs encoded polypeptides of more than one gliadin type, and the ones encoding α/β or γ gliadins were identified. These findings are consistent with the ideas on the mechanisms of evolution of the cereal genomes in that the prolamin storage protein genes originated from an ancestral gene by gene duplication (amplification) and divergence, by accumulation of mutations. Thus, one might expect to find such sequence homologies in all (wheat, rye, barley) but the most distantly related species (oats), as well as differences within homologous sequences. The presence of homologous sequences between two species (wheat, barley) that are absent in a third (rye) indicates deletion or repression of its expression in the latter species. The presence of unique sequences (wheat) indicates the evolution of a new set of transcription units since the species diverged (Bartels and Thompson 1983*a*, 1983*b*).

The nucleotide sequence of one of the wheat cDNAs hybridizing strongly to barley poly(A) RNA and to a B hordein cDNA, and encoding a γ gliadin, has been obtained (Bartels and Thompson 1983*a*). Comparison of this sequence with partial nucleotide sequences from the B hordein cDNA (Forde et al. 1981) confirms the good sequence homology (~82%). The sequence contains at least three near-perfect tandem repeats of 42 base pairs encoding 14 amino acid sequence repeats. Within these repeats are several groups of hydrophobic amino acids (tetraplets) bounded by glutamines, reminiscent of the repeating structural units in maize (Figure 8.1), and they may have a similar function in the gliadins. Physical measurements on gliadin solutions indicate that the proteins have a very compact structure involving some α helix and probably involving intramolecular disulfide bridges and noncovalent bonds (Kasarda et al. 1976; Payne and Rhodes 1982).

Such interactions are further illustrated by the aggregation of α gliadins under certain conditions into microfibrillar chains visible by electron microscopy (Kasarda et al. 1976). The 3′ noncoding region has, like the B hordein cDNAs (Rasmussen et al. 1983), three putative polyadenylation signals, though there is as yet no evidence that alternative signals are used. Preliminary evidence from Southern-blot analysis of genomic DNA has indicated that the gliadin genes are present in high copy numbers (50 copies or more per haploid genome) (Litts et al. 1983; R. Thompson and D. Bartels, personal communication).

cDNAs encoding the HMW glutenins have also been isolated and sequenced (Forde et al. 1984; Shewry et al. 1983*a*; Thompson et al. 1983). The possible correlation between the HMW glutenin polypeptides and good bread-making quality has been suggested (Payne et al. 1981). Estimates of the HMW glutenin gene copy number by Southern-blot hybridization with specific cDNA have indicated only 2–3 genes per type-1 chromosome (8–10 per cell) (Thompson et al. 1983; R. Thompson and D. Bartels, personal communication). The low copy number should allow increased gene dosage by appropriate engineering and transformation to improve on bread-making capabilities as will be discussed later. Nucleotide sequencing has revealed the presence of in-frame tandem repeats of six amino acid units with a consensus sequence GQQXGQ (R. Thompson, personal communication). Protein purification and *N*-terminal sequencing have shown at least four sequence types of HMW glutenins, with a high degree of homology, and each with a cysteine residue in position 10, but differing by amino acid substitutions or deletions. All are rich in glutamine (Glu + Gln) (34–39%), proline (13–17%), and glycine (14–20%) and contain 0.4–1.5% cysteine (Shewry et al. 1983*a*). A cDNA sequence predicts the presence of a single cysteine in the *C*-terminal region. The amino acid composition and sequence data indicate the presence of significant α helix content, especially at the *N* and *C* termini, possibly with proline or glutamine hinge regions and the high glycine content that confers chain flexibility. The cysteine in the *N*- and *C*-terminal regions could allow disulfide-linked linear polymers of glutenin molecules, as predicted by Ewart (1977), with additional intermolecular disulfide bonds forming between the other cysteines and those in the LMW glutenins (aggregated gliadins). Hydrophobic interactions involving prolines and hydrogen bonds involving glutamines might then be implicated in the interactions of the α, β, γ, and ω gliadins with the glutenin assembly (Shewry et al. 1983*a*, 1983*c*).

Soybean (*Glycine max* L.)

Soybean has been the most intensively studied of the legume crop species, mainly because of its agronomic importance. It has the highest protein content of any of the legumes (up to 50% of the seed dry weight) and produces high yields of good-quality oil. Many different types of proteins are found in relatively large amounts, including the major storage globulins glycinin (11S legumin) and β conglycinin (7S vicilin), a number of other globulins including the α and γ conglycinins and the albumins, soybean lectin, trypsin and chymotrypsin inhibitors, and a number of enzymes such as urease. A typical distribution of the globulin proteins is glycinin 40%, β conglycinin 30%, α conglycinin 15%, and γ conglycinins 3%, with the remainder largely uncharacterized (Koshiyama 1969). The ratio of glycinin to β conglycinin can vary from 3:1 to 1:1 in different cultivars (Nielsen 1984). Soybean proteins have been reviewed by Koshiyama (1983), Larkins (1981, 1983), Brown and associates (1982), Nielsen (1984), Mossé and Pernollet (1983), and Pernollet and Mossé (1983). Glycinin and β and γ conglycinins are all located in cotyledon protein bodies (Koshiyama 1972*a*, 1972*b*). Glycinin is the major storage protein, of M_r 360,000, consisting of six subunits of M_r ~60,000 each. Each subunit consists of an A (acidic pI) or α polypeptide of M_r ~37,000, disulfide-linked to a B (basic pI) or β polypeptide of M_r ~22,000 (Catsimpoolas et al. 1971). When dissociated and reduced, the subunits separate into A and B polypeptides; dissociation without reduction maintains disulfide bonds and holds A and B polypeptides together. This molecular behavior is typical of the legumin type of seed storage proteins (Derbyshire et al. 1976). Extensive size and charge heterogeneity has been demonstrated in the A and B polypeptides by one- or two-dimensional electrophoresis (Kitamura et al. 1980), and at least six A-type and five B-type polypeptides have been separated and purified (Moreira et al. 1979). Homologies and differences have been demonstrated by immunological methods (Moreira et al. 1981*a*) and by cyanogen bromide cleavage and sequencing (Moreira et al. 1981*b*), such that whereas the A-type polypeptides are distinct from the B types, all the A types show degrees of homology with each other, as do the B types. Sequence heterogeneity within purified subunits has also been demonstrated, suggesting that subunit types may be encoded by more than one gene (Nielsen 1984). Staswick and associates (1981) were able to show that subunits were composed of nonrandom pairs of polypeptides; a specific A type was always associated with a specific B type, consistent with the suggestions of Evans and associates (1979) and Croy and associates (1980*a*) that legumin subunits are synthesized as a precursor polypeptide containing a contiguous A-B polypeptide that subsequently undergoes posttranslational cleavage to give the A and B polypeptides (see the later section on *Pisum sativum*). One apparent exception was indicated by an A-type polypeptide (A_4) that did not appear to pair with any B type and that was later shown to be a proteolytic cleavage product from a larger A polypeptide, the short end of which (A_5) remained disulfide-linked to its specific B polypeptide (B_3) (Iyengar and Ravenstein 1981; Staswick and Nielsen 1983). Glycinin subunits have been classified into two distinct groups (I and II), within each of which there is good homology (85–90%), but between the two there is only about 50% homology. Group I subunits additionally have much more methionine, and it has been suggested that nutritional improvement might be effected by replacing group II legumin subunits with group I (Nielsen 1984). Close sequence homology has also been demonstrated between the *N*-terminal sequences of the β/B polypeptides of glycinin and broad bean legumin (Gilroy et al. 1979) and between the *N*-terminal sequences of the α/A polypeptides of glycinin and pea legumin (Casey et al. 1981*b*).

Glycinin is synthesized on membrane-bound polysomes (Sengupta et al. 1981; Tumer et al. 1981), and is translated in vitro from poly(A) RNA as nonreducible HMW (M_r ~60,000) precursors, each containing an A and a B polypeptide sequence analogous to pea legumin synthesis (Barton et al. 1982; Croy et al. 1981*a*; Tumer et al. 1981, 1982). At least three discrete size classes of glycinin precursors have been demonstrated, with M_r of 58,000 to 63,000 (Barton et al. 1982; Sengupta et al. 1981; Tumer et al. 1981, 1982). The precursors are synthesized as preproteins, with 10–20 amino acid leader sequences that, as in other systems, are cotranslationally removed in vivo (Barton et al. 1982), or in an oocyte translation system (Tumer et al. 1982). Pulse-chase experiments have confirmed the precursor-produce relationship between the M_r ~60,000 polypeptides and authentic A and B polypeptides of glycinin (Barton et al. 1982; Sengupta et al. 1981), and furthermore Barton and associates (1982) have suggested that the precursor polypeptides assemble initially into 7S trimeric glycinin "half-molecules," also analogous to the pea legumin system (Spencer and Higgins 1980). Additionally, Barton and associates (1982), on the basis of developmental studies, have suggested that final assembly of the glycinin half-molecule into 11S hexameric glycinin may not occur until the proteolytic cleavage of precursors to A and B polypeptides has occurred.

Three immunologically distinct conglycinins have been described (Catsimpoolas and Ekenstam 1969): α conglycinin is a LMW (2S) globulin and is probably a monomeric protein; γ conglycinin is a minor 7S glycoprotein globulin (M_r ~104,000) containing about 5% carbohydrate and consisting of three subunits of the same size (Koshiyama and Fukushima 1976); β conglycinin (soybean vicilin) is the major 7S globulin (M_r 140,000–210,000) consisting of trimeric molecules made from three major types of subunits, designated α (M_r 54,000–76,000), α′ (M_r 54,000–86,000), and β (M_r 40,500–53,000) subunits, which are all related, have similar amino acid compositions and acidic isoelectric points, are devoid of cysteine, and are glycosylated, containing 4–5% carbohydrate (Beachy et al. 1981; Thanh and Shibasaki 1977; Yamauchi and Yamagishi 1979). In these respects, β conglycinin closely resembles the vicilins of other legumes (Derbyshire et al. 1976), although it is antigenically unrelated to most of them (Dudman and Millerd 1975). Six of the trimeric molecular species out of the 10 possible combinations of the three subunits have been demonstrated (Thanh and Shibasaki 1978), and all are capable of reversible assembly into hexameric (9S) molecules (Thanh and Shibasaki 1979). Other minor β conglycinin subunits have been isolated, designated γ (1–3) and δ subunits, and microheterogeneity has been shown in the β subunits, indicating at least four species (β_1–β_4) (Beachy et al. 1983; Thanh and Shibasaki 1977). Thus, β conglycinin, like glycinin, is probably encoded by a mutigene family. Cotranslational and posttranslational processing of β conglycinin appears to be very complicated. β conglycinin subunits are also synthesized on membrane-bound polysomes, and their leader sequences are cotranslationally removed by the endoplasmic reticulum (Sengupta et al. 1981). Moreover, these workers have shown that in the α and α′ subunits an initial glycosylation event (core glycosylation) takes place cotranslationally or immediately after proteolytic removal of the leader sequence and completion of the polypeptide (i.e., within 5 min). This is then followed by further glycosylation events over a period of up to 90 min, and finally either proteolytic cleavage or deglycosylation over a long period of 1.5 to 6 hr to give the authentic α and α′ subunits (see Figure 7 in Sengupta et al. 1981). Similar cotranslational and posttranslational modifications have been observed subsequently in phaseolin of *Phaseolus vulgaris* (Bollini et al. 1983). A similar scheme of events presumably takes place in the synthesis and processing of the β subunit, which is synthesized only later in development (Gayler and Sykes 1981; Hill and Breidenbach 1974).

Glycinin and β-conglycinin-specific cDNAs have been cloned (Table 8.4) and used to estimate the sizes and quantities of mRNAs during seed development. Using northern-blot hybridizations, Barton and associates (1982) have shown that glycinin-specific mRNAs are about 2,000 nucleotides long, adequate to encode the largest glycinin precursor polypeptide (M_r 63,000). Similarly, cDNAs specific for the three β conglycinin subunits have been used to demonstrate the synthesis of β conglycinin mRNAs~2,500 nucleotides long, adequate to encode the α and α′ precursor subunits, and 1,700 nucleotides long, adequate to encode the β precursor subunitss (Beachy et al. 1981; Goldberg et al. 1981*a*; Schuler et al. 1982*a*).

Goldberg and associates (1981*a*, 1981*b*), using single-stranded or cloned cDNAs in kinetic hybridization studies to assess the abundance and sequence complexity of mRNAs transcribed during embryogenesis, have identified a set of at least 7 to 10 superabundant mRNA classes that are coordinately expressed and accumulated to about 50% of the total embryo mRNA at the midmaturation stage and are absent from leaf tissue. Kinetic hybridization of cloned cDNAs to genomic DNA gave an estimate of the reiteration frequency (copy number) of the genes that encode these mRNA classes, later identified by hybrid-release translation as glycinin (A28) (four copies per haploid genome), β conglycinin (A16) (five copies), Kunitz trypsin inhibitor (A37) (five copies), and an unidentified sequence (A36) encoding an M_r 15,000 polypeptide (two copies) (Goldberg et al. 1981*a*). These values closely agreed with those obtained from Southern-blot hybridizations with genomic DNA (Fischer and Goldberg 1983) and the frequency of gene sequences represented in a gene library (Goldberg et al. 1981*a*, 1981*b*). Furthermore, specific cDNAs have been used in northern-blot hybridizations to quantify β conglycinin and glycinin mRNAs during development and have shown that they are most abundant during active storage protein accumulation, supporting the conclusions of Goldberg and associates (1981*b*) that developmental controls of the storage protein genes operate primarily at the transcriptional level. Additionally, Meinke and associates (1981) have shown that β conglycinin mRNAs are transcribed earlier than those for glycinin, similar to the findings for pea, and both are detectable before the polypeptides are detected. Levels of specific mRNAs decay in the latter stages of seed development when dehydration commences. The apparent stability of these messages has been suggested as the primary factor controlling the dramatic accumulation of superabundant mRNA classes (Goldberg et al. 1981*a*).

At the time of submission of this review we are not aware of any published work on glycinin cDNA or gene sequences, although such data have been obtained (N. Nielsen and R. Goldberg, personal communication) and have been submitted for publication (cited by Nielsen 1984). Three of the nonallelic glycinin genes shown by Southern-blot analysis, designated G_1, G_2, and G_3, have been isolated and characterized by restriction and S1 mapping and by R-looping (Fischer and Goldberg 1983). These studies have shown the presence of at least one intron and possibly two introns. Genes G_1 and G_2 have been shown to be linked on a single genomic clone isolated from "gene-walking experiments" (Goldberg et al. 1983*a*). Furthermore, glycinin G_2 and G_3 genes, as well as a number of other seed protein genes, are linked to non-seed-protein genes that are expressed in different plant tissues but are inactive in seeds. Sequence information on the glycinin gene has shown the presence of consensus TATA and CAT box sequences; a leader sequence of 18 amino acids and three introns of 238, 292, and 624 base pairs long. The transcript contains 43 base pairs of 5′ and 226 base pairs of 3′ noncoding sequences, the 3′ region containing three potential polyadenylation signals, the last of which is known to function (Nielsen 1984).

Comparison of protein and mRNA (cDNA) predicted amino acid sequences has revealed the cleavage site(s) between the A and B polypeptides in the glycinin precursor. As proposed originally for pea legumin (Croy et al. 1982), the cleavage is suggested to occur at paired basic residues (Lys-Arg, Arg-Arg, etc.) analogous to animal prohormone processing (Nielsen 1984), as will be described later. Further processing involving the removal of a four-amino-acid "linker" sequence would be necessary to produce the observed *N* terminal of the B polypeptide. The removal of a *C*-terminal pentapeptide is also indicated by such comparisons, and the presence of paired basic residues at the *N* terminal of this peptide and the A_4 cleaved acidic fragment suggest the involvement of a similar processing enzyme in all three proteolytic cleavages (Nielsen 1984).

cDNAs encoding β conglycinin subunits have been classified into two major classes according to detailed restriction mapping and Southern-blot analysis. Sequencing and hybrid-release translations showed one class to encode α sequences and the other α′ sequences (Schuler et al. 1982*b*). The nucleotide sequences of cDNA and gene clones have been reported (Schuler et al. 1982*a*, 1982*b*). The cDNAs encode different numbers of α and α′ classes, as judged by nucleotide substitutions leading to amino acid changes accounting in part for some charge heterogeneity. Comparison of α and α′ sequences shows them to be highly homologous (~93% homology). Most of the nucleotide differences between α and α′ families occur in a "hot spot" encoding amino acid sequences close to the *C* terminal, but these changes do not appear to interfere with any protein secondary structures (Schuler et al. 1982*b*). The β subunit also shows some sequence homology with α and α′ subunits, as judged by hybrid-release experiments, but this is limited, and the β polypeptides are thought to be encoded by a separate gene family that shares little amino acid homology (Schuler et al. 1982*a*). One point of interest from these sequence studies was the discovery of a highly conserved (92–99% homology) sequence of 155 nucleotides, present in the α, α′, and probably β coding sequences, that was also present in an otherwise unrelated cDNA clone and caused cross-selection of mRNAs in hybrid-release experiments. This conserved sequence encodes ~52 amino acids in the *C*-terminal region of the α and α′ subunits, but is located in the 3′ noncoding region of the unrelated cDNA. These authors have suggested that this phenomenon indicates a selection pressure for mRNA primary or secondary structure rather than at the protein structural level (Schuler et al. 1982*a*). It also suggests the involvement of a transposition of DNA from an ancestral α/α′ gene to an unrelated gene. Sequence conservation between α and α′ sequences is higher in the 3′ noncoding region than in the coding region, further suggesting conservation of a functional sequence or control structure, and stable hairpin structures can be predicted within these sequences (Schuler et al. 1982*b*). The 3′ noncoding sequences contain double overlapping polyadenylation signals, reminiscent of those in pea legumin (Croy et al. 1982; Lycett et al. 1983*a*), and some cDNA evidence suggests alternative use of either signal. Goldberg and associates (1983*a*) have reported isolation of three different genomic clones, each of which contains two β conglycinin genes, showing the clustering of the genes in this family. Comparison of cDNA and gene sequences has revealed the presence of at least four small introns 85, 115, 132, and 40 nucleotides in length that show some similarities and some differences to consensus sequences for animal and other plant introns and splice sites (Schuler et al. 1982*b*). Schuler and associates (1983) have additionally compared the partial nucleotide sequences available for the β conglycinin and phaseolin genes (Sun et al. 1981) by dot-matrix comparisons and have shown that over comparable regions there is considerable homology between the two (~73% homology), despite the lack of immunological relatedness between the two proteins. Three of the introns are present at comparable positions in the two genes. The sequences diverge by single nucleotide mutations as well as by addition and deletion of long nucleotide sequences, and sequence divergence is greater within introns than in the coding regions. Evidence suggests that β conglycinin (M_r 83,000), which is much larger than phaseolin (M_r 50,000), may have evolved partly by duplication of a number of single-copy nucleotide sequences while maintaining the reading frame (Schuler et al. 1983). This gene homology has been shown to extend even further to pea vicilin (Croy et al. 1984*b*, in press; Lycett et al. 1983*b*), a species even less related to soybean.

French bean (*Phaseolus vulgaris* L.)

Various aspects of the storage proteins of *Phaseolus vulgaris* have been dealt with in reviews by Brown and associates (1982), Hall and associates (1979), Ma and Bliss (1978), Mossé and Pernollet (1983), and Ersland and associates (1983). Storage proteins in French beans are largely represented by a single 7S globulin protein class, variously termed phaseolin, glycoprotein II, G1, and vicilin (see the earlier section on seed protein nomenclature), that accounts for up to 50% of the protein in mature seeds. The protein has not been located in any other tissue. Phaseolin is glycosylated, containing 3–5% carbohydrate, and has an M_r of 140,000–160,000. Analysis by one-dimensional SDS electrophoresis shows phaseolin to consist of a small number of polypeptides (M_r 43,000–53,000) characterized by two, three, or four banded patterns, depending on the cultivar. The polypeptides have been designated as α, β, and γ subunits, in order of decreasing size (Brown et al. 1981). Pusztai and Stewart (1980) have presented evidence in support of a model that postulates that phaseolin molecules consist of trimeric combinations of these subunits, and this model has subsequently been confirmed by X-ray crystallography (Blagrove et al. 1981; Blagrove, personal communication) and by small-angle X-ray scattering (Pleitz et al. 1983*a*, 1983*c*). Phaseolin is capable of reversible association from its trimeric 7S state into an 18S (M_r ~600,000) form consisting of 12 subunits (tetrad of trimers) of largely unknown configuration (Blagrove et al. 1981; Pusztai and Stewart 1980; Pusztai and Watt 1970; Sun et al. 1974) and even larger aggregates (Pusztai and Stewart 1980). Whether subunit assemblies within 7S and 18S forms are random or specific has not been determined. Two-dimensional electrophoretic analysis has shown that the phaseolin subunits exhibit charge as well as size heterogeneity (Bollini and Vitale 1981; Brown et al. 1980, 1981). Screening of 150 cultivars has revealed only three different phaseolin subunit complements, exemplified as the T (tendergreen) type, S (sanilac) type, and C (contender) type, consisting of 5, 8, and 8 phaseolin polypeptides, respectively, and indicating a total of 14 different polypeptide types. Genetic analysis of the T, S, and C phaseolin types shows that the encoding genes are closely linked, and each group of genes is inherited as a codominant allelic alternative (Brown et al. 1981). Peptide mapping indicates considerable homologies within the polypeptides, though different levels of glycosylation and methionine content have been reported (Bollini et al. 1983; Hall et al. 1980; Romero et al. 1975).

Phaseolin subunits are synthesized on membrane-bound polysomes (Bollini and Chrispeels 1978; Bollini et al. 1983); their leader sequences are cotranslationally removed by the endoplasmic reticulum (Bollini et al. 1983), and the polypeptides are sequestered into the lumen (Baumgartner et al. 1980), where posttranslational modifications take place prior to transport to, and deposition in, protein bodies (Barker et al. 1976; Bollini and Chrispeels 1978; Pusztai et al. 1977). The complex cotranslational and posttranslational modifications that the phaseolin polypeptides undergo have been described in detail by Bollini and associates (1982, 1983) and include two glycosylation steps and possibly a deglycosylation step or proteolytic processing possibly involving the removal of a *C*-terminal peptide (R. Bollini, personal communication). This system appears to be analogous to the β conglycinin processing scheme in soybean (Sengupta et al. 1981).

cDNA clones encoding phaseolin polypeptides have been isolated and characterized by hybrid-release and hybrid-arrested translations (Table 8.4). These cDNAs showed cross-hybridizations with different mRNAs, further indicating homology between the different subunits (Hall et al. 1980; Sun et al. 1981). As in other seed protein gene systems, the synthesis of phaseolin is under strict developmental control (Sun et al. 1978); whereas the genes encoding certain of the phaseolin polypeptides appear to be coordinately regulated (Brown et al. 1982; Hall et al. 1980; Pusztai and Stewart 1980), differential accumulations are evident during development (Pusztai and Stewart 1980) and in different locations (Barker et al. 1976). The use of the phaseolin cDNAs in reassociation kinetic hybridizations and in Southern-blot analysis of genomic DNA has indicated a gene copy number of 7–14 copies per haploid genome (Hall et al. 1983*a*), and the Eco-RI genomic fragments containing these genes have been divided into three nonallelic classes on the basis of restriction mapping (Hall et al. 1983*a*). Phaseolin was the first legume storage protein for which genomic clones were obtained (Sun et al. 1981), and its encoding genes are thus well characterized. A cloned phaseolin gene was shown not to contain large introns, as judged by heteroduplex mapping (Hall et al. 1983*b*). However, comparison of the nucleotide sequences of the gene and a full-length cDNA clone revealed the presence of five short introns (IVS 1–5), 72, 88, 124, 128, and 108 base pairs long, within the coding sequence; IVS 1, 2, and 3 contain termination codons. These introns all conform to the GT/AG intron boundary rule of Breathnach and associates (1978) and show some similarities and differences from consensus boundary sequences (Hall et al. 1983*a*; Slightom et al. 1983; Sun et al. 1981). The gene sequence consists of 1,990 base pairs from cap site to poly(A) addition site: 77 base pairs of 5′ noncoding sequence, 1,263 base pairs of coding sequence, 515 base pairs of introns, and 135 base pairs of 3′ noncoding sequence. Three possible TATA box sequences and two CAT box sequences are located upstream from the transcription start, and Hall and associates (1983*a*) have suggested that these may form multiple promoter regions that might function more efficiently by their combined effects, as demonstrated for one of the zein genes (Langridge and Feix 1983). As in other storage protein genes, multiple polyadenylation signals are present in the 3′ noncoding region, and one of the sequenced cDNAs has provided evidence for the use of alternative signals (Hall et al. 1983*a*). Comparison of a variety of cDNA sequences illustrates some of the possible underlying reasons for the size and charge heterogeneity in the phaseolin polypeptides. Two pairs of tandem repeats, one 15-base-pair repeat and one 27-base-pair repeat, are present in the α-type (larger) cDNAs, but only one copy is present in the β (smaller) cDNAs, corresponding to a difference of about M_r 1,600 within the encoded polypeptides. A number of other small deletions, usually involving single-codon deletions, are also evident. Whereas a number of the single-base changes observed cause no change in the encoded amino acid, others give rise to changes that would affect the charge on the encoded polypeptide, giving rise to the charge heterogeneity (Hall et al. 1983*a*). The homology between phaseolin and β conglycinin genes at the nucleotide level illustrated by dot-matrix comparisons of cDNA sequences was mentioned previously (Schuler et al. 1983). This homology is now known to extend to pea vicilin genes (Croy et al. 1984*b*, in press).

The first successful genetic engineering experiments involving the functional transfer of a seed protein gene have recently been reported, using a cloned phaseolin gene (Murai et al. 1983). The phaseolin gene, either fused to the 5′ end of the octopine synthase gene or complete with its own promoter, was inserted into the T region of the Ti plasmids pTi15955 and pTiA66 and transferred into *Agrobacterium tumefaciens* via a shuttle vector system (Barton and Chilton 1983; Leemans et al. 1982) and used to transform sunflower tissue. RNA isolated from transformed tissue contained transcripts of the expected sizes, as detected by northern-blot hybridizations with phaseolin cDNA. Poly(A) RNA isolated from transformed tissue and translated in vitro produced immunoprecipitable phaseolin polypeptides of the approximate size of authentic phaseolin (from tissue transformed with the phaseolin gene construct) or larger than phaseolin (from tissue transformed with the phaseolin-octopine synthase gene fusion). The results suggest successful transfer and transcription of the phaseolin gene from either the octopine synthase promoter or its own promoter, although the latter was found to be relatively inefficient, possibly reflecting its developmental regulation. Because the phaseolin gene contains introns, it would appear that correct transcript processing, as well as polyadenylation and transport, occurs in the transformed tissue. However, translation in vivo showed that most of the phaseolin was cleaved to smaller peptides (M_r 28,000, 26,000, and 14,000), indicating that the protein synthetic machinery in the undifferentiated tissue is not receptive to the synthesis of foreign proteins. This emphasizes the importance of consideration of constraints on gene controls in the logistics of manipulating and transferring seed protein genes, as will be described later.

Pea (*Pisum sativum* L.)

Within the legumes, the storage proteins of the Faboideae exemplified by pea and broad bean (*Vicia faba* L.) represent one of the most complex protein systems, mainly because of their extensive posttranslational modifications. The properties, synthesis, and genetics of pea storage proteins have been reviewed by Gatehouse and associates (1984*b*), Ersland and associates (1983), Casey and Domoney (1984), Brown and associates (1981), Spencer (1984), Higgins (1984), and Mossé and Pernollet (1983). As in soybean, pea storage proteins consist mainly of two immunologically distinct protein classes: 11S, legumin, and 7S, vicilin globulins, which account for 60–70% of the total seed protein. A third distinct globulin class, termed convicilin, which is immunologically related to vicilin, has recently been isolated and described (Casey and Sanger 1980; Croy et al. 1980*c*). The ratio of vicilin to legumin varies considerably between cultivars (0.5–4.0) (Casey et al. 1982; Gatehouse et al. 1984*b*; Müller 1983), and in some convicilin, usually a minor storage protein, is present as a major component (Casey and Domoney 1984). All three globulin protein classes are located in cotyledon protein bodies.

Legumin, like its homologue, glycinin, is a hexameric protein of M_r 380,000–410,000 consisting of six subunits (M_r ~60,000), each containing a disulfide-linked α (acidic) (M_r ~38,000) and β (basic) (M_r ~21,000) polypeptide pair, homologous to the A and B polypeptides in glycinin (Casey et al. 1981*b*, Croy et al. 1979). Such disulfide-linked polypeptides separate completely on reduction and dissociation. One- and two-dimensional gel electrophoresis of pea legumin shows that the α polypeptides exhibit considerable size and charge heterogeneity, and the β polypeptides, while displaying charge heterogeneity, show much smaller variations in size. (Casey 1979*a*, 1979*b*; Croy et al. 1979; Krishna et al. 1979; Przybylska et al. 1981; Thomson et al. 1978). On this basis, legumin polypeptides have been classified into major and minor types, depending on their relative abundances (Casey 1979*a*). This system was extended by Matta and associates (1981) to include two different minor classes with larger α polypeptides (big legumin, M_r 40,000–43,000) and smaller α polypeptides (small legumin, M_r 24,000–25,000), each class pairing with specific β polypeptides. A total of 22 α and 11 β polypeptides can be identified that in their α-β pairs constitute at least three hexameric legumin types (Gatehouse et al. 1984*b*). Casey (1979*a*) has demonstrated that the genes encoding the α^M (major acidic) polypeptides are closely linked in a single genetic locus, and, additionally, Matta and associates (1981) showed that big-legumin and small-legumin α polypeptides segregated independent of each other and the major polypeptides. Thus, three genetic loci, Lg-1, Lg-2, and Lg-3, have been proposed for the major big-legumin and small-legumin genes. Lg-1 has been mapped to a position 17 map units from the round/wrinkled locus on chromosome 7 (Davies 1980; Matta and Gatehouse 1982).

Early studies of in vitro synthesis of legumin and glycinin failed to detect any synthesis of the separate α and β polypeptides. Evans and associates (1979) and Croy and associates (1980*a*) were able to demonstrate in vitro and in vivo synthesis of non-reducible pea legumin subunits (M_r 60,000), which they proposed was a precursor of the α and β polypeptides. These observations led to the precursor-product model proposed for the biosynthesis of all legumin-type proteins, including glycinin, in which legumin subunit pairs are synthesized as single polypeptide chains containing both α and β polypeptides and are subsequently cleaved by posttranslational proteolysis but remain held together by disulfide and noncovalent bonds. This model was subsequently confirmed by Spencer and Higgins (1980) and Chrispeels and associates (1982*a*, 1982*b*) using pulse-chase experiments in which labeled amino acids incorporated initially into unreducible M_r 60,000 polypeptides could be chased with "cold" amino acids into the separate α and β polypeptides. Additionally, they showed in vivo synthesis of several different size classes of pea legumin precursor and their partial assembly in vivo to a 7S trimeric intermediate legumin molecule. They estimated the half-life of these legumin precursors to be about 1–2 hr, suggesting that the processing enzyme is located in the protein bodies. Chrispeels and associates (1982*b*) have extended these findings to show that the assembly of newly synthesized precursors into the intermediate trimeric form takes place in the endoplasmic reticulum, from which they are transported to the protein bodies, where rapid proteolytic processing takes place. The final assembly of processed legumin to its hexameric form then occurs very slowly within the protein bodies. Although legumin is synthesized on membrane-bound polysomes, the presence of leader sequences removed by endoplasmic reticulum (ER) from the precursor polypeptides has not yet been clearly demonstrated. Synthesis of similar legumin precursors has also been demonstrated in broad bean (*Vicia faba*) (Bassüner et al. 1983*b*; Croy et al. 1980*a*; Weber et al. 1981), in soybean (*Glycine max*) (Barton et al. 1982; Sengupta et al. 1981; Tumer et al. 1981, 1982), and, rather surprisingly, in two nonlegume species: oats (Brinegar and Peterson 1982; Matlashewski et al. 1982, 1983) and rice (Yamagata et al. 1982).

Although purified pea vicilin appears highly homogeneous by ultracentrifugation or gel-filtration analysis, giving M_r value of 145,000–170,000 (Gatehouse et al. 1981; R. Croy and A. Pusztai, unpublished results), analysis by SDS electrophoresis presents a very complex polypeptide pattern. Thus, major polypeptides of M_r 50,000, 33,000, 19,000, 16,000, 13,500, and 12,500 and minor components of M_r 35,000 and 31,000 are discernible (Gatehouse et al. 1981, 1984*b*; Thomson et al. 1978). Depending on the methods of purification, most vicilin preparations also contain variable amounts of convicilin, which contains subunits of M_r 70,000 and has a molecular weight of about 210,000–280,000, corresponding to a trimeric or tetrameric protein (Croy et al. 1980*c*). The M_r values of vicilin and convicilin polypeptides reported by different groups vary somewhat, although close scrutiny of the SDS gel patterns confirms the identity of equivalent components. To avoid confusion, we have used only those values deduced by our own group. Thus, the large number of constituent "subunits" in vicilin has made the elucidation of a molecular structure difficult. Vicilin synthesized in vitro, in comparison, showed a much simpler pattern of polypeptides, approximating in size the M_r 70,000 convicilin and M_r 50,000 vicilin components, but none corresponding to the lower-M_r polypeptides (Croy et al. 1980*b*). Vicilin synthesized in vivo in early development similarly showed only M_r ~50,000 subunits, with little or none of the smaller polypeptides (Croy et al. 1980*b*; Gatehouse et al. 1981). Pulse-chase experiments similarly showed incorporation of radioactive amino acids into M_r ~50,000 vicilin subunits initially, but then the label could be chased with "cold" amino acids into the M_r 33,000 and the other lower-M_r vicilin polypeptides (Chrispeels et al. 1982*a*, 1982*b*). Concomitantly, at least one of the labeled M_r ~50,000 vicilin subunits (M_r 47,000) disappeared, clearly implying a precursor-product type of relationship. On the basis of these results, Gatehouse and associates (1981) advanced a model for the structure of vicilin suggesting that the molecule was essentially a trimer of structurally similar M_r 50,000 subunits. Some of these subunits contained sites susceptible to posttranslational proteolytic cleavage or "nicking," so that although the overall structure and size of the molecule were maintained, some of the polypeptides had breaks in them, but were still held together by the noncovalent forces supporting their secondary and tertiary structures. Other polypeptides with no apparent cleavage sites remained intact, with M_r of 50,000. Vicilin molecules assembled from different combinations of precursors and subunits would thus give rise to the observed diversity of vicilin molecular species (Gatehouse et al. 1981, 1984*b*). More extensive pulse-chase experiments have confirmed this model and additionally have shown that the newly synthesized vicilin precursors and subunits are assembled into 7S trimers in the endoplasmic reticulum and transported to the protein bodies, where very slow processing of the susceptible polypeptides occurs within 6–12 hr. Such a model conforms with the demonstrated structures of other vicilin molecules (phaseolin and β conglycinin), neither of which undergoes such extensive proteolytic processing. The immunological similarity of convicilin coupled with the difficulty in separating it cleanly from vicilin, as shown by the early fractionation studies of Thomson and associates (1980), tend to suggest that hybrid molecules containing both convicilin and vicilin subunits may be possible (Croy et al. 1980*c*; Gatehouse et al. 1981).

Two-dimensional gel analysis of vicilin showed considerable charge heterogeneity, especially in the M_r 50,000 subunits, with six or more polypeptides present, whereas the lower-M_r polypeptides and convicilin showed less heterogeneity (Gatehouse et al. 1981).

Vicilin polypeptides, like those of legumin, are synthesized on rough endoplasmic reticulum, and cotranslational removal of leader sequences has been clearly demonstrated (Gatehouse et al. 1981; Higgins and Spencer 1981). Like other vicilins, pea vicilin is glycosylated. Chrispeels and associates (1982*a*) have shown that this glycosylation occurs in the endoplasmic reticulum, not the Golgi apparatus, and they and Badenoch-Jones and associates (1981) and Davey and Dudman (1979) have shown that polypeptides of M_r 50,000 and M_r 16,000 are the main glycosylated components of mature vicilin. Homologies between M_r 16,000 and M_r 12,500 polypeptides, indicated by immunological cross-reactions and peptide mapping, suggest that they may differ only by one being glycosylated, namely M_r 16,000 (Davey et al. 1981).

Qualitative and quantitative studies on the protein accumulations during pea seed development have indicated a differential synthesis of polypeptides such that vicilin appears early in development, followed two to three days later by legumin and convicilin (Croy et al. 1980*c*; Gatehouse et al. 1981, 1982*a*, 1984*b*). In peas, as observed in other seeds, transcriptional activity increases rapidly to a maximum during early embryogenesis and then decreases (Cullis 1976; Gatehouse et al. 1982*a*; Millerd and Spencer 1974; Morton et al. 1983). During this period the sequence complexity of the mRNA population decreases dramatically, and small numbers of mRNAs become very abundant, as judged by cDNA-mRNA hybridization kinetic analysis (Morton et al. 1983). These abundant mRNAs encode storage proteins, as inferred by the predominance of these polypeptides translated in vitro from mRNAs and polysomes isolated during the developmental period (Gatehouse et al. 1981, 1982*a*). Additionally, cloning and isolation of pea-seed-protein-specific cDNAs (Table 8.4) has allowed direct quantitation of specific mRNAs during development using northern-blot and dot-blot hybridizations. Such studies have confirmed that the levels of legumin and vicilin mRNAs increased and decreased in agreement with the estimated rates of synthesis and accumulation of the encoded polypeptides (Gatehouse et al. 1982*a*, 1984*b*) and confirmed that the mRNAs are relatively long-lived ($T_{1/2} > 10$ hr). The recent northern-blot analyses of Chandler and associates (in press) (cited in Higgins 1984), illustrated in Figure 8.5, using a number of different cDNAs, have confirmed the results for vicilin and legumin and have shown similar developmental responses in several other abundant seed protein mRNAs, including convicilin, pea major albumin, and pea lectin. Moreover, Figure 8.5 clearly shows the differential expression of vicilin, legumin, and convicilin genes.

It is clear from this evidence and that described earlier for soybean that control of storage protein synthesis is effected by modulating mRNA levels, most probably at the level of transcription. Direct evidence of changes in the expression of specific storage protein genes during seed development has been obtained by transcription assays on isolated pea cotyledon nuclei (Evans et al. 1984); however, despite the observation that relative transcriptional activities, mRNA accumulations, and protein synthesis rates for different storage protein genes through seed development were all consistent with each other, it is possible that other factors such as differential heteronuclear RNA (hnRNA) processing and translational controls also play roles in the control of gene expression (Gatehouse et al. 1982*a*, 1983).

Several of the legumin and vicilin cDNAs have been sequenced (Croy et al. 1982; Gatehouse et al. 1982*b*, 1983; Lycett et al. 1983*a*, 1984*b*; Spencer et al. 1984). Detailed comparisons of the amino acid sequences predicted by the cDNAs and those obtained from complete or partial polypeptide sequencing have confirmed that both protein types are synthesized as precursors and have allowed the amino acid sequence specificity of the putative processing sites to be deduced. Thus, Croy and associates (1982) showed that the cDNA sequences encoding the legumin α and β polypeptides are contiguous, because the predicted amino acid sequence continues upstream from the known *N*-terminal sequence of the β polypeptide, into the *C*-terminal sequences of the α polypeptide, unbroken by any initiation or termination codons. It was also apparent from this work that the α polypeptide region is synthesized first, being encoded at the 5′ end of the mRNA. Initial amino acid sequence data supported the suggestion that the paired basic residues (Arg-Arg), located in the probable processing region, might form one of at least two cleavage sites, analogous to similar sites involved in the processing of animal hormone precursors (Croy et al. 1982, 1984*b*; Gatehouse et al. 1984*b*). However, more recent *C*-terminal peptide data now indicate that, at least in pea legumin, there may be only a single cleavage site adjacent to the *N*-terminal amino acid of the β polypeptide (Gly) comprising an Asn-Gly peptide bond (Lycett et al. 1984*b*), and no removal of a linking pentapeptide sequence (Croy et al. 1984*b*). Such a single site is simpler, requiring only one processing enzyme, and is more consistent with the putative processing sites of vicilin, pea lectin (Higgins et al. 1983*a*), and a number of other seed proteins, all of which have an exposed Asn-*X* peptide bond. The hydrophilicity profile for the legumin precursor constructed from the cDNA predicted amino acid sequence (Figure 8.3A) shows that the region containing the processing site probably lies on the protein surface.

Subsequent legumin cDNA sequence data have revealed the presence of three long tandem repeats of about 54 base pairs each that encode three highly homologous polar (acidic) amino acid repeats 18 residues long, near the *C* terminal or the legumin α polypeptide (Croy et al. 1984*b*; Lycett et al. 1984*b*). These repeats were not found in some sequenced cDNAs (Croy et al. 1982; Lycett et al. 1984*b*) and thus may account for some of the size heterogeneity of the legumin α polypeptides. Structural predictions in this region (Figure 8.3A) show that the repeats probably lie on or near the protein surface and consist largely of α helix, possibly accounting for absent or poor immunological cross-reactions with legumin homologues from other species. Overall, the α polypeptide structure appears to be somewhat flexible or less rigid than the β polypeptide, containing relatively few α helical regions ($< 30\%$, mainly in the repeat regions) compared with the much tighter β polypeptide, which contains almost 50% of its sequence as α helices; this, together with the predominantly negative hydrophilicity index and the order in which the polypeptides are synthesized, supports our contention that the β polypeptide region is largely internal within the subunits and protein. Such an arrangement might account for the greater degree of size heterogeneity in the α polypeptides – the β regions being apparently much more conserved structurally.

All of the 3′ noncoding regions of the legumin cDNAs examined thus far have multiple polyadenylation signals, notably double overlapping signals of the form AATAAATAAA that have also been found in a small number of other plant seed protein cDNAs (Lycett et al. 1983*a*). The cDNAs have also been used to investigate and isolate legumin genomic restriction fragments. Thus, Croy and associates (1982) showed the presence of at least four Eco-RI genomic fragments hybridizing to one cDNA on Southern-blot analysis, although additional, apparently less homologous genes were indicated on lower-stringency hybridizations or prolonged exposure of autoradiographs (Croy et al. 1984*a*, 1984*b*). Four different legumin genes (Leg A, B, C, D) have been isolated from λ gene banks and characterized by restriction mapping and Southern-blot analysis (Croy et al. 1984*a*; Shirsat 1984) (Table 8.6). Comparison of sizes of unique restriction fragments in legumin cDNAs and in three of the legumin genes indicates the presence of at least one intron about 100 bp long in the β polypeptide coding region (Croy et al. 1984*a*, 1984*b*). Subsequent gene sequencing has confirmed this (IVS 3 = 99 bp) and shown the presence of two other introns (IVS 1 and 2) in the α polypeptide coding region of 88 bp each. All three introns conform to the plant intron consensus boundary sequences (Lycett et al. 1984*a*). The positions of the pea legumin introns appear to correspond to those in a soybean glycinin gene, though the sizes of the pea gene introns are much smaller than those of the soybean gene and cannot be detected by heteroduplex mapping (Croy et al. 1984*b*; Nielsen 1984). The perfect homology of the Leg A gene sequence with one of the legumin cDNA sequences indicates that Leg A is transcriptionally active, and examination of its 5′ noncoding sequence shows all the putative transcription control consensus sequences, including a TATA box (position −66), a CAT box (position −126), and additionally a possible AGGA box (Lycett et al. 1984*a*; Messing et al. 1983). The transcriptional start of the legumin message has been located by S1 mapping to a CATC sequence 25 bp downstream from the TATA box, implicating these sequences in the promoter region for the Leg A gene (Lycett et al. 1984*a*). Interestingly, at position −186, 55 bp upstream from the CAT box, is a sequence of 10 bp showing 80% homology with the adenovirus enhancer sequence; the significance of this sequence has not been elucidated, although the Leg A gene is transcribed in a HeLa-cell-free transcription system (I. Evans, personal communication). In the 3′ noncoding sequence are three sets of polyadenylation subsequences, only two of which have been found in the cDNAs, that are polyadenylated 19–20 bp downstream from the second signal. This signal lies in a region of considerable potential secondary structure; computer calculations indicate a stable stem-loop structure very similar to that observed in the equivalent position in the soybean lectin mRNA (Lycett et al. 1984*a*; Vodkin et al. 1983).

Domoney and Casey (1984) have recently isolated a different legumin cDNA that preferentially selects mRNA encoding an M_r 80,000 polypeptide in hybrid-release experiments; this minor polypeptide is synthesized and processed in vivo and is structurally closely related to the major pea legumins. It has been suggested that there may be a relationship between this precursor and the minor small-legumin subunits, described by Matta and Gatehouse (1982), for which a precursor has not been demonstrated (Domoney and Casey 1984).

Complete and partial amino acid sequences of the different vicilin polypeptides (M_r ~50,000) show strong homology to different M_r 50,000 polypeptide regions predicted by individual vicilin cDNA molecules (Gatehouse et al. 1982*b*; Hirano et al. 1982). Thus, these polypeptides appear to be derived from M_r 50,000 precursor encoded in certain mRNAs, by proteolysis, consistent with the model proposed earlier (Gatehouse et al. 1981). The generalized vicilin precursor proposed, designated αβγ, has two potential sites at which proteolytic processing may occur; cleavage at both sites produces three polypeptide products of M_r 19,000 (α), 13,500 (β), and 12,500 (or 16,000 if glycosylated) (γ); cleavage at only one site produces two products of M_r 33,000 (αβ) (homologous to the M_r 19,000 and 13,500 polypeptides) and 12,500 (or 16,000 if glycosylated) (γ); cleavage at the other site only also produces two products of M_r 19,000 (α) and 31,000 (βγ), the latter being homologous to the M_r 13,500 and 12,500/16,000 polypeptides; the absence of both of the susceptible sites in a large proportion of the M_r 50,000 polypeptides accounts for the abundance of uncleaved M_r 50,000 polypeptides (αβγ) in vicilin (Boulter 1984; Croy et al. 1984*b*; Gatehouse et al. 1982*b*).

Close examination of the amino acid sequences at the putative cleavage sites predicted by the cDNA sequences and the *N*- and *C*-terminal amino acid sequences of the individual vicilin polypeptides has allowed Gatehouse and associates (1983) to propose that cleavage is determined by amino acid sequence specificity, located in polar (surface) regions of the vicilin polypeptide; thus, sequences such as Gly-Lys-Glu-Asn- immediately preceding the cleavage site allow proteolysis, whereas less hydrophilic sequences such as Glu-Gly-Leu-Arg- do not. The possible importance of asparagine at the cut site is noteworthy. The hydrophilicity profile for the cDNA predicted amino acid sequence of the vicilin precursor shown in Figure 8.3B clearly illustrates that the sequence allowing cleavage lies at or near the surface of the protein. Recent work by Spencer and associates (1984) using serological studies and comparison of *N* terminal with vicilin cDNA predicted amino acid sequences has further confirmed this model. The secondary structures predicted for the vicilin and legumin precursor polypeptides (Figures 8.3A and 8.3B) indicate that potential proteolytic processing occurs within regions of irregular peptide conformation or β turns, usually linked with one or two regions of more rigidly defined structure (α helix, β sheet). The nearly complete vicilin cDNA sequence reported by Lycett and associates (1983*b*) predicts the presence of a leader sequence, although of unknown length, and shows only a single, simple polyadenylation site (AATAAA) (Lycett et al. 1983*a*). The cDNA nucleotide and predicted amino acid sequences have been compared with the published coding and amino acid sequences of phaseolin and β conglycinin by computer and visualized by dot matrices (Croy et al. 1984*b*, in press). Although these confirm the homology of pea vicilin to that of French bean and soybean, they also show that the regions of sequence divergence in all three species occur mainly around the positions of the two potential proteolytic processing sites in vicilin, though neither phaseolin nor β conglycinin is cleaved. The homolo-

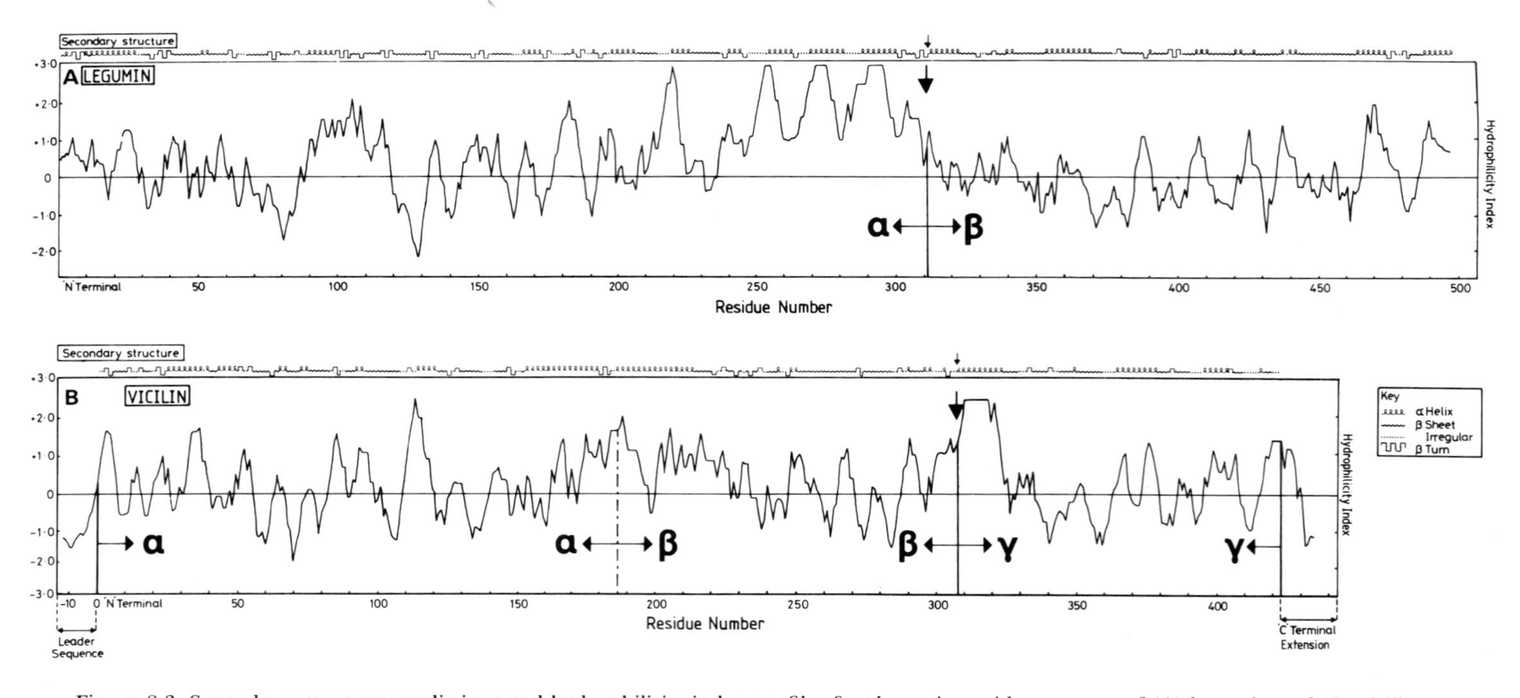

Figure 8.3. Secondary structure predictions and hydrophilicity index profiles for the amino acid sequences of (A) legumin and (B) vicilin, precursors predicted by cDNA nucleotide sequences. The computer program used for the secondary structure predictions was based on the analyses of Garnier et al. (1978). Hydrophilicity index profiles were constructed by a program adapted directly from that of Hopp and Wood (1981). The key to the predicted structural features is shown in the box inset.

gous polypeptide regions may therefore represent sequences important for conservation of structural features.

Vicilin genes have been quantified by Southern-blot analyses (Gatehouse et al. 1983, 1984*b*) using different vicilin cDNAs. The results indicate five to seven gene copies per haploid genome. One vicilin sequence known to encode a processed precursor is present in only two or three copies (Gatehouse et al. 1983).

Domoney and Casey (1983) have recently isolated a cDNA clone specifically encoding a convicilin polypeptide (M_r 70,000) that hybridizes to an mRNA appreciably longer than vicilin mRNAs. Preliminary nucleotide sequence information (R. Casey, personal communication) confirms the close homology of convicilin and vicilin but does not elucidate the nature of the extra polypeptide sequence to account for the size difference between the two (~200 amino acids).

Conclusions

It has not been possible in this review to describe all of the work on seed proteins using genetic engineering techniques; further studies have included those on cottonseed (*Gossypium* sp.) (Dure et al. 1983), broad bean (*Vicia faba*) (Bassüner et al. 1984; R. Bassüner and U. Wobus, personal communication), oats (*Avena sativa*) (Matlashewski et al. 1983; I. Altosaar, personal communication), castor bean (*Ricinus communis*) (J. Lord, personal communication), and rapeseed (*Brassica napus*) (Crouch et al. 1983). Moreover, work involving a number of other seed protein genes that may be of interest in future applications of genetic engineering has not been fully covered, including that on pea seed lectin (Higgins et al. 1983*a*, 1983*b*), French bean lectin (Hoffman et al. 1982), soybean lectin (Goldberg et al. 1983*b*), soybean Kunitz trypsin inhibitor (Goldberg et al. 1983*a*), and soybean urease (Polacco et al. 1979) (see also the following section). Certain general conclusions can be drawn from the present knowledge of the different seed protein gene and protein systems. The structural features common to many, if not all, seed proteins and their genes have been summarized in Figure 8.4 in the form of an idealized seed protein gene system. All seed protein genes thus far studied are under strict developmental control; their transcription and the accumulation of the translated seed proteins are confined to well-defined stages of development of seed tissue. This, coupled with mRNA longevity, accounts for the rapid accumulation of abun-

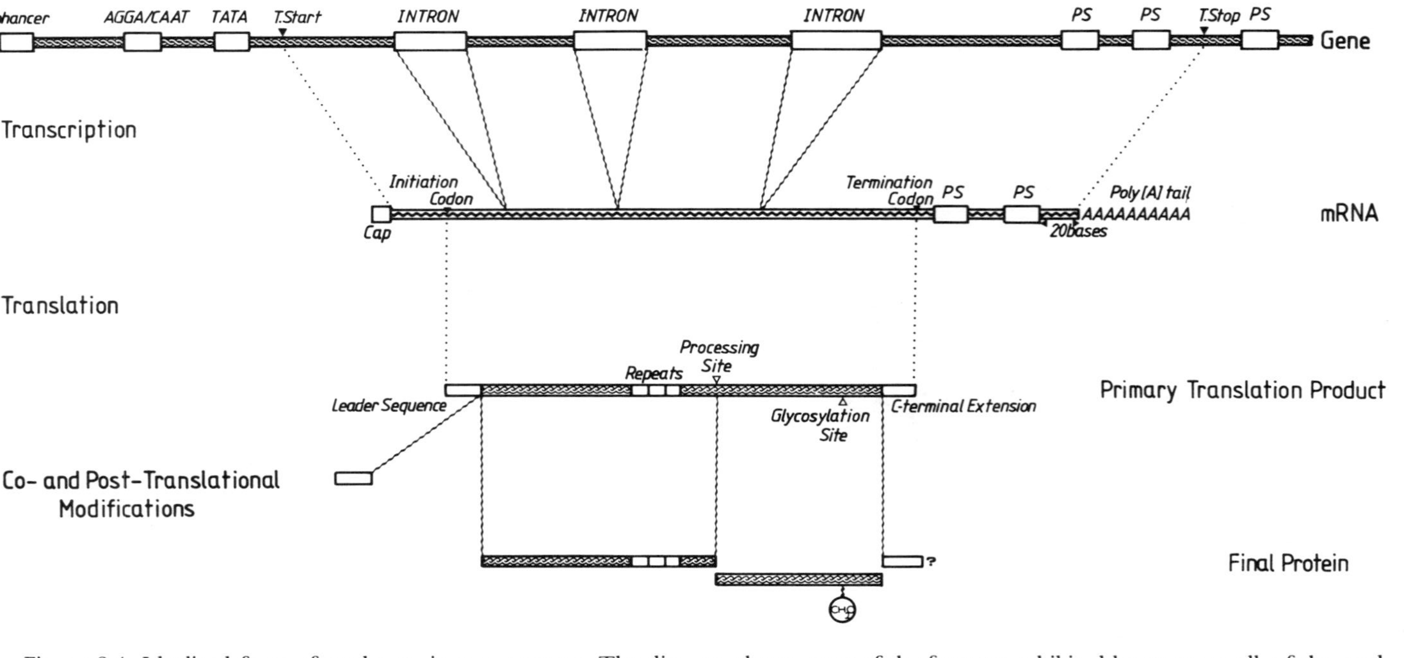

Figure 8.4. Idealized form of seed protein gene systems. The diagram shows many of the features exhibited by some, or all, of the seed protein genes and their transcripts and translation products: TATA and AGGA/CAT consensus sequences or boxes represent putative 5′ short-range controls: enhancer or tissue-specific regions represent additional sequences that may modify expression during development or in particular plant tissues; introns have been demonstrated present in at least three seed protein systems and absent in one (zein); CAP or 5′-terminal cap structure present on most eukaryotic mRNAs, and possibly also on seed mRNAs (Hall 1979); C-terminal extension cleavage has been suspected in several seed proteins, though it is not certain if the fragment is lost or retained on the protein; leader sequence – all storage proteins synthesized initially with these signal peptides, which are cotranslationally removed; processing and glycosylation sites – specific amino acid sequences on the protein surface recognized by proteolytic or glycosylating enzymes that either cut or add on sugars (mannose + others); repeat sequences – demonstrated present on several storage protein genes and translated into homologous amino acid sequence repeats (often in tandem).

dant seed proteins. Posttranscriptional processing may also be a factor in regulating mRNA levels, at least in early seed development (Evans et al. 1984). The storage protein genes are present as multigene families, with between 4 copies (pea legumin) and 100 copies (zein) per haploid genome, though the extent to which this feature contributes to the high level of transcription of storage protein mRNAs has not been assessed. The genes all have the 5′ consensus control sequences in one form or another, and recent evidence points to additional putative tissue-specific control or "enhancer" sequences (Lycett et al. 1984*a*). Complex double-promoter systems have also been suggested in some storage protein genes; 3′ sequences often contain mutiple polyadenylation signals, and some evidence suggests alternative usage of these signals. Stable secondary stem-loop structures can be predicted within many 3′ and 5′ noncoding regions, and although they are of unknown significance, they may contribute to mRNA stability and efficient expression.

The coding sequences commonly contain short repeat sequences, often in tandem, that encode similar repeated units of amino acid sequence. Such repeats imply an evolutionary mechanism whereby the plant has multiplied certain regions of coding information, enabling more of certain amino acids to be accumulated in the storage protein nitrogen store. Such a mechanism must inevitably reflect a fair degree of flexibility in the final protein structure in order to accommodate such extra sequences and yet still conserve the functional properties of a storage protein. In this respect the high-molecular-weight variants of certain storage protein polypeptides are of considerable interest. The presence of multigene families, many of which are closely linked or clustered in the plant genome, suggests that these families have arisen through a mechanism of evolution involving gene duplication or amplification followed by divergence of certain genes by accumulation of mutations (point mutations, deletions, repeat sequences, etc.), giving rise to the observed charge and size heterogeneity in the storage protein polypeptides. Further duplications could also take place to produce related gene subfamilies at more distinct loci through transposition and/or translocation mechanisms.

All of the storage proteins thus far investigated are synthesized by membrane-bound polysomes and therefore contain leader sequences that are cotranslationally removed. The significance of the various posttranslational modifications, including glycosylation, proteolysis, and assembly, is as yet unknown.

Potential applications of genetic engineering of seed protein genes

Constraints and considerations

Many of the characteristics desired by plant breeders in conventional breeding programs for seed crop improvement are ones that can also be achieved or facilitated by genetic engineering. However, the latter has the major advantages of speed, accuracy (in that only the required gene sequences need be manipulated), and wide variability (because genetic engineering is not necessarily restricted by interspecific barriers, even between animals and plants). Although conventional breeding techniques will continue to be extensively employed (Miflin and Lea 1984), genetic engineering of crop plants is likely to be used in producing changes difficult or impossible to produce conventionally. Such engineering is likely to be of two types: (1) gene alteration, in which a seed protein gene already present and expressed in the seeds is isolated and its coding sequence changed by addition, substitution, or deletion using modifying enzymes or oligonucleotides or by site-directed mutagenesis; (2) gene addition, in which new capabilities or qualities are conferred on the seed by addition of new genes.

Structural constraints

We know a great deal about the biosynthesis, posttranslational modifications, and structures of the seed proteins, and we have detailed descriptions of the ultrastructural events during seed development and germination. However, we are largely ignorant of how seed protein structures relate to their functional properties in connection with seed viability and vigor, or of what the molecular arrangements (packing) are for storage protein molecules within the protein bodies. While constraints at the protein structure level are widely accepted, the possibility of structural limitations on new genetic combinations at the mRNA level has also been suggested (Barton and Brill 1983; Schuler et al. 1982*a*).

Protein sequence and structure. Storage protein genes have a higher potential for changes in amino acid sequences that do not significantly affect their physiological role than do metabolically active proteins, in which possible sequence changes are limited by the necessity of maintaining a catalytic or other active function. However, although the function of storage proteins is apparently limited to supplying amino acids for the developing seedling, constraints on their structures and sequences are present (Boulter 1976; Brown et al. 1982). These can be summarized as (1) maintenance of the sequence and structure necessary for correct posttranslational transport to the protein bodies, (2) maintenance of the sequence and structure necessary for deposition (packing) in the protein bodies (e.g., Barton and Brill 1983), (3) maintenance of the sequence and structure necessary for correct utilization (proteolysis) of the storage protein on germination, and (4) maintenance of the correct overall amino acid composition for synthesis and utilization of the storage proteins. Whereas the various posttranslational modifications of storage proteins are probably relevant only to constraints 1 and 2, the conservation of essential three-dimensional structural configurations and surface regions (domains) in the protein will have importance for all constraints; see Hall and associates (1983*a*).

To attempt to understand the nature of the functional constraints on storage protein sequences we have available a number of techniques. Established sequences can be used to predict three-dimensional structures using various computer modeling techniques such as that shown for pea storage proteins (Figure 8.3). These techniques will find increasing use in attempting to predict the effects of sequence changes on protein structure and their functional properties. Further information aiding these predictions may be provided by close analysis of existing variants of the proteins, either by examining sequence variation within a line or by comparison between lines (genetic variation) or between species. Ongoing studies on such structural variants will almost certainly help define the essential structural features of these proteins and show which regions may be used for subsequent manipulation. Conversely, in vivo and in vitro studies on proteins expressed from modified storage protein genes will perhaps help elucidate the functions of different parts of the molecule and of the various posttranslational modifications in transport, assembly, deposition, and dehydration in seed development, and rehydration and disassembly in seed germination.

Are posttranslational modification sites necessary? It has been suggested that some post translational modifications, such as the endoproteolytic processing of certain storage protein polypeptides and the "tagging" of some subunits with sugar moieties (glycosylation), have no physiological function. However, the presence of processing sites and "nicks" in these proteins may in some subtle way alter the utilization of storage protein nitrogen as compared with unprocessed proteins, such as by providing increased latent protein nitrogen by increasing access to the internal regions of the molecule. Such a function would not necessarily be obvious under normal germination and growth, but it could have a significant survival advantage under stress conditions such as poor mineral nutrition or drought. Assessment of the performance of seeds genetically engineered to alter or remove such processing sites could elucidate such functions and be used to enhance the suitability of legumes for adverse environments.

Although the significant levels of carbohydrate present in some seed proteins may play a role in the nutrition of the seedling, perhaps as a nascent carbohydrate source, it is hard to conceive of such a function for the extremely low levels present in other seed proteins. It is clearly suggested from the work of Badenoch-Jones and associates (1981), Chrispeels and associates (1982*a*), and Bassüner and associates (1983*a*) using glycosylation inhibitors (tunicomycin) that glycosylation is not essential for synthesis, assembly, processing, or secretion in vivo or in oocytes. However, specific transport mechanisms involving membrane systems other than at the site of protein synthesis, including that of the protein bodies and possibly involving a role in systematic deposition, cannot be completely ruled out. Seed glycoproteins have alternatively been proposed to play a role as "receptors" in recognition functions within the seed, possibly in conjunction with seed lectins, and may aid in specific subunit assembly or deposition. A further suggested role is as "hydration centers," in which the glycosylated proteins may be transported by different routes and deposited in specific locations within protein bodies, thereby forming hydrophilic regions or "hydration channels." Such channels might then function in facilitating rapid and uniform rehydration of the protein bodies, which might in turn allow access by de novo synthesized hydrolytic enzymes, activation of indigenous enzymes, and egress of stored proteins, hydrolysis products, and other materials. Genetic engineering could be used to remove or replace the glycosylation sites and thereby allow accumulation of nonglycosylated storage proteins; this would allow conclusions to be made about the function of this type of modification in seed viability and vigor and would determine whether or not such modifications are essential or desirable in engineered proteins.

Which sequence regions can be manipulated? One of the first objectives in manipulating the seed protein genes will be to identify those regions of the protein that, in light of the foregoing considerations, could be used to insert, delete, or substitute amino acid sequences with the minimum amount of disruption of essential structure. Four types of sequence regions can be considered: (1) Nonessential surface regions. It is obvious that surface areas of proteins are much less likely to disrupt the protein configuration than internal regions. Thus, hydrophilic regions, which tend to lie on or near the surface of the protein, might be considered for alterations. (2) Repeat sequences. The exact functional status of repeat sequences within storage protein genes is for the most part obscure. Whereas it appears that in at least one case, zein, repeat sequences may be important in maintaining structure, the natural variation in repeated sequence lengths and types can be exploited to allow modifications or additions within such repeats. (3) Sequences linking structural repeats. If repeat sequences have an essential structural role, this may require strict conservation of amino acid sequences. However, the regions linking such repeat sequences may tolerate a certain amount of flexibility. (4) Terminal sequences. If most of the primary sequence of the protein is apparently essential for the integrity of the molecule, there remains the possibility of using the *N*- or *C*-terminal regions for the addition of extra amino acid sequences. The effects of significant (possibly homopolymeric) terminal additions on the folding and subsequent properties of storage proteins are as yet unknown. In this case, however, studies of the synthesis and structure of fusion proteins from bacterial-expression-vector/cDNA hybrids might help in testing the feasibility of this approach. Already, structural studies involving nucleic acid sequencing of cDNAs encoding variant polypeptides of certain seed storage proteins have suggested regions that might tolerate some alteration (Messing et al. 1983).

Regulation and control of efficiency of expression

Besides consideration of how the proteins can themselves be altered, genetic engineering of such proteins will be limited by problems of ensuring that the genes are expressed in the desired manner. Knowledge of factors controlling the expression of seed protein genes is limited, but such controls may be divided into those inherent in the development of the plant and those produced by pertubations in its environment.

Inherent controls of gene expression

Temporal (programmed) effects. The use of the developing seed as a model system for study of the control of plant gene expression has been put forward by many authors over the past few years. Under defined conditions, seeds develop in a highly reproducible "programmed" manner such that a number of specific sequential events take place at predictable stages during the developmental period. Such programmed development involves dramatic changes in the metabolic and morphologic status of certain plant tissues, which in turn leads to gene activation and deactivation. Further, during seed development there is temporal control of seed protein gene expression, as shown by numerous reports to be cited later. In summary, the types of studies carried out with seed materials from different stages of development include the following: analysis of protein accumulation (Carasco et al. 1978; Guldager 1978; Manteuffel et al. 1976; Miflin et al. 1983*a,* 1983*b;* Millerd et al. 1978 and protein subunit accumulation (Croy et al. 1980*c;* Gatehouse et al. 1981;) Pusztai et al. 1983; Sengupta et al. 1981; Sun et al. 1978); analysis of nascent protein synthesis by pulse-chase experiments (Chrispeels et al. 1982*a,* 1982*b;* Gatehouse et al. 1981; Sengupta et al. 1981; Spencer and Higgins 1980); quantitation of total mRNAs by hybridization kinetics (Dure and Galau 1981; Galau and Dure 1981; Goldberg et al. 1981*a;* Morton et al. 1983; Viotti et al. 1979) and of specific mRNAs by in vitro translation systems (Bassüner et al. 1983*b;* Sun et al. 1978) and northern-blot analysis (Barton et al. 1982; Chandler et al. in press; Gatehouse et al. 1982*b;* Goldberg et al. 1981*b*) or by northern-dot-blot hybridizations (Gatehouse et al. 1982*a;* Larkins et al. 1983); specific nascent nuclear transcription (Evans et al. 1984; R. Goldberg, personal communication).

The northern-blot analyses of Chandler and associates (in press) shown in Figure 8.5 clearly illustrate the differential and sequential control of transcription of several seed protein gene sets, using cloned cDNAs encoding different protein families to measure the relative abundance of their specific mRNAs at different stages of seed development.

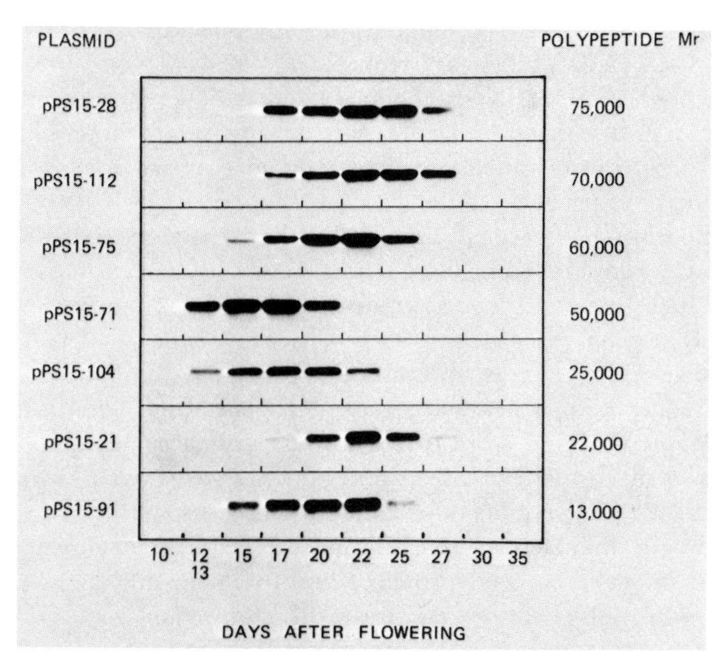

Figure 8.5. Developmental regulation of pea seed protein gene expression. Northern blots of seed RNA isolated from different stages of development and fractionated by gel electrophoresis, hybridized with cDNA specific for polypeptides of M_r 75,000 (convicilin), M_r 70,000, M_r 60,000 (legumin), 50,000 (vicilin), M_r 25,000 (pea major albumin, PMA), M_r 22,000 (pea lectin), and M_r 13,000 (pea LMW albumin, Psa-L_A). The figure illustrates the overall developmental regulation of seven individual gene sets plus the apparent differential control of expression of these genes. Adapted from Chandler et al. (1984) with permission.

Despite the accumulated body of evidence for control of gene expression at the transcriptional level, nothing is known about the actual molecular interactions and mechanisms ("gene switches") that effect these controls. Such an understanding is likely to be essential for practical application of genetic engineering to the seed proteins in order that inserted genes can be expressed in the correct, developmentally regulated manner, or in any other manner required. It is likely that such gene controls could be effected at the following two levels (Boulter 1981; Freelings and Woodman 1979; Gatehouse et al. 1984*b*).

Short-range controls involving the sequences immediately upstream and downstream from the 5′ and 3′ ends of the gene. The 5′ regions of many plant genes contain consensus sequences such as TATA and CAAT/AGGA boxes (Lycett et al. 1983*a;* Messing et al. 1983) that may form part of promoter sequences. They may, in addition, contain enhancer-type elements further upstream, as found in mammalian viruses or the mammalian genome, that may further modify promoter efficiency; see the review by Banerji and Schaffner (1983). Comparison of storage protein gene flanking sequences may identify specific, temporal control sequences. Initial sequence comparison of two such genes, expressed coincidentally in soybean, has unfortunately not yet revealed any such consensus sequences (Fischer and Goldberg 1983). Lycett and associates (1984*a*) have reported a possible sequence in a pea storage protein gene. Information will also undoubtedly come from detailed studies of these "temporal" controls in in vivo and in vitro transcription systems (e.g., Evans et al. 1984). The 3V regions also contain consensus sequences, such as polyadenylation signals, that will need to be maintained or incorporated into altered seed protein genes to obtain mRNA processing and stable, efficient expression. In this connection, a search for sequences capable of Z-DNA formation and for proteins binding to these regions would seem particularly appropriate (Kolata 1983).

Long-range controls, as yet unidentified, that may involve remote DNA sequences, genes, or other structural features within the chromosomes (Davidson et al. 1977; Thompson and Murray 1981). These may directly modify the activity of the genes by affecting the access of polymerase proteins to certain gene sets by conformational changes in the chromatin or indirectly by regulatory gene products that may act on the short-range control systems activating gene transcription. Whereas short-range control sequences can be relatively easily incorporated into modified or new genes, as will be described later, the possible need for selected site insertion within the host plant genome may be needed to provide the necessary long-range controls to obtain optimum gene expression. The ideal situation in this respect for a new or modified storage protein gene would be to insert it near or alongside genes under the desired control system, so that it would presumably act under the same controls. Arranging such transfers is not yet a practical proposition and may be difficult to develop. However, vectors that contain significant amounts of sequence complementary to the desired insertion region and that might also contain genes encoding the enzymes necessary for promoting DNA integration (integrase) – site-directed, homology mediated transfer vectors – will no doubt be developed for plant systems. Although it is not the intention of this review to elucidate specific developments of potential vector systems, those based on *Agrobacterium* Ti plasmids (Shaw et al. 1983) and eukaryotic transposon-type elements (Federoff et al. 1983*a*, 1983*b*; Rubin and Spradling 1982) are obvious candidates.

Whereas the resident storage protein gene controls of the host plant are likely to be the system of choice in controlling any seed protein genes engineered for nutritional improvement, seed-specific expression may not be the system desired for all applications of manipulated seed protein genes, especially those with a potential role in insect and fungal resistance. In these cases, consideration of controls that operate to modulate gene expression at different periods in development, or in different plant tissues or continuously throughout the plant (constitutive expression), will be necessary to produce the seed protein at a desired site and time. In addition, the possibility of modulation of gene expression by posttranscriptional processing of hnRNA or by translation-level controls may need to be considered. Initial attempts to secure controlled expression of introduced genes in plants will undoubtedly take place on a trial-and-error basis.

Spatial effects. The synthesis of most seed proteins is limited to seed tissues (Guldager 1978; Millerd 1975), and this has been demonstrated to be due to tissue-specific expression of the encoding genes. However, seeds consist of different tissue types (Croy 1977; Fulcher and Wong 1980; Payne and Rhodes 1982), and there have been several reports indicating that close examination of seed proteins within these different tissues reveals important quantitative as well as qualitative differences in their accumulation. Perhaps the most extreme case of this differential distribution is in those cereals such as wheat, barley, maize, and sorghum in which the prolamins are the major storage proteins and are deposited only in the starchy endosperm (Payne and Rhodes 1982), whereas the albumin and globulin proteins, including enzymes, lectins, and enzyme inhibitors, as well as storage proteins, tend to be largely concentrated in the embryo and aleurone tissue of these seeds (Dierks-Ventling and Ventling 1982; Payne and Rhodes 1982). The wheat lectin, for example, has been located specifically in the surface cell layers of the embryonic radicle, coleoptile, and scutellum (Mishkind et al. 1982).

Similarly, although albumins and globulins are found throughout the legume seed, the outer cell layers of the abaxial cotyledon surfaces of the French bean (*Phaseolus vulgaris*) (Barker et al. 1976; Zimmerman et al. 1967); and cowpea (*Vigna unguiculata*) (Gatehouse and Boulter 1983) have been shown to contain higher levels of albumin proteins, notably the trypsin inhibitors. Meinke and associates (1981) have reported several differences between the storage proteins synthesized in embryonic axes and in cotyledons of soybeans. Relative to the cotyledons, the axes contained very little glycinin protein and altered proportions and types of β conglycinin protein subunits and contained high levels of the enzyme urease. Manickam and Carlier (1980) have described a LMW albumin protein, apparently a storage protein (M_r 12,000), present almost exclusively in the axis tissue of mung beans (*Vigna radiata*). It is reasonable to expect to find such differential expression of certain genes in different tissues, because subsequently, on germination, each tissue will fulfill a different function and thus exhibit different metabolic rates and pathways. Such differential spatial (tissue-specific) accumulations of related and unrelated seed proteins may provide important clues into the nature of the regulation of these genes and a possible means of ensuring expression of introduced genes only at specific sites. Comparison with genes expressed specifically in other tissue types (leaves or roots) will add to this information, further aiding the genetic engineer in this respect.

Genetic effects. Large numbers of genetic variants and mutants have been described and shown to affect the rates of synthesis and levels of storage protein accumulation. However, as with the hormone effects to be discussed in a later section, little is known about the mechanisms whereby such effects are mediated. There are significant natural quantitative variations in the levels and rates of synthesis of storage proteins among different cultivars, which may indicate the involvement of regulatory loci, but in some cases may only reflect differential responses and metabolic efficiencies to a given set of environmental conditions (Casey et al. 1982; Cullis 1976; Gatehouse et al. 1982*a;* Manzocchi et al. 1980; Martensson 1980; Miflin et al. 1980; Mutschler and Bliss 1977; Saio et al. 1969*a*, 1969*b;* Schroeder 1982; Thomson and Schroeder 1978; Thomson and Doll 1979). A great deal of work has been done on the putative regulatory loci in maize that affect the synthesis of zein polypeptides. These include the mutant alleles opaque-2 (O2), opaque-6 (O6), opaque-7 (O7), and floury-2 (FL2), all of which depress the synthesis of zein; see Soave and Salamini (1983, 1984) for reviews. The detailed genetics, gene dosage effects, and interactions of these mutant alleles have been described (Di Fonzo et al. 1980; Soave and Salamini 1983). Decreased zein synthesis in O2 mutants is accompanied by reciprocal increases in other seed proteins, particularly the globulins, but also albumins, and it has been suggested that this may be a response possibly genetically linked to zein gene suppression (Dierks-Ventling 1981). Burr and Burr (1982), using cDNA probes, confirmed that the levels of M_r 22,000 zein mRNAs are depressed in O2 mutants, but they eliminated gene deletion as a cause. As mentioned earlier in this review, the expression of zein genes during development is in some way synchronized, despite the fact that the genes are located on at least three different chromosomes. Such coordinated control must be mediated by a diffusible gene product. Moreover, the O2 locus on the short arm of chromosome 7 and the O7 locus on the long arm of chromosome 10 preferentially depress zein gene classes mainly located on other chromosomes, namely, M_r 22,000 genes on chromosomes 4 and M_r 20,000 genes on chromosomes 7, 4, and 10, respectively. This also implies the involvement of a diffusible component. Toward the identification of such a component, Soave and associates (1981) and Galante and associates (1983) have described two potential protein candidates, designated b-32 and b-70, associated with the O2/O6 loci and FL2 locus, respectively, as indicated by their absence or overproduction in the mutant lines. However, these proteins are found in quantities higher than might be expected for a nuclear gene effector, and the b-70 protein at least has been located in the cytoplasm and associated with protein body and ER membranes, suggesting a post-transcriptional role possibly interfering with zein translation. Similar systems may exist in the high-lysine mutant lines of other cereals, such as the Risø, Notch, and hys series of barleys; see also Miflin and Shewry (1981), Doll (1983), and Payne (1983). Risø 1508, for example, carries an effector locus, Lys 3A, on chromosome 7 that results in dramatic and differential decreases in the polypeptides of the B and C hordeins, with compensatory increases in the D hordeins. These effects are mirrored by the decreases and increases in the levels of the corresponding mRNAs transcribed from the genes on chromosome 5 (Miflin et al. 1984). This system may be analogous to the effects of the O2 and O7 mutations in maize. The absence of the major B hordeins in the γ-radiation-induced mutant Risø 56 has recently been shown to be due to a deletion of at least 80 kb at the Hor-2 locus, containing all the major B hordein genes (Kreis et al. 1983*b;* Miflin et al. 1984).

Similar situations have been described in legumes. Different alleles of genetic loci that are linked with storage protein genes have been associated with the suppression of these genes. Thus, in *Pisum sativum* (pea) the structural genes for legumin are linked with the round/wrinkled locus on chromosome 7, and in wrinkled phenotypes (r_ar_a), homozygous for the recessive r_a locus, the legumin levels are much lower than in the round phenotypes (R_aR_a) (Davies 1980; Domoney and Davies 1983). However, demonstration of direct control of legumin genes by the r_a locus has not been possible, and other effects on pea starch and carbohydrate metabolism have been associated with this locus (Davies and Domoney 1983). Genetic loci have also been described in *Phaseolus vulgaris* (French bean) that enhance or depress phaseolin synthesis (Bliss and Brown 1982; Sullivan 1981). A closely related species, *Phaseolus coccineus,* contains no detectable phaseolin (Brown et al. 1982).

The unequivocal association of these or any other factors with a regulatory role, directly at the target gene level, remains to be established. However, it is certain that with the continuing studies of gene and chromatin structure in relation to gene expression, especially in characterized mutant lines, the molecular mechanisms responsible for storage protein gene control will be elucidated.

Influences on gene expression by environmental factors. Although tissue development is one parameter that influences the expression of seed protein genes, various other potentially exploitable experimental situations have been described that also modify this expression. These systems may directly affect gene activity or alternatively may perturb the seed metabolism or "hormone" balance in a way that modifies programmed seed development and gene expression and may putatively be used to effect the expression of introduced genes. Although we shall briefly describe some of these systems here, more detailed accounts are presented in other reviews (Brown et al. 1982; Higgins 1984).

Nutritional effects. Plants grown under certain nutritional deficiencies exhibit, in addition to physiologic effects, dramatically altered proportions of their seed storage proteins. Studies have examined the effects of deficiencies in the major elements nitrogen and sulfur on crop plants such as peas (Chandler et al. 1983, 1984; Randall et al.

1979; Schroeder 1982), barley (Kirkman et al. 1982; Koie et al. 1976; Shewry et al. 1983*b*), and wheat (Blagrove et al. 1976; Gillespie et al. 1978; Wrigley et al. 1980); see also Byers and associates (1978). Perhaps the best characterized of these systems is that of sulfur deficiency. As might be predicted under these conditions, the synthesis of sulfur-rich proteins is selectively and markedly depressed as compared with that for other proteins. For example, in peas, sulfur deficiency causes reductions in the levels of legumin and several of the albumin proteins, including pea major albumins (PMA) (Croy et al. 1984*a;* Schroeder 1984*a*), LMW albumins (Psa-L_a) (Gatehouse et al. 1985; Schroeder 1984*a*), and trypsin inhibitors (R. Croy, unpublished observations). In contrast, the level of pea vicilin, which is devoid of sulfur amino acids (Croy et al. 1980*c*), is maintained or increased (Randall et al. 1979). Similar changes have been reported within the storage protein groups of other species subjected to sulfur deficiency, including lupins (Blagrove et al. 1976; Gillespie et al. 1978), barley (Rahman et al. 1983; Shewry et al. 1983*b*), and wheat (Wrigley et al. 1980). The nature of the reduction in sulfur-rich protein synthesis in peas has been studied in detail by Chandler and associates (1983, 1984) using quantitative northern blots hybridized with specific cDNAs. Reduction in protein synthesis has been correlated to a reduction in the accumulation of the corresponding mRNAs, potentially due to a reduction in the levels of specific transcription. The results for legumin shown in Figure 8.6 illustrate the almost complete absence of legumin mRNAs in cotyledons from seeds deprived of sulfur nutrient, compared with seeds supplied with normal levels of sulfur, and, more important, show the apparent induction of legumin mRNA transcription within two days of resumption of normal sulfur nutrition (Chandler et al. 1983). Vicilin mRNA transcription in deprived seeds appears, in contrast, to be increased and extended well beyond the usual period of expression of the vicilin genes, indicating the apparent absence of a gene switch-off control. The results were similar using seeds from whole plants or from pods grown in culture. Thus, this system should provide a useful and convenient tool for detailed study of gene controls.

Osmotic effects. Domoney and associates (1980) have suggested that a high osmotic strength in tissue culture medium could be used to initiate legumin synthesis in cultured immature pea cotyledons, which otherwise do not accumulate this protein (Millerd et al. 1975). Similarly, Crouch and Sussex (1981) reported that high levels of

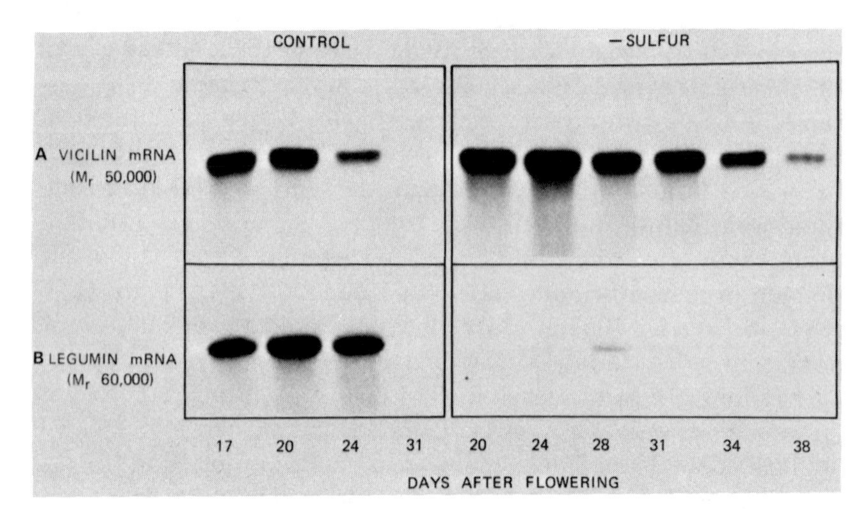

Figure 8.6. Environmental control of expression of pea seed protein genes. Northern-blot analyses of seed RNA isolated at different stages of development from sulfur-fed (control) and sulfur-deprived (−sulfur) pea plants, hybridized with vicilin-specific (sulfur-poor) cDNA and legumin-specific (sulfur-rich) cDNA. The figure shows the near total absence of the legumin mRNA under conditions of limiting sulfur, indicating a possible transcriptional control. Adapted from Chandler et al. (1984) with permission (Higgins 1984).

sucrose (12%) or mannitol in culture media promoted the synthesis of rapeseed (*Brassica napus*) legumin (12S) proteins. These results suggest that the influence of osmotica in cultures of developing seed embryos may be significant in the activation or modulation of seed protein gene expression and may take place in vivo as seed development proceeds (Crouch and Sussex 1981; Miflin and Shewry 1981; Smith 1973). Tsai and associates (1978) have suggested that elevated sucrose concentrations may be implicated in the decreased synthesis of storage proteins in certain maize mutant lines. It is well established that immature seed embryos actively accumulate solutes such as sucrose, mineral salts, and amino acids (Smith 1973; Thorne 1982), creating a particular osmotic environment and establishing a unique pattern of gene expression; their development ends in desiccation, which plays a role in permanently suppressing developmental gene expression and inducing germination-specific gene expression (Dasgupta and Bewley 1982). Whether this phenomenon or any of the systems described earlier acts directly on the genome or is the cause or effect of changes in the hormone levels remains to be evaluated.

Hormonal effects. Although it is apparent that hormones or growth substances (Trewavas 1982) can play a role in the modulation of storage protein gene expression, it is by no means certain if any of the many growth substances play specific roles, because many parameters, such as osmotic pressure, desiccation, and developmental time (metabolic status), may influence the endogenous levels of growth substances such as abscisic acid (Ackerson 1984*a*). The difficulty in demonstrating a direct causative relationship between hormone and protein synthesis is likely to complicate the elucidation of any specific roles and primary mechanism of action, despite numerous reports of the involvement of plant hormones in seed development and germination (Dasgupta and Bewley 1982; Ho 1979; Sussex and Dale 1979; Trewavas 1982; Walton 1980).

Sussex and Dale (1979), using cultured *Phaseolus vulgaris* embryos, and Crouch and Sussex (1981), using rapeseed embryos in culture, reported stimulation of storage protein synthesis by abscisic acid, although gene expression had already been "switched on" in both systems. In contrast, Davies and Bedford (1982), using pea embryos, and Dure and Galau (1981), using cotton embryos, could show no specific stimulation of storage protein synthesis. Exogenous growth substances may not be necessary to switch on and maintain storage protein synthesis, because Shimamoto and associates (1983) have recently reported the onset and sustained expression of zein genes in maize endosperm cell culture in which the endosperm tissue used to start the culture was excised prior to commencement of zein synthesis. However, a report by Schroeder (1984*b*), cited by Higgins (1984) and Spencer (1984), indicates that various growth effectors applied to developing pea seeds on the intact plant do have differential effects on the expression of different seed protein genes, again suggesting that these growth substances are implicated in some way in the programmed embryogenesis and subsequent seed germination. In this way they should prove a valuable tool for the study of gene control (Ackerson 1984*b*).

Other considerations

Genetic background. In the transfer of modified storage protein genes, especially those designed to produce proteins of a higher nutritional quality or of better functional properties for food processing, one major difficulty is likely to be the expression of the modified gene above a substantial background of homologous and nonhomologous storage protein gene expression. In many cases, therefore, even assuming equal degrees of expression, the desired effects are likely to be diluted or obscured, and even recognition of the new products may be difficult (Thomson and Doll 1979). Attention may have to be given to ensuring (1) that inserted genes have expression at least as good as, and preferably more efficient than, that of the background genes or (2) that multiple copies of the gene are inserted (Hartley and Gregori 1981) or (3) that mutant plant lines depleted of particular storage protein families are used as hosts (Kreis et al. 1983*b*) or (4) that site-directed mutagenic inactivation or deletion of specific genes can be carried out. Although systems able to carry out requirement 4 have not yet been developed, vectors similar to those proposed for site-directed insertions could be used for site-directed, homology-mediated deletion or insertional inactivation. Obviously, such systems must be highly specific for the genes to be inactivated, and in the case of insertional inactivation the inserted sequence must be stable against reversion. As discussed previously, natural insertional inactivations of several plant gene systems (including one seed protein gene) (Goldberg et al. 1983*b*) by transposon-like elements have already been demonstrated and are thus, in theory at least, feasible. Such vectors would, of course, need to recognize not just single genes but perhaps (in the case of polymorphic proteins) all the members of a gene family for this strategy to be completely effective. Furthermore, such gene alterations would also need to be completely stable, and thus deletions and frame-shift insertions might be more desirable. Recent reports by Goldberg and associates (1983*b*) and Vodkin and associates (1983) have shown a "naturally" occurring insertional inactivation of the lectin gene in one soybean cultivar, conferring on it a lectin − phenotype. As illustrated in Figure 8.7, the inactivation is caused by an insertion into the coding sequence of the lectin gene. The insert is a DNA element some 3.4 kb long that is structurally analogous to prokaryotic and other eukaryotic insertional elements in having inverted terminal repeats and sequence duplications in the region of the target site. However, the most interesting feature of the soybean lectin insertion element is the symmetry of inverted repeats around the target site, which the authors describe as "striking and suggestive of site specificity." Although no information on the activity or stability of this insertion is available, similar or identical elements have been demon-

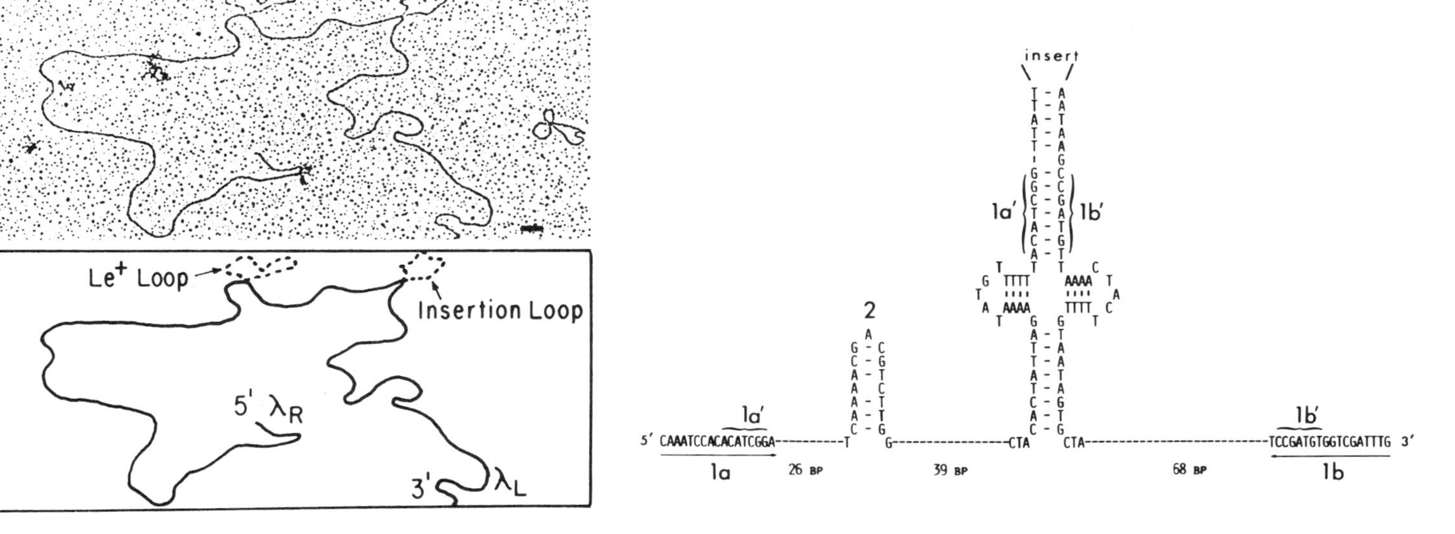

Figure 8.7. Insertional inactivation of soybean seed lectin gene. (A) Heteroduplex structure formed on hybridizing an Le⁺ ("normal" lectin-positive line) genomic clone with an Le⁻ (mutant lectin-lacking line) genomic clone visualized by electron microscopy. The Le⁺ loop (~4.8 kb) arises from the additional 5′ region present in the Le⁺ clone but absent from the Le⁻ clone. The insertion loop (~3.6 kb) in the Le⁻ clone lies within the lectin gene, dividing it into a 0.65 kbp 5′ region and a 0.35-kbp 3′ region. (B) Inverted repeat sequences at the termini of the soybean insertion element reminiscent of transposon-like structures. Adapted from Goldberg et al. (1983*b*) and Vodkin et al. (1983) with permission. Original figure copyrights held by M.I.T.

strated in several other lectin − soybean lines, and thus such an element may form the basis for future work on site-specific insertion or deletion vectors. Other eukaryotic insertional elements such as the P elements in *Drosophila* (Rubin and Spradling 1982), the Ac-Ds elements of maize (Federoff et al. 1983*a*, 1983*b*), the Mu element in maize (Strommer et al. 1982), and the Tam 1 and Tam 2 elements that inactivate the chalcone synthase gene in *Antirrhinum majus* (snapdragon) (Saedler et al. 1983; H. Saedler, personal communication) should all provide information on the necessary features required to construct specific gene inactivating systems. Once the molecular mechanisms of gene controls have been fully elucidated, it may be possible to effectively remove whole sets of genes by a single site-specific insertion or deletion, possibly by altering the long-range gene controls.

The problem of background genetic expression will obviously be more serious in those gene families, such as cereal storage proteins, in which there are relatively large numbers of members (of the order of 100). In this case it is likely to be very difficult to "silence" the endogenous genes enough for a homologous introduced gene to be expressed in significant amounts, if required (Doll 1984; Thomson and Doll 1979). In this respect, legumes are likely to be better candidates for genetic engineering, as their storage proteins are encoded by small gene families (<10 members). However, the foregoing option 3, the use of deletion mutant cereal lines such as Risø 56 of barley that have lost specific blocks of storage protein genes and therefore exhibit low background expression, should prove useful for adding back engineered genes (Kreis et al. 1983*b*).

Amino acid composition and codon usage. It is now well established that plants have their own preferred amino acid codon usage in cases in which there is a multiple codon choice. Thus, usually only two of the six possible arginine codons are frequently used in plant storage proteins (Lycett et al. 1983*a*). The maintenance of such preferred codon usage in manipulated storage protein genes is likely to be very important for efficient expression in the utilization of the established tRNA pools within the host plant seed. Furthermore, many of the proposed manipulations of storage proteins will intentionally introduce significant changes to their amino acid composition. Such alterations are likely to place new demands on the amino acid pool within the seed that must be accommodated if the new protein is to be efficiently synthesized.

Specific applications of genetic engineering of seed proteins

Nutritional quality

One of the immediate potential applications of genetic engineering of storage proteins will be to improve the nutritional balance of essential amino acids within the major storage proteins, thereby correcting deficiencies in the total protein complement (Payne 1983).

Utilization of genes encoding the major storage protein classes. Gene transfers leading to efficient, controlled expression seem more likely to succeed using the same plant species to supply the genes for manipulation (donor) and also to accept the altered genes (host). In such homologous systems inserted genes would continue to recognize the indigenous control systems and should be efficiently expressed, as will be discussed later. Thus, one strategy will be to isolate a particular storage protein gene, preferably one known to be efficiently expressed, and then to alter its coding sequence by appropriate methods, within the constraints discussed earlier. The altered gene will then be inserted into the same or closely related species by appropriate means. Similar strategies have been proposed by other authors (Miflin and Shewry 1981). The amino acid deficiencies of legume seeds might therefore be alleviated by manipulating the globulin genes so that the proteins contain much more methionine, theonine, tryptophan, or valine, depending on the limiting amino acids in the host plant species (Burr 1975). Similarly, deficiencies in cereal seed storage proteins might also be corrected by manipulating prolamin or glutelin genes to encode more lysine, tryptophan, or isoleucine, as appropriate (Payne 1983). It is clear that more than one amino acid type will be required to make all of the modified protein complement available as a dietary protein. Thus, it has been estimated that beans would require twice the present average content of tryptophan and three times the level of sulfur amino acids to allow complete utilization of the seed protein (Evans and Gridley 1980).

Genes encoding legume vicilins and legumins have already been isolated and fully characterized (Croy et al. 1984*c;* Hall et al. 1983*a;* Lycett et al. 1984*a;* Nielsen 1984). Furthermore, one method for gene transfer with demonstrable expression of a vicilin gene (phaseolin) has been reported (Murai et al. 1983), as described earlier, and will undoubtedly be developed into a routine and efficient procedure within the next few years. Thus, the basic requirements for this and the following strategies are, at least for the legumes, already available, although the problem of ensuring reproducible and tissue-specific gene expression is as yet unsolved.

Utilization of selected storage protein genes. An alternative strategy, which obviates the need for remodeling proteins, might be to exploit the existing natural variation in limiting amino acid contents within certain groups of related storage proteins. Where such a variation can be demonstrated, those gene variants encoding the best amino acid compositions would be selected for addition, preferably as multiple copies (Hartley and Gregori 1981), to the existing complement of homologous genes. Within the sulfur-containing legumin proteins in peas, broad beans, and soybeans, certain legumin subunits are present that contain little or no methionine and therefore serve only to exacerbate methionine deficiency in these species (Casey and Short 1981; Horstmann 1983; Larkins 1983; Staswick et al. 1981). In contrast, other legumin subunits have relatively high levels of methionine, and it has been suggested (Lycett et al. 1984*a*) that these genes, which have now been isolated, could be used for amplification (Boulter 1984). Similar variations exist in a number of other protein types, including the convicilin-vicilin protein family in pea (Casey and Sanger 1980; Casey et al. 1982; Croy et al. 1980*c;* Domoney and Casey 1983) and *Phaseolus vulgaris* (Hall et al. 1983*a*, 1983*b;* Romero et al. 1975). The variations in lysine content for different classes of prolamins in cereals (Miflin and Shewry 1981) might be exploited in a similar manner by adding back multiple copies of high-lysine genes to the host species. Thus, for example, increasing the number and type of B hordein genes in barley would bring about increased lysine content (Kreis et al. 1983*b;* Miflin and Shewry 1981; Shewry et al. 1980).

Utilization of genes encoding other seed protein classes. Several authors have proposed conventional breeding programs within a species to increase the levels of different classes of seed proteins that, although present in lower amounts than the major storage proteins, are nutritionally more desirable (Bliss and Hall 1977;

Boulter 1982; Davies 1976; Doll 1984; Rhodes and Jenkins 1978). Such suggestions might also be exploited by genetic engineering, either by modifying the controls of the new gene sets to enhance their expression or by selecting specific genes from these sets and providing them with appropriate control sequences from the major storage protein genes, thus avoiding the disadvantages of conventional breeding/mutation programs. Thus, the enhanced levels of lysine in the Hiproly varieties of barley have been attributed in part to a small number of high-lysine protein components in the albumin and globulin fractions (Doll 1984; Ingversen 1983; Payne 1983). These have been identified and designated as SPII albumin (Jonassen 1980; Svendsen et al. 1980), β-amylase, chymotrypsin inhibitors I and II, and protein Z (Hejgaard and Boisen 1980). Protein Z contains 2 moles of cysteine and 20 moles of lysine per mole and contributes to foam stability and haze formation in beer (Hejgaard 1982). Similarly, the high-lysine maize mutation opaque-2, while displaying decreased accumulation of the low-lysine prolamin proteins (zeins) (Jones et al. 1977*a*, 1977*b*), additionally shows elevated lysine levels in other protein fractions (Deutscher 1974; Dierks-Ventling 1981; Payne 1983), particularly in the globulin fraction (Dierks-Ventling 1982), indicating that in maize, too, genes encoding high-lysine proteins are likely to exist. It is highly likely that several of the other high-lysine mutants of maize (Denić 1983) and other cereals (Doll 1984; Hejgaard and Boisen 1980) will also show the presence of such potentially manipulable high-lysine genes.

In legumes, nutritional deficiencies attributable largely to low levels of sulfur amino acids in the major storage globulins may be corrected by increasing the levels of proteins from the albumin fraction, which is comparatively rich in sulfur amino acids, but may represent only 20% of the total seed protein (Croy et al. 1984*a;* Croy 1977). A number of relatively sulfur-rich proteins have been isolated from the albumin fractions of several legumes, including pea (Croy et al. 1984*a;* Jakubek and Przybylska 1982; Murray 1979; Schroeder 1984*a*), that do not appear to be toxic and therefore are unlikely to confer undesirable properties. Genes encoding such high-sulfur proteins would therefore be of great value. Furthermore, because some of these proteins are small, consisting of as few as 50 amino acids (Gatehouse et al. 1985), the possibility exists for use of the whole or part of the coding sequences from their genes to add onto or into storage protein genes, as in the foregoing strategy 1. Such hybrid genes would have the advantages of improved amino acid composition and maintenance of storage protein gene controls and therefore potentially good expression efficiency. Polacco and associates (1979) have suggested genetic manipulation of urease genes to increase its levels in seeds and thus to improve on the nutritive value of legumes such as soybean.

Utilization of genes from unrelated species. It is possible that in many crops the nutritionally desirable alternative proteins described in the previous section will not be found. For example, high-lysine proteins have not been discovered in wheat, possibly because the polyploid nature of this cereal makes it difficult to detect recessive high-lysine mutants and obscures the presence of such proteins (Doll 1984). It may additionally be the case that engineering of the resident storage protein genes in particular species (strategy 1) cannot conveniently be done. In these situations it may be necessary to adopt a strategy of creating "unnatural" or interspecific combinations of storage proteins. Thus, the genes encoding the relatively lysine-rich legume globulins (Gatehouse et al. 1984*b*) may be transferred into the lysine-poor cereals. Conversely, any source of genes encoding methionine-rich proteins (zein, for example) could be usefully transferred into legumes.

Perhaps because most nutritional comparisons of seed proteins use animal proteins, such as egg ovalbumin and casein, as high-quality standards, it should not be put outside the bounds of speculation to consider transferring genes encoding animal secretory proteins for seed storage protein production, although the structural constraints discussed earlier may prevent such extreme changes.

The choice of strategy obviously will depend on how many of limiting amino acids are required by a particular host species and how many of these could be supplied by the insertion of engineered genes, selected existing genes, or suitable foreign genes.

Deletion of undesirable gene sequences. As discussed in the introduction, an additional problem in the use of many seeds, especially in legumes, for human and animal food is the presence of several toxic or antimetabolic proteins; see reviews by Liener (1978), Gatehouse (1984), and Eggum and Beames (1983). The major toxic constituent of *Phaseolus vulgaris* seeds has been unequivocally identified as the lectin proteins (King et al. 1980; Pusztai et al. 1979*a*, 1979*b*).

Seed lectins and phytohemagglutinins are extremely widespread throughout the grain legumes, cereals, and other species, and close homologies between certain lectin groups have been demonstrated by Toms and Western (1971), Pusztai and associates (1983), Stinissen and associates (1983), Grant and associates (1983), and Gatehouse (1984). The levels of lectins in different species and cultivars also vary widely: from zero or near zero (Pusztai et al. 1983) up to as much as 20% of the total seed protein in different *Phaseolus* varieties (Pusztai et al. 1979*b*, 1983). Their toxicity in different animal types and strains also varies considerably and additionally depends on the proportion of lectin in the diet, the duration of exposure to lectin in the diet, and the source and type of lectin. Although lectins are heat-denaturable, it is questionable if the presence of any amount of lectin is desirable in unprocessed seed products for human or non-ruminant animal feeds.

Of at least equally widespread occurrence in seed-bearing plant species are enzyme inhibitors, notably the trypsin, chymotrypsin, and amylase inhibitors that have also been implicated in the antimetabolic performance and poor digestibility of seed proteins (Gatehouse et al. 1984*a;* Liener 1978). Like the lectins, these proteins are largely albumin proteins. Although by no means have all of these proteins been positively identified as antimetabolites, it has been suggested that the poor growth of animals fed on raw legume diets may result partly from interference by protease inhibitors in the digestion of storage proteins and partly from indigestibility of the inhibitors themselves and thus the loss of availability of their appreciable sulfur amino acid content (Phillips et al. 1981), exacerbating the sulfur deficiencies of the legumes. It is also worthy of note that certain of these inhibitors are highly heat-stable and thus cannot be removed efficiently by heat processing.

The foregoing types of antimetabolites may not be so important in cereals, because they are likely to be degraded in ruminant animals, and cereal flours are invariably heat-processed prior to human consumption (Payne and Rhodes 1982). However, the cereal prolamin storage proteins themselves, notably those from wheat, have been implicated in causing human celiac disease or gluten-sensitive enteropathy in sensitive individuals (Kasarda et al. 1976). It is apparent that peptides from the digestive breakdown of α gliadins are the active toxic factors (Patey et al. 1975).

It is clear from these few examples that certain valuable crop species or varieties could be improved by removal or substitution of all or part of the genes encoding such undesirable components. Poor nutritional quality of cereals and legumes is largely due to the amino acid deficiencies in the major storage proteins: prolamins in cereals, globulins in the legumes. Deletion of some of these genes might (1) bring about a better nutritional composition, possibly by compensatory increases in other proteins, similar to the high-lysine maize mutants, and perhaps without the disadvantages, and (2) provide deletion lines suitable for further engineering similar to the Risø 56 barley mutant. As discussed earlier, deletion of specific genes is likely to be highly problematical, requiring the development site-specific insertions or deletions to inactivate the undesirable gene types.

Insect and pathogen resistance

Although some antimetabolic proteins in seeds may be nutritionally undesirable for human or animal consumption, it is apparent that many of these proteins may form the basis of resistance in certain seeds to insect or pathogen attack. Gatehouse and associates (1979) were able to demonstrate that the resistance shown by seeds of one variety of cowpea to the larvae of the bruchid seed weevil, which is a natural pest of cowpea, was specifically due to the effects of elevated levels of trypsin inhibitors. Although this was one of the first reports of elucidation of the biochemical basis of plant resistance against a natural insect pest, it is apparent that such antimetabolic seed proteins can exert toxic effects on insects that do not naturally feed on these seeds. Thus, Steffens and associates (1978) showed that soybean trypsin inhibitors were toxic to corn borer larvae, whereas the natural host inhibitors were ineffective. Gatehouse and Boulter (1983) have shown that both soybean and lima bean trypsin inhibitors also have toxic effects on bruchid beetle larvae, though neither was as effective as the cowpea inhibitors. Lectins have also been shown to be effective toxins against a number of insect pests. The mechanism of the toxicity of kidney bean lectins in rats has been established (King et al. 1980; Pusztai et al. 1979*a*, 1979*b*) as due to a primary binding of the lectins to sites on the surface of the proximal region of the small intestine, causing subsequent extensive disruption of the microvilli and secondary effects leading to death. Gatehouse and associates (1984*a*) have similarly demonstrated that kidney bean lectins are highly toxic to bruchid beetle larvae, confirming and extending earlier findings of Janzen and associates (1976). As shown in Figure 8.8, using the same

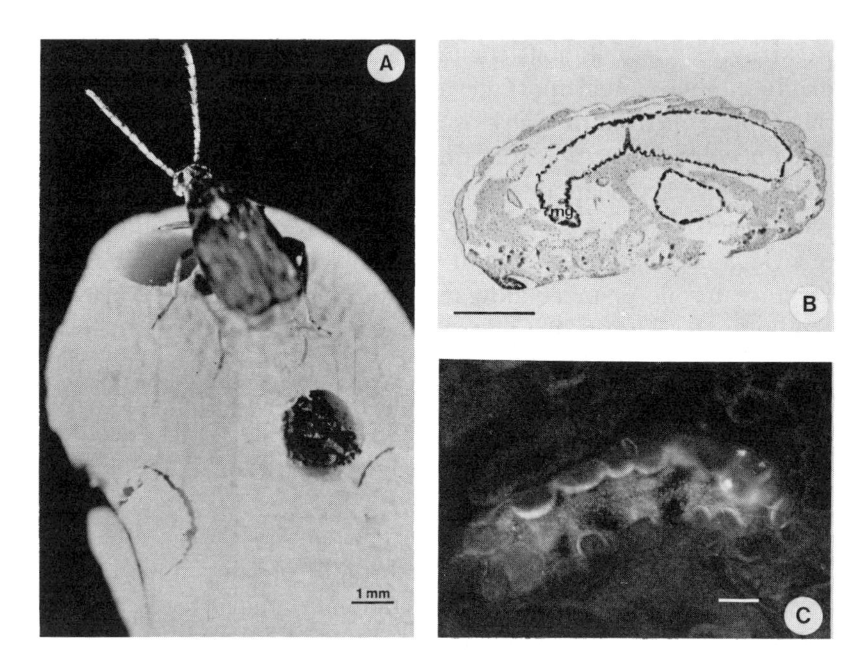

Figure 8.8. Toxic effects of French bean (*Phaseolus vulgaris*) lectins on bruchid beetles (*Callosobruchus maculatus*), a major storage pest of legumes. (A) Adult bruchid beetle newly emerged from an infested seed that contains no lectin. (B) Longitudinal section (L/S) of a normal beetle larva. Bar represents 100 μm; mg = midgut region. (C) Fluorescence microscopy of an L/S of midgut region of a bruchid larva, fed on a diet containing 2% lectin for 48 hr, incubated with rabbit antilectin antibodies, followed by fluorescein-isothiocyanate-conjugated antirabbit antibodies. Figure shows lectin immunfluorescence specifically associated with the cell surfaces adjacent to the midgut lumen. It has been suggested that the lectin exerts its toxic effect by binding to the midgut epithelial cell surfaces, causing disruption of nutrient transport and cell structure (Gatehouse et al. 1984*a*). Bar represents 10 μm. Unpublished photographs reproduced with kind permission of Dr. A. Gatehouse.

immunologic localization techniques as King and associates (1980), the authors have proposed that the mechanism of toxicity is analogous to that in the rat, because lectin-fed larvae showed specific binding of lectin protein to the epithelial cell surface of the midgut apparently interfering with the normal digestive and absorptive processes.

Information on the effects of other seed proteins on pathogens is somewhat sparse. Wheat germ lectin (WGA) has been implicated as a natural resistance factor against fungal attacks on the grain during dormancy, germination, and early growth (Mirelman et al. 1975).

The lectin specifically binds to surface chitin on the growing tips of fungal hyphae and inhibits their extension and growth, and additionally inhibits fungal spore germination (Mirelman et al. 1975). It is also apparent from the work of Stinissen and associates (1983) that lectins homologous to wheat lectin are found in many cereal species and may have a similar function.

Such seed proteins are likely to be extremely valuable introductions via genetic engineering into susceptible crop plants, conferring resistance to one or more indigenous pests. In this respect, cDNAs encoding several lectins have already been cloned, including soybean (Goldberg et al. 1983*b*), French bean (Hoffman et al. 1982), pea (Higgins et al. 1983*a,* 1983*b*), and castor bean (Lamb 1984; J. Lord, L. Roberts, and I. Lamb, personal communication). Genes encoding soybean lectin have also been isolated and characterized (Goldberg et al. 1983*b;* Vodkin et al. 1983).

Although certain of these proteins would have toxic effects if consumed directly as food, they would still be of value in food crops for cooking or food processing, in ruminant animal feeds, and in nonfood crops. The variability in the types and effects of these proteins is such that expedient choice of an effective inhibitor that is nontoxic to the consumer may be possible. Other seed proteins that have not been fully investigated but that have a toxic effect on certain organisms or an inhibitory effect on their digestive enzymes and thus may be potentially useful in such a strategy include amylase inhibitors (De Ponte et al. 1976; Powers and Culbertson 1983), purothionins (Ohtani et al. 1977; Ozaki et al. 1980), and a number of proteolytic enzyme inhibitors, as reviewed by Richardson (1977), Ramshaw (1982), and Gatehouse and associates (1984*a*).

A more potent antimetabolic activity is shown by a number of seed proteins ("cytotoxins") that are distributed widely in different plant species, often in high concentrations. They have been shown to be very toxic and to exert a very potent inhibitory action on eukaryotic protein synthesis in plant and animal cells. Examples of these toxins include ricin from *Ricinus communis* (castor bean) seeds (Montanaro et al. 1975), abrin from *Abrus precatorious* (rosary pea) seeds (Olsnes and Pihl (1977), and modeccin from *Adenia digitata* (Gasperi-Campani et al. 1980; Olsnes et al. 1978). The toxin molecules consist of two disulfide-linked polypeptides: one, the "effectomer" or toxic component (A chain) that exerts the inhibitory action by enzymatically destroying part of the 60S ribosome subunits; second, the "haptomer" (B chain), which has lectin-like properties. The lectin moiety binds the toxin to specific sites on cell surfaces and mediates the entry of the A chain, possibly facilitated by hydrophobic domains (Uchida et al. 1980), into the cell cytoplasm, where it exerts its toxic effects. These toxins are extremely potent,and it has been estimated that only a single absorbed toxin molecule is sufficient to kill a cell, irreversibly inactivating ribosomes and polysomes at a rate of 200 ribosomes per minutes (Stirpe et al. 1980). Recently, however, a second type of seed protein has been found that is at least as effective in inhibiting protein synthesis, but shows little or no toxicity to whole cells or animals (Gasperi-Campani et al. 1977, 1978, 1980). Structurally this second type appears to consist only of the A-chain-type polypeptides, which, lacking the lectin-type moiety, cannot efficiently enter into intact cells. Such proteins have been isolated from many sources, including tritin from wheat germ (Coleman and Roberts 1981; Roberts and Stewart 1979), momordin from *Momordica charantia* seeds (Barbieri et al. 1980), gelonin from *Gelonium multiflorum* seeds (Stirpe et al. 1980), crotin from *Croton tiglium* seeds (Stirpe et al. 1976), curcin from the seeds of *Jatropha curcas* (Sperti et al. 1976), and PAP-S, pokeweed antiviral protein from *Phytolacca americana* seeds (Barbieri et al. 1982). Similar proteins can be extracted from the roots and leaves of other plants (Irvin 1975; Stirpe et al. 1981). Possibly the most important feature of these proteins is their demonstrated antiviral properties, as exemplified by the leaf proteins, pokeweed antiviral protein (PAP) (Irvin 1975) and the dianthins from *Dianthus caryophyllus* (carnation) (Stirpe et al. 1981), which exert potent inhibition of viral infections of crops such as tobacco, tomato, potato, and cucumber, an effect presumed to be due to inhibition of protein synthesis within the infected host cells. The transfer of gene constructs encoding inactive precursors of these nontoxic inhibitors, which might then subsequently be activated in appropriate tissue sites (possibly extracellular), might provide a means of conferring general viral resistance to susceptible crop plants. The demonstration that the nontoxic inhibitors can be converted into toxins, effective on intact cells, by in vitro coupling to seed lectins (Stirpe et al. 1980; Uchida et al. 1980) or to antibodies (Musuho et al. 1982; Thorpe et al. 1981) carries tremendous potential for construction of efficient, specific, and biological antiviral or antipathogenic agents. Such "hybrid" toxins have been constructed from ricin A chain and *Wisteria floribunda* lectin (Uchida et al. 1980) and from gelonin and concanavalin A lectin (Stirpe et al. 1980) and demonstrated to be toxic to intact animal cells. Genetic engineering makes it a feasible proposition to produce such hybrids in vivo in important crops; they would specifically recognize and bind to receptors on the surfaces of major indigenous disease pathogens, mediated by a specific lectin or other suitable carrier moiety. Subsequent entry into the cell along with the pathogen would lead to specific inactivation of the infected cell, effectively preventing further spread of the infection. Such a strategy would involve the transfer of "linked" genes or a "fused" gene construct, encoding a suitable A-chain-type inhibitor, such as gelonin (Stirpe et al 1980), and a specific lectin or other haptomer known to bind to the appropriate pathogen. Fused gene constructs would produce products analogous to the ricin precursor that contains both A and B chains synthesized as a continuous polypeptide (Butterworth and Lord 1983; J. Lord, personal communication), which might then be transported and processed to the active toxin by the host plant processing systems.

Remodeling storage proteins for food processing

Apart from the nutritional considerations discussed in the previous sections, the functional role of vegetable and seed proteins in food processing is to provide the required physical properties to the food material either during processing or in the final product. The physical properties of the starting materials (protein extracts, isolates, concentrates, or flours) and products are determined by the level of protein present, the proportions of different protein types (Saio et al. 1969*a*, 1969*b*), and the presence of nonprotein components (Lillford 1978; Wright and Bumstead 1984), and such properties are likely to manifest themselves in different ways, depending on the processing procedures used (Kinsella 1976). The types of functional properties sought in proteins are manyfold and include those responsible for emulsification, for foam formation and stabilization, and for texturing (Graveland et al. 1980; Kinsella 1976; Kinsella and Shetty 1979; Wright and Bumstead 1984). Whereas many functional properties rely on maintenance of the native configuration of the proteins, several others arise through complete or partial denaturation of the proteins, followed by rearrangement of the polypeptide chains and formation of new intramolecular and intermolecular bonds. In all cases, however, the behavior of a particular protein type depends ultimately on its intrinsic primary structure (amino acid sequence) encoded by the genes. Thus, the proportions and functional properties of certain seed proteins could be manipulated by genetic engineering to suit particular applications in food processing.

For example, it has been shown that the legumin proteins in soybean (glycinins) have a profound effect on the physical properties of foods made from soybean flour (Mori et al. 1982*a*). The hexameric legumin proteins of soybean (*Glycine max*) (Moreira et al. 1979, 1981*b;* Utsumi et al. 1981), broad bean (*Vicia faba*) (Horstmann 1983; Utsumi and Mori 1980, 1981), and pea (*Pisum sativum*) (Casey et al. 1981*a,* 1981*b;* Croy et al. 1979; Krishna et al. 1979; Matta et al. 1981; Thomson et al. 1978) are all heterogeneous in terms of size, charge, and composition – a direct reflection of the variations in the amino acid sequences of the component subunits. Variations in the functional properties of glycinins (such as heat gelling time, gel hardness, etc.) isolated from a number of soybean cultivars exhibiting different α and β subunit compositions have been related to the presence of particular subunits or subunit pairs (Nakamura et al. 1984). Thus, short gelling time was related to the presence of the AS IV subunit (Nakamura et al. 1984), and high gel mechanical strength was proportional to the amount of the AS III subunit (Mori et al. 1982*b;* Nakamura et al. 1984). Furthermore, it was suggested that differences in gel turbidity (clarity) correlated to the number of surface cysteine residues on the glycinin molecules, possibly on the α subunits (Mori et al. 1982*a*), implying that the positions of sulfhydryl groups in the primary sequence affect their activity in cross-linking polypeptides during processing (Nakamura et al. 1984). Similar properties have been assigned to broad bean legumin subunits (Utsumi et al 1983). Although it is apparent that most, if not all, legumin α and β subunit combinations are preselected by virtue of their being part of the same contiguous polypeptide precursor (Barton et al. 1982; Croy et al. 1980*b,* 1982, 1984*c;* Horstmann 1983; Nielsen 1984), their three-dimensional structures and cysteine positions in different α and β subunits are sufficiently similar for subunit interchange to take place in in vitro dissociation-reconstruction systems (Croy et al. 1980*b;* Mori et al. 1982*b*). Thus, artificial combinations of subunits can be made that self-assemble into hexameric pseudoglycinins (Mori et al. 1982*a,* 1982*b*) pseudolegumins (Utsumi and Mori 1983; C. Horstmann personal communication). Furthermore, similarities in certain legumin subunits from different species are close enough to allow the formation of interspecific hybrid $\alpha = \beta$ subunit combinations and in some cases reconstitution of new hexameric 11S proteins (Utsumi et al. 1983). Thus, hybrid 11S globulins have been formed from combinations of native subunits from glycinin and broad bean legumin (Mori 1984; Utsumi et al 1980) and glycinin and sesame seed globulins (Mori et al. 1979). Studies of the functional properties of pseudoglycinins (Mori et al. 1982*a*, 1982*b*) and hybrid 11S globulins (Mori 1984; Utsumi et al. 1983) have demonstrated that proteins assembled from certain new combinations of subunits have distinctive and desirable functional properties. As illustrated in Figure 8.9 genetic enginering could enable any desired combinations of compatible subunits to be synthesized in vivo from hybrid legumin genes (T. Mori, personal communication). The genes encoding several legumin proteins have already been cloned, including those from pea (Croy et al. 1984*b*, 1984*c*) and soybean (Nielsen 1984), making this strategy a viable proposition.

Elucidation of the amino acid sequences in different types of subunits that confer the unique functional properties would furthermore allow direct incorporation of such sequences into any legumin proteins by addition of appropriate coding sequences into the resident legumin genes. In both cases, maintenance of the assembly of the subunit pairs into the 11S hexameric structure would be essential for expression of certain functional properties (T. Mori, personal communication). The potential exists for similar investigations involving other multisubunit seed proteins such as the vicilin (7S) proteins or lectins, especially in crops such as *Phaseolus* and *Vigna,* whose seeds contain large amounts of such proteins and little or no legumin (11S) proteins (Derbyshire et al. 1976). Subunit interchange in cereal lectins and soybean vicilins is already known to occur (Koshiyama 1983; Peumans et al. 1982). Altering the proportions of particular storage protein classes to increase the amount of a desirable component for processing is another feasible application for genetic engineering, as was described previously (Wright and Bumstead 1984).

Another example of the importance of protein functional properties is in baking quality. Payne and Rhodes (1982) have described a number of different baked products and the different qualities of wheat grain and wheat protein required for each. An important part of the basis of baking quality lies in the composition of the gluten protein fraction in wheat flour. Gluten is the water-insoluble viscoelastic protein mass left after soluble proteins, starch, and other non-

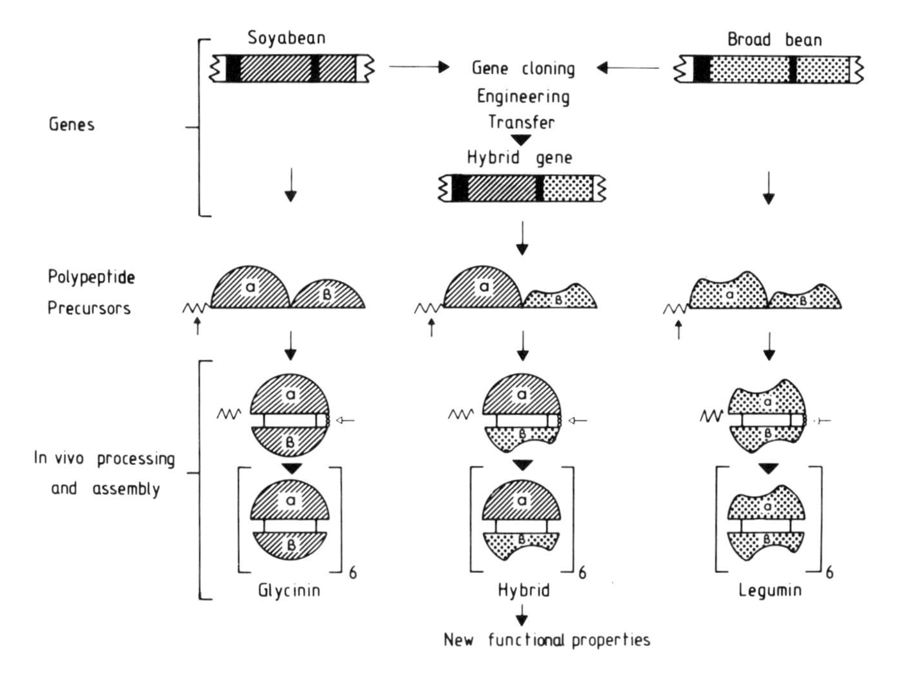

Figure 8.9. Genetic engineering of legumin storage proteins for improved functional properties (T. Mori, personal communication). In this strategy the genes encoding two legumin proteins, each with a desirable feature, are cloned, and the required sequences from each gene are fused together to form a hybrid gene. Such a hybrid gene, after transfer, will then synthesize a hybrid legumin precursor that undergoes endogenous processing and assembly into the legumin hexameric configuration. In this example the gene sequence encoding an α/A polypeptide of soybean is fused to that encoding a β/B polypeptide of broad bean legumin to produce a hybrid legumin with new functional properties. To avoid confusion, the glycinin polypeptides have been termed α and β instead of the more usual A and B. Wavy lines indicate leader sequences; black-tip arrows indicate leader-sequence removal; white-tip arrows indicate legumin processing system; lines linking two halves indicate disulfide bonds.

protein materials have been washed and processed out of the flour. Gluten is composed mainly of hydrated forms of the two major wheat protein fractions, namely the gliadins (wheat prolamins) and the glutenins (wheat glutelins), as described in the earlier section on nomenclature (Khan and Bushuk 1979; Sarkki 1980). These proteins contribute in different ways to the properties of the flour during processing and in the final product (Daniels and Frazier 1978). Thus, the gliadins provide viscosity and extensibility to bread dough, whereas the glutenins provide the elasticity that is all-important in dough stability and in the structure and texture of bread (Khan and Bushuk 1979; Pomeranz 1980). Differences in the proportions and properties of these two protein fractions determine whether a particular wheat variety has good bread-making properties or is more suitable for the production of other products, such as pasta or biscuits (Payne and Rhodes 1982; Payne et al. 1984). A molecular basis for these properties has been proposed generally, based on the properties of the purified components and the known amino acid composition, or more recently on the primary structures of certain of the wheat proteins (Miflin et al. 1983*a,* 1984; Shewry et al. 1984). However, the components responsible for good bread-making quality have only recently been tentatively identified. Such protein quality has been demonstrated to reside in the glutenin fraction, notably the HMW glutenin proteins (Payne 1983; Payne and Corfield 1979; Payne et al. 1980*a*). Payne and associates (1979, 1981) have shown a positive correlation between the molecular weight of native glutenin and the amount of HMW glutenin subunits and have suggested that interactions of these subunits with other polypeptides are important in stabilizing the glutenin in structure. Furthermore, Payne and associates (1981), using natural variations in these HMW glutenin subunits to study the inheritance of bread-making quality, have concluded that this allelic variation does correlate with good or poor baking quality, though other factors, possibly other wheat proteins, may also be involved (Miflin et al. 1983*a;* Payne 1983). Thus, certain of the gliadin polypeptides have also been implicated in baking quality and dough strength (Sosinov and Poperelya 1980; Wrigley 1980). The positive identification and cloning of the genes encoding polypeptides that contribute to good bread-making quality could potentially allow the transfer of this trait to poorer-quality wheats that carry other desirable characteristics. Additionally, it might be possible to manipulate the functional properties of the gluten for purposes other than bread making (Sarkki 1980) by transferring multiple copies of certain genes, thereby altering the proportion of the HMW glutenins, for example. This would seem a feasible proposition, because the HMW glutenin genes appear to be present in low copy numbers, in contrast to the higher copy numbers of the gliadin genes (R. Thompson and D. Bartels, personal communication; Thompson et al. 1983). In this context, cDNAs encoding HMW glutenin polypeptides have already been cloned and sequenced

(Forde et al. 1984; Shewry et al. 1983*a;* Thompson et al. 1983) as a preliminary to isolation of the genes themselves. Such sequencing studies should help elucidate the nature of the molecular interactions of the proteins in bread-making quality and may indicate further scope for alteration and improvement of intrinsic properties by genetic engineering (Miflin et al. 1984). Furthermore, it should be possible to resolve the question of definite assignment of quality to specific polypeptides, as opposed to a mere genetic linkage to the "true quality genes" (Miflin et al. 1983*a;* Porceddu et al. 1983). In this way, genetic engineering could be used to aid in the construction of wheat and other cereal varieties with predetermined functional properties designed for a precise processing purpose, such as bread-making, breakfast foods, meat analogues, and hydrolyzed products, as well as for nonfood applications (Porceddu et al. 1983; Sarkki 1980).

Conclusions and future prospects

Seeds represent the single greatest protein-synthesizing capacity of any natural system. Bacterial and yeast fermentation systems are capable of producing large amounts of concentrates containing up to 50% protein (Worgan 1978), but these require a great deal of preparation, maintenance, regulation, and processing, as well as defined growth conditions. Perhaps most important, these systems also require considerable energy input. Plants, in contrast, have been able to exploit most habitats, maximizing the use of available resources and utilizing solar energy to drive their synthetic processes, thus requiring no direct input of energy. Furthermore, many crop plants, notably the legumes, are nitrogen-fixing and are therefore also potentially self-sufficient in terms of nitrogen nutrition. Seed tissues have evolved to become the repositories of significant nitrogen stores and thus have been exploited as the main protein source for human and animal nutrition.

The ultimate goal, and perhaps the most popular notion of genetic engineering, is to transfer any desirable characteristic from one species of organism into a totally unrelated species and to obtain phenotypic expression of the characteristic in this host species. Although such manipulations are only now being successfully developed to commercial viability in bacterial and yeast host species, the encouraging signs for success in endeavors to carry out genetic engineering of plants suggest that these systems may also be used to produce desirable proteins and other materials. This concept of "green fermenters" is not as outlandish as it may appear at first sight, because many drugs and other materials are already extracted from plant tissues where they happen to occur as a result of natural "genetic engineering." Would a field full of plants that accumulate large amounts of, for example, insulin, interferon, or growth hormone in their seeds be competitive with production of the material in a bacterial fermenter? Further speculations in this direction are best left to the reader, but it should be stressed that all such projects are dependent not only on having a gene or genes to transfer and a usable transfer system but also on making the transferred gene work in the correct way after it has been integrated into the host genome.

As opposed to the production of materials from foreign genes incorporated into plants, most present ideas revolve around the goal of making plants better in what they do at present, in other words, increasing or improving desirable characteristics such as yield, protein content and composition, or disease resistance. In many cases the desired result is not likely to prove easy to achieve, either because the genes responsible might be difficult to identify or clone (e.g., crop yield) or because the system as a whole is too complex (e.g., nitrogen fixation). Those systems in which success is most likely to be achieved are those in which a single identified gene conferring the desirable properties is transferred into a host species that has little or no endogenous expression of similar genes; for example, transferring a high-methionine-containing protein gene into a legume with low-methionine storage proteins or transfer of a gene encoding a protein conferring insect resistance might fall into this category. Again, the tremendous commercial potential for genetic engineering of plants will depend on the successful transfer of a genomic sequence such that high controlled levels of expression are achieved in the host. In this respect, as Miflin and Lea (1984) stress, genetic engineering should be regarded as a tool to create new genetic combinations for use in conventional plant breeding.

Finally, what advances in scientific knowledge are likely to result from genetic engineering of seed proteins? Although a much better understanding of the relationships between protein sequences and functional properties is likely to be a valuable gain, perhaps most important is the possibility of being able to explain tissue-specific gene expression at the molecular level. Plants are good materials in which to study tissue-specific expression, as the range of cell types is comparatively limited, and even a differentiated plant cell is often still capable of dedifferentiation and regeneration of whole plants. Studies of events at the molecular level in tissues in which the genes are or are not being expressed, and reintroductions of genes with altered putative control sequences, are likely to result in considerable advances in knowledge of gene control mechanisms. Reintroduction of genes will probably be performed by using *Agrobacterium* Ti plasmid transfer systems, which have been improved to an extent allowing them to be considered routine for gene transfer in plants, at least to those host species in which infection and subsequent tissue culture can be readily carried out. Alternatively, various microinjection techniques are being developed for transfer to those species that cannot be infected by *Agrobacterium,* or where regeneration is not possible. Further valuable information on control of plant gene expression may come from the development of other transfer systems, such as that employing transposable elements, which may allow exploration of the relationship between the location of a gene on the genome and its activity.

The last five years have seen remarkable advances in our knowledge of plant genes and transfer systems, and it is not an unreasonable hope that the next 10 years will see not only the first commercially viable plant varieties produced by genetic engineering but also the gaining of enough understanding of how plant genes work to enable a "second generation" of genetically engineered plants to be produced.

Acknowledgments

We would like to thank all those scientists who supplied us with preprints of articles before publication. Our thanks are particularly due to Dr. T.J.V. Higgins, who supplied Figures 8.5 and 8.6, and Dr. A.M.R. Gatehouse, who supplied Figure 8.8. We would also like to thank our friends and colleagues in the Botany Department at Durham University for their helpful discussions and suggestions during the preparation of this review and particularly Professor D. Boulter for support and encouragement. Finally, we would like to thank Mrs. E. Ellis, Mrs. M. Raine, and Mrs. E. Croy for invaluable help in preparing the manuscript.

References

Ackerson, R. C. (1984*a*). *J. Exp. Bot.* **35,** 403–13.
(1984*b*). *J. Exp. Bot.* **35,** 414–21.
Alt, F. W., Kellens, R. E., Bertino, J. R., and Schimke, R. T. (1977). *J. Biol. Chem.* **253,** 1357–70.
Alwine, J. C., Kemp, D. J., Parker, B. A., Reiver, J., Renart, J., Stark, G. R., and Wahl, G. M. (1979). *Methods Enzymol.* **68,** 220–44.
Alwine, J. C., Kemp, D. J., and Stark, G. R. (1977) *Proc. Natl. Acad. Sci. USA* **74,** 5350–4.
Argos, P., Pedersen, K., Marks, M. D., and Larkins, B. A. (1982). *J. Biol. Chem.* **257,** 9984–90.
Badenoch-Jones, J., Spencer, D., Higgins, T. J. V., and Millerd, A. (1981). *Planta* **153,** 201–9
Bailey, C. J., and Boulter, D. (1970). *Eur. J. Biochem.* **17,** 460–6.
(1972). *Phytochem.* **11,** 59–64.
Banerji, J., and Schaffner, W. (1983). in *Genetic Engineering: Principles and Methods, Vol. 5,* ed. J. K. Setlow and A. Hollaender, pp. 19–32. Plenum Press, New York.
Barbieri, L., Aron, G. M., Irvin, J. D., and Stirpe, F. (1982). *Biochem. J.* **203,** 55–9.
Barbieri, L., Famboni, M., Lorenzoni, E., Montanaro, L., Sperti, S., and Stirpe, F. (1980). *Biochem. J.* **186,** 443–52.
Barker, R. D. J., Derbyshire, E., Yarwood, A., and Boulter, D. (1976). *Phytochem.* **15,** 751–7.
Bartels, D., and Thompson, R. (1983*a*). *Nucl. Acids Res.* **11,** 2961–77.
(1983*b*). *Theor. Appl. Genet.* **64,** 269–73.
Barton, K. A., and Brill, W. J. (1983). *Science* **219,** 671–6.
Barton, K. A., and Chilton, M.-D. (1983). *Methods Enzymol.* **101C,** 527–39.
Barton, K. A., Thompson, J. F., Madison, J. T., Rosenthal, R., Jarvis, N. P., and Beachy, R. N. (1982). *J. Biol. Chem.* **257,** 6089–95.
Bassüner, R., Huth, A., Manteuffel, R., and Rapoport, T. A. (1983*a*). *Eur. J. Biochem.* **133,** 321–6.
Bassüner, R., Manteuffel, R., Muntz, K., Puchel, M., Schmidt, P., and Weber, E. (1983*b*). *Biochem. Physiol. Pflanzen.* **178,** 665–84.
Bassüner, R., Wobus, U., and Rapoport, T. A. (1984). *FEBS Lett.* **166,** 314–20.
Bathurst, I. C., Craig, R. K., Herries, D. G., and Campbell, P. N. (1980). *Eur. J. Biochem.* **109,** 183–91.
Baumgartner, B., Tokuyasu, K. T., and Chrispeels, M. J. (1980). *Planta* **150,** 419–25.
Beachy, R. N., Doyle, J. J., Ladin, B. F., and Schuler, M. A. (1983). in *Structure and Function of Plant Genomes,* ed. O. Ciferri and L. Dure, pp. 101–12. Plenum Press, New York.
Beachy, R. N., Jarvin, N. P., and Barton, K. A. (1981). *J. Mol. Appl. Genet.* **1,** 19–27.

Bendich, A. J., Anderson, R. S., and Ward, B. L. (1980). in *Genome Organization and Expression in Plants,* ed. C. J. Leaver, pp. 31–3. Plenum Press, New York.
Benton, W. D., and Davies, R. W. (1977). *Science* **196,** 180–2.
Beone, B. J., and Pecora, R. (1976). *Dynamic Light Scattering.* Wiley, New York.
Berk, A. J., and Sharp, P. A. (1977). *Cell* **12,** 721–32.
Biggin, M. D., Gibson, J. G., and Hong, G. F. (1983). *Proc. Natl. Acad. Sci. USA* **80,** 3963–5.
Bishop, J. O., and Davies, J. A. (1980). *Mol. Gen. Genet.* **179,** 573–80.
Blagrove, R. J., Colman, P. M., Lilley, G. G., Van Dohkelaar, A., and Vargnese, J. N. (1981). *Australian Biochem. Soc. Abst.* [cited in Ersland et al. (1983)].
Blagrove, R. J., Gillespie, J. M., and Randall, P. J. (1976). *Aust. J. Plant Physiol.* **3,** 173–84.
Blagrove, R. J., and Lilley, G. G. (1980). *Eur. J. Biochem.* **103,** 577–84.
Blattner, F. R., Blechl, A. E., Thompson, K. D., Faber, H. E., Richards, J. E., Slightom, J. L., Tucker, P. W., and Smithies, L. (1978). *Science* **202,** 1279–84.
Blattner, F. R., Williams, B. G., Blechl, A. E., Thompson, K. D., Faber, H. E., Furlong, L. A., Grunwald, D. J., Kiefer, D. O., Moore, D. D., Schumm, J. W., Sheldon, E. L., and Smithies, O. (1977). *Science* **196,** 141–69.
Bliss, F. A., and Brown, J. W. S. (1982). *Qual. Plant. Plant Fd. Hum. Nutr.* **31,** 269–79.
Bliss, F. A., and Hall, T. C. (1977). *Cereal Foods World* **22,** 106–13.
Blobel, G., and Dobberstein, R. (1975). *J. Cell Biol.* **67,** 835–51.
Bolivar, F. (1978). *Gene* **4,** 121–36.
Bolivar, F., Rodriguez, R. L., Betlach, M. C., and Boyer, H. W. (1977*a*). *Gene* **2,** 75–93.
Bolivar, F., Rodriguez, R., Greene, P., Betlach, M., Heyneker, H., Boyer, H., Crosa, J., and Falkow, S. (1977*b*). *Gene* **2,** 95–113.
Bollini R., and Chrispeels, M. J. (1978). *Planta* **142,** 291–8.
Bollini, R., Van der Wilden, W., and Chrispeels, M. J. (1982). *Physiol. Plant* **55,** 82–92.
Bollini, R., and Vitale, A. (1981). *Physiol. Plant* **55,** 82–92.
Bollini, R., Vitale, A., and Chrispeels, M. J. (1983). *J. Cell Biol.* **96,** 999–1007.
Bonner, W. M., West, M. H. P., and Stedman, J. D. (1980). *Eur. J. Biochem.* **109,** 17–23.
Buolter, D. (1976). in *Genetic Improvement of Seed Protein,* pp. 231–50. National Academy of Sciences, Washington, D.C.
(1981). in *Advances in Botanical Research, Vol. 9,* ed. H. W. Woolhouse, pp. 1–31. Academic Press, New York.
(1982). *Proc. Nutr. Soc.* **41,** 1–6.
(1984). *Phil. Trans. R. Soc. Lond.* [*Biol.*] **304,** 323–32.
Brammar, W. J. (1982). in *Genetic Engineering, Vol. 3,* ed. Williamson, R., pp. 53–83. Academic Press, New York.

Brandt, A. (1979). *Carlsberg Res. Commun.* **44,** 255–67.
Brandt, A., and Ingversen, J. (1978). *Carlsberg Res. Commun.* **43,** 451–69.
Breathnach, R., Benoist, C., O'Hare, K., Gannon, F., and Chambron, P. (1978). *Proc. Natl. Acad. Sci. U.S.A.* **75,** 4853–7.
Brinegar, A. C., and Peterson, D. M. (1982). *Plant Physiol.* **70,** 1767–9.
Brown, J. W. S., Bliss, F. A., and Hall, T. C. (1980). *Plant Physiol.* **66,** 838–40.
(1981). *Theoret. Appl. Genet.* **60,** 251–9.
Brown, J. W. S., Ersland, D. R., and Hall, T. C. (1982). in *The Physiology and Biochemistry of Seed Development, Dormancy and Germination,* ed. A. A. Khan, pp. 3–42. Elsevier, The Hague.
Burgess, S. R., Shewry, P. R., Matlashewski, G. J., Altosaar, I., and Miflin, B. J. (1983). *J. Exp. Bot.* **34,** 1320–2.
Burr, B., and Burr, F. A. (1976). *Proc. Natl. Acad. Sci. U.S.A.* **73,** 515–19.
Burr, B., Burr, F. A., Rubenstein, I., and Simon, M. N. (1978). *Proc. Natl. Acad. Sci. U.S.A.* **75,** 696–700.
Burr, B., Burr, F. A., St. John, T. P., Thomas, M., and Davis, R. W. (1982). *J. Mol. Biol.* **154,** 33–49.
Burr, F. A., and Burr, B. (1980). in *Genome Organization and Expression in Plants,* ed. C. J. Leaver, pp. 227–31. Plenum Press, New York.
(1981). *J. Cell Biol.* **90,** 427–34.
(1982). *J. Cell Biol.* **94,** 201–6.
Burr, H. K. (1975). in *Protein Nutritional Quality of Foods and Feeds, Vol. 1,* ed. M. Friedman, pp. 119–34. Marcel Dekker, New York.
Butterworth, A. G., and Lord, J. M. (1983). *Eur. J. Biochem.* **137,** 57–65.
Byers, M., Kirkman, M. A., and Miflin, B. J. (1978). in *Plant Proteins,* ed. G. Morton, pp. 227–43. Butterworth, London.
Caldwell, K. A. (1983). *J. Exp. Bot.* **34,** 1411–20.
Cameron-Mills, V., Brandt, A., and Ingversen, J. (1980). in *Cereals for Food and Beverages–Recent Progress in Cereal Chemistry,* ed. G. E. Inglett and L. Munck, pp. 339–64. Academic Press, New York.
Cameron-Mills, V., Ingversen, J., and Brandt, A. (1978). *Carlsberg Res. Commun.* **43,** 91–102.
Carasco, J. F., Croy, R. R. D., Derbyshire, E., and Boulter, D. (1978). *J. Exp. Bot.* **29,** 309–23.
Casey, R. (1979*a*). *Heredity* **43,** 265–72.
(1979*b*). *Biochem. J.* **177,** 509–20.
Casey, R., and Domoney, C. (1984). *Phil. Trans. R. Soc. Lond.* [*Biol.*] **304,** 349–58.
Casey, R., March, J. F., and Sanger, E. (1981*a*). *Phytochem.* **20,** 161–3.
Casey, R., March, J. F., Sharman, J. E., and Short, M. N. (1981*b*). *Biochim. Biophys. Acta* **670,** 428–32.
Casey, R., and Sanger, E. (1980). *Biochem. Soc. Trans.* **8,** 657–8.
Casey, R., Sharman, J. E., Wright, D. J., Bacon, J. R., and Guldager, P. (1982). *Qual. Plant. Plant Fd. Hum. Nutr.* **31,** 333–46.
Casey, R., and Short, M. (1981). *Phytochem.* 20, 21–3.
Catsimpoolas, N., and Ekenstam, C. (1969). *Arch. Biochem. Biophys.* **129,** 490–7.

Catsimpoolas, N., Kenney, J. A., Meyer, E. W., and Szuhaj, B. F. (1971). *J. Sci. Fd. Agric.* **22,** 448–50.
Chandler, P. M. (1982). *Anal. Biochem.* **127,** 9–16.
Chandler, P. M., Higgins, T. J. V., Randall, P. J., and Spencer, D. (1983). *Plant Physiol.* **71,** 47–54.
Chandler, P. M., Spencer, D., Randall, P. J., and Higgins, T. J. V. (1984). *Plant Physiol.* **75,** 651–7.
Chen, Y. H., Yang, J. K., and Chan, K. H. (1974). *Biochemistry* **13,** 3350–9.
Chirguin, J. M., Przybyla, A. E., Macdonald, R. J., and Rutter, W. J. (1979). *Biochemistry* **18,** 5294–9.
Chlan, C. A., and Dure, L. (1983). *Mol. Cell. Biochem.* **55,** 5–15.
Chou, P. Y., and Fassman, G. D. (1974). *Biochemistry* **13,** 222–34.
Chrispeels, M. J., Higgins, T. J. V., Craig, S., and Spencer, D. (1982*a*). *J. Cell Biol.* **93,** 5–14.
Chrispeels, M. J., Higgins, T. J. V., and Spencer, D. (1982*b*). *J. Cell Biol.* **93,** 306–13.
Clarke, L., and Carbon, J. (1976). *Proc. Natl. Acad. Sci. U.S.A.* **72,** 4361–5.
Coleman, W. H., and Roberts, W. K. (1981). *Biochim. Biophys. Acta* **654,** 57–66.
Crouch, M. L., and Sussex, I. M. (1981). *Planta* **153,** 64–74.
Crouch, M. L., Tenbarge, K. M., Simon, A. E., and Ferl, R. (1983). *J. Mol. Appl. Genet.* **2,** 273–83.
Croy, R. R. D. (1977). Ph.D. thesis, University of Aberdeen.
Croy R. R. D., Derbyshire, E., Krishna, T. G., and Boulter, D. (1979). *New Phytol.* **83,** 29–35.
Croy, R. R. D., Hoque, M. S., Gatehouse, J. A., and Boulter, D. (1984*a*). *Biochem. J.* **218,** 795–803.
Croy, R. R. D., Lycett, G. W., Cottrell, J., and Boulter, D. (in press).
Croy, R. R. D., Lycett, G. W., Gatehouse, J. A., and boulter, D. (1984*b*). *Kulturpflanze* **32.**
Croy, R. R. D., Lycett, G. W., Gatehouse, J. A., Yarwood, J. N., and Boulter, D. (1982). *Nature* **295,** 76–9.
Croy, R. R. D., Gatehouse, J. A., Evans, I. M., and Boulter, D. (1980*a*). *Planta* **148,** 49–56.
(1980*b*). *Planta* 148, 57–63.
Croy, R. R. D., Gatehouse, J. A., Tyler, M., and Boulter, D. (1980*c*). *Biochem, J.* **191,** 509–16.
Croy, R. R. D., Shirsat, A., and Boulter, D. (1984*c*). *Nuc. Acids Res.*
Cullis, C. (1976). *Planta* **131,** 293–8.
Cunningham, B. A., Hemperly, J. J., Hopp, T. P., and Edelman, G. M. (1979). *Proc. Natl. Acad. Sci U.S.A.* **76,** 3218–22.
Damaschum, G., Miller, J. J., and Beilka, H. (1979). *Methods Enzymol.* **59,** 706–50.
Daniels, N. W. R., and Frazier, P. J. (1978). in *Plant Proteins,* ed. G. Norton, pp. 299–315. Butterworth, London.
Darby, G. K., Jones, A. S., Kennedy, J. F., and Walker, R. T. (1970). *J. Bacteriol.* **103,** 159–65.

Dasgupta, J., and Bewley, J. D. (1982). *Plant Physiol.* **70,** 1224–7.
Davey, R. A., and Dudman, W. F. (1979). *Aust. J. Plant Physiol.* **6,** 435–7.
Davey, R. A., Higgins, T. J. V., and Spencer, D. (1981). *Biochem. Int.* **3,** 595–602.
Davidson, E. H., Klein, W. H., and Britten, R. J. (1977). *Dev. Biol.* **55,** 69–84.
Davies, D. R. (1976). *Euphytica* **25,** 717–24.
(1980). *Biochem. Genet.* **18,** 1207–19.
Davies, D. R., and Bedford, I. D. (1982). *Plant Sci. Lett.* **27,** 227–43.
Davis, B. J. (1964). *Ann. N.Y. Acad. Sci.* **121,** 404–27.
Denić, M. (1983). in *Seed Proteins: Biochemistry, Genetics, Nutritive Value,* ed. W. Gottschalk and H. P. Müller, pp. 245–68. Martinus Nijhoff, The Hague.
De Ponte, R., Parlamenti, R., Petrucci, T., Silano, V., and Tomasi, M. (1976). *Cereal Chem.* **53,** 805–20.
Derbyshire, E., Wright, D. J., and Boulter, D. (1976). *Phytochem.* **15,** 3–24.
Deutscher, D. (1974). in *Nutritional Improvement of Food and Feed Proteins,* ed. M. Friedman, pp. 281–300. Plenum Press, New York.
Dierks-Ventling, C. (1981). *Eur. J. Biochem.* **120,** 177–82.
(1982). in *Embryonic Development, Part B: Cellular Aspects,* pp. 545–53. Alan R. Liss, New York.
Dierks-Ventling, C., and Cozens, K. (1982). *FEBS Lett.* **142,** 147–50.
Dierks-Ventling, C., and Ventling, D. (1982). *FEBS Lett.* **144,** 167–72.
Di Fonzo, N., Fornasar, E., Salamini, F., Reggiani, R., and Soave, C. (1980). *J. Heredity* **71,** 397–402.
Doll, H. (1983). in *Seed Proteins: Biochemistry, Genetics, Nutritive Value,* ed. W. Gottschalk and H. P. Müller, pp. 207–23. Martinus Nijhoff, The Hague.
(1984). *Phil. Trans. R. Soc. Lond.* [*Biol.*] **304,** 373–82.
Domoney, C., and Casey, R. (1983). *Planta* **159,** 446–53.
(1984). *Eur. J. Biochem.* **139,** 321–7.
Domoney, C., and Davies, D. R. (1983). in *Perspective for Peas and Lupins as Protein Crops,* ed. R. Thompson and R. Casey, pp. 256–71. Martinus Nijhoff, The Hague.
Domoney, C., Davies, P. R., and Casey, R. (1980). *Planta* **149,** 454–60.
Donovan, G. R., Lee, J. W., and Longhurst, T. J. (1982). *Aust. J. Plant Physiol.* **9,** 59–68.
Dudman, W. F., and Millerd, A. (1975). *Biochem. Syst. Ecol.* **3,** 25–33.
Dure, L., and Chlan, C. (1981). *Plant Physiol.* **68,** 180–6.
Dure, L., Chlan, C., and Galau, G. A. (1983). in *Structure and Function of Plant Genomes,* ed. O. Ciferri and L. Dure, pp. 113–21. Plenum Press, New York.
Dure, L. S., and Galau, G. A. (1981). *Plant Physiol.* **68,** 187–97.
Edens, L., Heslinga, L., Klok, R., Ledeboer, A. M., Maat, J., Tooner, M. Y., Visser, C., and Verrips, C. T. (1982). *Gene* **18,** 1–12.
Efstratiadis, A., Posakong, J. W., Maniatis, T., Lawn, R. M., O'Connell, C., Spritz, R. A., De Riel, J. K., Forget, B. G., Weissman, S. M., Slightom,

J. L., Blechl, A. E., Smithies, O., Baralle, F. E., Shoulders, C. C., and Proudfoot, N. J. (1980). *Cell* **21,** 653–68.
Eggum, B. O., and Beames, R. M. (1983). in *Seed Proteins: Biochemistry, Genetics, Nutritive Value,* ed. W. Gottschalk and H. P. Müller, pp. 499–531. Martinus Nijhoff, The Hague.
Ehrlich, H. A., Cohen, S. N., and MacDevitt, H. O. (1979). *Methods Enzymol.* **68,** 443–53.
Elleman, T. C. (1977). *Aust. J. Biol. Sci.* **30,** 33–45.
Ersland, D. R., Brown, J. W. S., Casey, R., and Hall, T. C. (1983). in *Seed Proteins: Biochemistry, Genetics, Nutritive Value,* ed. W. Gottschalk and H. P. Müller, pp. 355–76. Martinus Nijhoff, The Hague.
Esen, A., Bietz, J. A., Paulis, J. W., and Wall, J. S. (1982). *Nature* **296,** 678–9.
Evans, I. M., Croy, R. R. D., Brown, P., and Boulter, D. (1980). *Biochim. Biophys. Acta* **610,** 81–95.
Evans, I. M., Croy, R. R. D., Hutchinson, P., Boulter, D., Payne, P. I., and Gordon, M. E. (1979). *Planta* **144,** 455–63.
Evans, I. M., Gatehouse, J. A., Croy, R. R. D., and Boulter, D. (1984). *Planta* **160,** 559–68.
Evans, A. M., and Gridley, H. E. (1980). in *Commentaries in Plant Science, Vol. 2,* ed. H. Smith, pp. 143–59. Pergamon, New York.
Ewart, J. A. D. (1977). *J. Sci. Fd. Agric.* **28,** 191–9.
Faulks, A. J., Shewry, P. R., and Miflin, B. J. (1981). *Biochem. Genet.* **19,** 841–58.
Federoff, N. (1983). *Plant Molecular Biology Reporter* **1,** 27–9.
Federoff, N., Chaleff, D., Courage-Tebbe, U., Doring, H.-P., Geiser, M., Starlinger, P., Tillman, E., Weck, E., and Werr, W. (1983*a*). in *Structure and Function of Plant Genomes,* ed. O. Ciferri and L. Dure, pp. 61–71. Plenum Press, New York.
Federoff, N., Wessler, S., and Shure, M. (1983*b*). *Cell* **35,** 235–42.
Feix, G., Langridge, P., and Weinand, U. (1981). in *Genetic Engineering in the Plant Sciences,* ed. N. J. Panopoulas, pp. 73–84. Praeger, New York.
Fischer, R. L., and Goldberg, R. B. (1983). *Cell* **29,** 651–60.
Flashman, S. M., and Levings, C. S. (1981). in *The Biochemistry of Plants – A Comprehensive Treatise, Vol. 6, Proteins and Nucleic Acids,* ed. P. K. Stumpf and E. E. Conn, pp. 83–109. Academic Press, New York.
Fliegerova, O., Salvetova, A., Ticha, M., and Kocouret, J. (1974). *Biochim. Biophys. Acta* **351,** 416–26.
Forde, B. G. (1983). in *Techniques in Molecular Biology,* ed. J. M. Walker and W. Gaastra, pp. 167–84, 221–38. Croom Helm, London.
Forde, B. G., Kreis, M., Bahramian, M. B., Matthews, J. A., and Miflin, B. J., Thompson, R. D., Bartels, D., and Flavell, R. B. (1981). *Nucl. Acids Res.* **9,** 6689–707.
Forde, J., Forde, B. G., Fry, R., Kreis, M., Shewry, P. R., and Miflin, B. J. (1984). *Nucl. Acids Res.*
Freeling, M., and Woodman, J. C. (1979). in *The Plant Seed – Development, Preservation and Germination,* ed. I. Rubenstein, R. L. Phillips, C. E. Green, and B. G. Gengenbach. Academic Press, New York.

Frischauf, A.-M., Lehrach, H., Poustka, A., and Murray, N. (1983). *J. Mol. Biol.* **170,** 827–42.

Fukushima, D., and Koshiyama, I. (1976). *Phytochem.* **15,** 161–4.

Fulcher, R. G., and Wong, S. I. (1980). in *Cereals for Food and Beverages – Recent Progress in Cereal Chemistry and Technology,* ed. G. E. Inglett and L. Munck, pp. 1–26. Academic Press, New York.

Gaastra, W., and Oudega, B. (1983). in *Techniques in Molecular Biology,* ed. J. M. Walker and W. Gaastra, pp. 287–307. Croom Helm, London.

Gabriel, O. (1971). *Methods Enzymol.* **22,** 565–78.

Galante, E., Vitale, A., Manzocchi, L., Soave, C., and Salamini, F. (1983). *Mol. Gen. Genet.* **192,** 316–21.

Galau, G., and Dure, L. (1981). *Biochemistry* **20,** 4169–78.

Gallagher, T. F., and Ellis, R. J. (1982). *EMBO Journal* **1,** 1493–8.

Garnier, J., Osguthorpe, D. J., and Robson, P. (1978). *J. Mol. Biol.* **120,** 97–120.

Gasperi-Campani, A., Barbieri, L., Lorenzoni, E., Montanaro, L., Sperti, S., Bonetti, E., and Stirpe, F. (1978). *Biochem. J.* **174,** 491–6.

Gasperi-Campani, A., Barbieri, L., Lorenzoni, E., and Stirpe, F. (1977). *FEBS Lett.* **76,** 173–6.

Gasperi-Campani, A., Barbieri, L., Morelli, P., and Stirpe, F. (1980). *Biochem. J.* **186,** 439–41.

Gatehouse, A. M. R. (1984). in *Developments in Food Proteins, Vol. 13,* ed. B. J. F. Hudson, pp. 245–94. Elsevier Applied Science Publishers, London.

Gatehouse, A. M. R., and Boulter, D. (1983). *J. Sci. Fd. Agric.* **34,** 345–50.

Gatehouse, A. M. R., Gatehouse, J. A., Dobie, P., Kilminster, A. M., and Boulter, D. (1979). *J. Sci. Fd. Agric.* **30,** 948–58.

Gatehouse, A. M. R., Dewey, F. M., Dove, J., Fenton, K. A., and Pusztai, A. (1984*a*). *J. Sci. Fd. Agric.* **35,** 373–80.

Gatehouse, J. A., Croy, R. R. D., and Boulter, D. (1984*b*). in *Critical Reviews in Plant Science, Vol. I,* pp. 287–314. CRC Press, Boca Raton, Fla.

Gatehouse, J. A., Hoque, M. S., Gilroy, J., and Croy, R. R. D. (1985). *Biochem. J.* **225,** 239–47.

Gatehouse, J. A., Croy, R. R. D., Morton, H., Tyler, M., and Boulter, D. (1981). *Eur. J. Biochem.* **118,** 627–33.

Gatehouse, J. A., Evans, I. M., Bown, D., Croy, R. R. D., and Boulter, D. (1982*a*). *Biochem. J.* **208,** 119–27.

Gatehouse, J. A., Lycett, G. W., Croy, R. R. D., and Boulter, D. (1982*b*). *Biochem. J.* **207,** 629–32.

Gatehouse, J. A., Lycett, G. W., Delauney, A. J., Croy, R. R. D., and Boulter, D. (1983). *Biochem. J.* **212,** 427–32.

Gayler, K. R., and Sykes, G. E. (1981). *Plant Physiol.* **67,** 958–61.

Geraghty, D. E., Messing, J., and Rubenstein, I. (1982). *EMBO Journal* **1,** 1329–35.

Geraghty, D., Peifer, M. A., Rubenstein, I., and Messing, J. (1981). *Nucl. Acids Res.* **9,** 5163–74.

Gillespie, J. M., and Blagrove, R. J. (1978). *Aust. J. Plant Physiol.* **5,** 357–69.

Gillespie, J. M., Blagrove, R. J., and Randall, P. J. (1978). *Aust, J. Plant Physiol.* **5,** 641–50.

Gilroy, J., Wright, D. J., and Boulter, D. (1979). *Phytochem.* **18,** 315–16.

Goding, L. A., Bhatty, R. S., and Finlayson, A. J. (1970). *Can. J. Biochem.* **48,** 1096–101.

Goldberg, R. B., Fischer, R. L., Harada, J. J., Jofuku, D., and Okamuro, J. K. (1983*a*). in *Structure and Function of Plant Genomes,* ed. O. Ciferri and L. Dure, pp. 37–45. Plenum Press, New York.

Goldberg, R. B., Hoschek, G., Ditta, G. S., and Breidenbach, R. W. (1981*a*). *Dev. Biol.* **83,** 218–31.

Goldberg, R. B., Hoschek, G., Tam, S. H., Ditta, G. S., and Breidenbach, R. W. (1981*b*). *Dev. Biol.* **83,** 201–17.

Goldberg, R. B., Hoschek, G., and Vodkin, L. O. (1983*b*). *Cell* **33,** 465–75.

Gorecki, M., and Rozenblatt, S. (1980). *Proc. Natl. Acad. Sci. U.S.A.* **77,** 3686–90.

Graham, D. E. (1978). *Anal. Biochem.* **85,** 609–13.

Grant, G., More, L. J., Mckenzie, N. H., Stewart, J. C., and Pusztai, A. (1983). *Brit. J. Nutr.* **50,** 207–14.

Graveland, A., Bosveld, P., and Lichtendonk, W. J. (1980). in *Cereals for Food and Beverages–Recent Progress in Cereal Chemistry,* ed. G. E. Inglet and L. Munck, pp. 171–81. Academic Press, New York.

Greene, F. C. (1981). *Plant Physiol.* **68,** 778–83.

(1983). *Plant Physiol.* **71,** 40–6.

Grierson, D., and Spiers, J. (1983). in *Techniques in Molecular Biology,* ed. J. M. Walker and W. Gaastra, pp. 135–58. Croom Helm, London.

Grunstein, M., and Hogness, D. (1975). *Proc. Natl. Acad. Sci. U.S.A.* **72,** 3961–5.

Guldager, P. (1978). *Theor. Appl. Genet.* **53,** 241–50.

Hagen, G., and Rubenstein, I. (1980). *Plant Sci. Lett.* **19,** 217–23.

(1981). *Gene* **13,** 239–49.

Hall, T. C. (1979). in *Nucleic Acids in Plants,* ed. T. C. Hall and T. W. Davis, pp. 217–51. CRC Press, Boca Raton, Fla.

Hall, T. C., Ma, Y., Buchbinder, B. U., Pyne, J. W., Sun, S. M., and Bliss, F. A. (1978). *Proc. Natl. Acad. Sci. U.S.A.* **75,** 3196–200.

Hall, T. C., Slightom, J. L., Ersland, D. R., Murray, M. G., Hoffman, L. M., Adang, M. J., Brown, J. W. S., Ma, Y., Mathews, J. A., Cramer, J. H., Barker, R. F., Sutton, D. W., and Kemp, J. D. (1983*a*). in *Structure and Function of Plant Genomes,* ed. O. Ciferri and L. Dure, pp. 123–42. Plenum Press, New York.

Hall, T. C., Slightom, J. L., Ersland, D. R., Scharf, P., Barker, R. F., Murray, M. G., Brown, J. W. S., and Kemp, J. D. (1983*b*).· *Proceedings Miami 15th Winter Symposium.*

Hall, T. C., Sun, S. M., Buchbinder, B. U., Pyne, J. W., Bliss, F. A., and Kemp, J. D. (1980). in *Genome Organization and Expression in Plants,* ed. C. J. Leaver, pp. 259–72. Plenum Press, New York.

Hall, T. C., Sun, S. M., Ma, Y., McLeester, R. C., Pyne, J. W., Bliss, F. A., and Buchbinder, B. U. (1979). in *The Plant Seed–Development, Preservation, and Germination,* ed. I. Rubenstein, R. L. Phillips, C. E. Green, and B. G. Gengenbach, pp. 3–26. Academic Press, New York.

Hanahan, D., and Meselson, M. (1980). *Gene* **10,** 63–7.
Hartley, J. L., and Gregori, T. J. (1981). *Gene* **13,** 347–53.
Heidecker, G., and Messing, J. (1983). *Nucl. Acids Res.* **11,** 4891–906.
Hejgaard, J. (1982). *Physiol. Plant* **54,** 174–82.
Hejgaard, J., and Boisen, S. (1980). *Hereditas* **93,** 311–20.
Higgins, T. J. V. (1984). *Annu. Rev. Plant Physiol.* **35,** 191–221.
Higgins, T. J. V., Chandler, P. M., Zurawski, G., Button, S. C., and Spencer, C. (1983*a*). *J. Biol. Chem.* **258,** 9544–9.
Higgins, T. J. V., Chrispeels, M. J., Chandler, P. M., and Spencer, D. (1983*b*). *J. Biol. Chem.* **258,** 9550–2.
Higgins, T. J. V., and Spencer, D. (1981). *Plant Physiol.* **67,** 205–11.
Hill, J. E., and Breidenbach, R. W. (1974). *Plant Physiol.* **53,** 747–51.
Hirano, H., Gatehouse, J. A., and Boulter, D. (1982). *FEBS Lett.* **145,** 99–102.
Ho, D. T.-H. (1979). in *Molecular Biology of Plants,* ed. I. Rubenstein, R. L. Phillips, C. E. Green, and B. G. Gengenbach, pp. 217–40. Academic Press, New York.
Hoffman, L. M., Ma, Y., and Barker, R. F. (1982). *Nucl. Acids Res.* **10,** 7819–28.
Holder, A. A., and Ingversen, J. (1978). *Carlsberg Res. Commun.* **43,** 177–84.
Hopp, T. R., and Wood, K. R. (1981). *Proc. Natl. Acad. Sci. U.S.A.* **78,** 3824–8.
Horstmann, C. (1983). *Phytochem.* **22,** 1861–6.
(1984). *Kulturpflanze* **32.**
Hu, N.-T., Peifer, M. A., Heidecker, G., Messing, J., and Rubenstein, I. (1982). *EMBO Journal* **1,** 1337–42.
Hurkman, W. J., Smith, L. D., Richter, J., and Larkins, B. A. (1981). *J. Cell Biol.* **89,** 292–9.
Hynes, M. J. (1968). *Aust. J. Biol. Sci.* **21,** 827–9.
Ingversen, J. (1983). in *Seed Proteins,* ed. J. Daussant, J. Mosse, and J. Vaughan, pp. 193–204. Academic Press, New York.
Irvin, J. D. (1975). *Arch. Biochem. Biophys.* **169,** 522–8.
Iyengar, R. B., and Ravenstein, P. (1981). *Cereal Chem.* **53,** 258–78.
Jackson, E. A., Holt, L. M., and Payne, P. I. (1983). *Theor. Appl. Genet.* **66,** 29–37.
Jakubek, M. F., and Przybylska, J. (1982). *Pisum Newsl.* **14,** 26–8.
Janzen, D. H., Juster, H. B., and Liener, I. E. (1976). *Science* **192,** 795–6.
Jonassen, I. (1980). *Carlsberg Res. Commun.* **45,** 47–68.
Jones, R. A., Larkins, B. A., and Tsai, C. Y. (1976). *Biochem. Biophys. Res. Commun.* **69,** 404–10.
(1977*a*). *Plant Physiol.* **59,** 525–9.
(1977*b*). *Plant Physiol.* **59,** 733–7.
Kaback, D. B., Angerer, L. M., and Davidson, N. (1979). *Nucl. Acids Res.* **6,** 2499–517.
Kafatos, F. C., Jones, C. W., and Efstratiadis, A. (1979). *Nucl. Acids Res.* **7,** 1541–52.
Kamiya, N., Sakabe, K., Sakabe, N., Sasaki, K., Sakakibara, M., and Naguchi, H. (1983). *Agric. Biol. Chem.* **47,** 2091–8.

Karn, J., Brenner, S., and Barnett, L. (1983). *Methods Enzymol.* **101,** 3–19.
Karn, J., Brenner, S., Barnett, L., and Cesareni, G. (1980). *Proc. Natl. Acad. Sci. U.S.A.* **77,** 5172–7.
Kasarda, D. D. (1980). *Ann. Technol. Agric.* **29,** 151–73.
Kasarda, D. D., Bernardin, J. E., and Nimmo, C. C. (1976). in *Advances in Cereal Science and Technology, Vol. 1,* ed. Y. Pomeranz, pp. 158–236. American Association of Cereal Chemistry, St. Paul, Minn.
Khan, K., and Bushuk, W. (1979). in *Functionality and Protein Structure,* ed. A. Pour-El, pp. 191–206. American Chemical Society, St. Paul, Minn.
Khan, M. R. I., Gatehouse, J. A., and Boulter, D. (1980). *J. Exp. Bot.* **31,** 1599–611.
Khoo, U., and Wolf, M. J. (1970). *Amer. J. Bot.* **57,** 1042–50.
Kim, S. I., Charbonnier, L., and Mossé, J. (1978). *Biochim. Biophys. Acta* **537,** 22–30.
King, C. R., Udell, D. S., and Deeley, R. G. (1979). *J. Biol. Chem.* **254,** 6781–7.
King, T. P., Pusztai, A., and Clarke, E. M. W. (1980). *Histochem. J.* **12,** 201–8.
Kinniburgh, A. J., Mertz, J. E., and Ross, J. (1978). *Cell* **14,** 681–9.
Kinsella, J. E. (1976). *Crit. Rev. Fd. Sci. Nutr.* **7,** 219–80.
Kinsella, J. E., and Shetty, K. J. (1979). in *Functionality and Protein Structure,* ed. A. Pour-El, pp. 37–63. American Chemical Society, St. Paul Minn.
Kirkman, M. A., Shewry, P. R., and Miflin, B. J. (1982). *J. Sci. Food Agric.* **33,** 115–27.
Kislev, N., and Rubenstein, I. (1980). *Plant Physiol.* **66,** 1140–3.
Kitamura, K., Toyokawa, T., and Harada, K. (1980). *Phytochem.* **19,** 1841–3.
Koie, B., Ingversen, J., Anderson, A. J., Doll, H., and Eggum, B. O. (1976). in *Evaluation of Seed Protein Alterations by Mutation Breeding,* pp. 55–9. IAEA, Vienna.
Kolata, G. (1983). *Science* **222,** 495–6.
Koshiyama, I. (1969). *Agric. Biol. Chem.* **33,** 281–4.
(1972*a*). *Agric. Biol. Chem.* **36,** 62–7.
(1972*b*). *Agric. Biol. Chem.* **36,** 2255–7.
(1983). in *Seed Proteins: Biochemistry, Genetics, Nutritive Value,* ed. W. Gottschalk and H. P. Müller, pp. 427–50. Martinus Nijhoff, The Hague.
Koshiyama, I., and Fukushima, D. (1976). *Int. J. Peptide Protein Res.* **8,** 283–9.
Kreis, M., Rahman, S., Forde, B. G., Pywell, J., Shewry, P. R., and Miflin, B. J. (1983*a*). *Mol. Gen. Genet.* **191,** 201–6.
Kreis, M., Shewry, P. R., Forde, B. G., Rahman, S., and Miflin, B. J. (1983*b*). *Cell* **34,** 161–7.
Krishna, T. G., Croy, R. R. D., and Boulter, D. (1979). *Phytochem.* **18,** 1879–80.
Laemmli, U. K. (1970). *Nature* **238,** 680–5.
Lamb, I. (1984). Ph.D. thesis, University of Warwick.
Land, H., Grez, M., Hansen, H., Lindenmaier, W., and Schutz, G. (1981). *Nucl. Acids Res.* **9,** 2251–66.
Langridge, P., and Feix, G. (1983). *Cell* **34,** 1015–22.

Langridge, P., Pintor-Toro, J. A., and Feix, G. (1982). *Mol. Gen. Genet.* **187,** 432–8.

Larkins, B. A. (1981). in *The Biochemistry of Plants – A Comprehensive Treatise, Vol. 6, Proteins and Nucleic Acids,* ed. P. K. Stumpf and E. E. Conn, pp. 449–89. Academic Press, New York.

(1983). in *Genetic Engineering of Plants,* ed. T. Kosuge, C. P. Meredith, and A. Hollaender, pp. 93–118. Plenum Press, New York.

Larkins, B. A., and Hurkman, W. J. (1978). *Plant Physiol.* **62,** 256–63.

Larkins, B. A., Pedersen, K., Handa, A. K., Hurkman, W. J., and Smith, I. D. (1979). *Proc. Natl. Acad. Sci. U.S.A.* **76,** 6448–52.

Larkins, B. A., Pedersen, K., Marks, M. D., Wilson, D. R., and Argos, P. (1983). in *Structure and Function of Plant Genomes,* ed. O. Ciferri and L. Dure, pp. 73–83. Plenum Press, New York.

Lathe, R. F., and Lecocq, J.-P. (1977). *Virology* **83,** 204–6.

Lathe, R. F., Lecocq, J.-P., and Everett, R. (1983). in *Genetic Engineering, Vol. 4,* ed. R. Williamson, pp. 2–57. Academic Press, New York.

Law, C. N. (1983). in *Genetic Engineering: Principles and Methods, Vol. 5,* ed. J. K. Setlow and A. Hollaender, pp. 157–72. Plenum Press, New York.

Leder, P., Tiemeier, D., and Enquist, L. (1977). *Science* **196,** 175–7.

Lee, K. H., Jones, R. A., Dalby, A., and Tsai, C. Y. (1976). *Biochem. Genet.* **14,** 641–50.

Leemans, J., De Greve, H., Henalsteens, J. P., Shaw, C., Wilmitzer, L., Otten, L., Van Montagu, M., and Schell, J. (1982). in *Molecular Biology of Plant Tumours,* ed. G. Kahl and J. Schell, pp. 537–50. Academic Press, New York.

Lehrach, H., Diamond, D., Wozney, J. M., and Boedther, H. (1977). *Biochemistry* **16,** 4743–51.

LeMeur, M., Glanville, N., Mandel, J. L., Gerlinger, P., Palmiter, R., and Chambon, P. (1981). *Cell* **23,** 561–71.

Lewis, E. D., Hagen, G., Mullins, J. I., Mascia, P. N., Park, W. D., Benton, W. D., and Rubenstein, I. (1981). *Gene* **14,** 205–15.

Liener, I. E. (1978). in *Plant Proteins,* ed. G. Norton, pp. 117–40. Butterworth, London.

Lillford, P. J. (1978). in *Plant Proteins,* ed. G. Norton, pp. 289–98. Butterworth, London.

Litts, J. C., Anderson, O. D., Okita, T. W., and Greene, F. C. (1983). *Plant Physiol.* [*Suppl. 2*] **72.**

Loenen, W., and Brammar, W. F. (1980). *Gene* **20,** 249–59.

Londerdal, B., and Janson, J.-C. (1972). *Biochim. Biophys. Acta* **278,** 175–83.

Lycett, G. W., Croy, R. R. D., Shirsat, A., and Boulter, D. (1984*a*). *Nucl. Acids Res.* **12,** 4493–506.

Lycett, G. W., Delauney, A. J., and Croy, R. R. D. (1983*a*). *FEBS Lett.* **153,** 43–6.

Lycett, G. W., Delauney, A. J., Gatehouse, J. A., Gilroy, J., Croy, R. R. D., and Boulter, D. (1983*b*). *Nucl. Acids Res.* **11,** 2367–80.

Lycett, G. W., Delauney, A. J., Zhao, W., Gatehouse, J. A., Croy, R. R. D., and Boulter, D. (1984*b*). *Plant Mol. Biol.* **3,** 91–6.

Ma, Y., and Bliss, F. A. (1978). *Crop Sci.* **17,** 431–7.
McMaster, G. K., and Carmichael, G. G. (1977). *Proc. Natl. Acad. Sci. U.S.A.* **77,** 5201–5.
McPherson, A. (1980). *J. Biol. Chem.* **255,** 10472–80.
Maniatis, T., Fritsch, E. F., and Sambrook, J. (1982). *Molecular Cloning–A Laboratory Manual.* Cold Spring Harbor Laboratory, New York.
Manickam, A., and Carlier, A. R. (1980). *Planta* **149,** 234–40.
Manley, J. L., Fire, A., Cano, A., Sharp, P. A., and Gefter, M. L. (1980). *Proc. Natl. Acad. Sci. U.S.A.* **77,** 3855–9.
Manteuffel, R., Muntz, K., Puchel, M., and Scholz, G. (1976). *Biochem. Physiol. Pflanzen.* **169,** 595–605.
Manzocchi, L. A., Daminati, M. G., and Gentinetta, E. (1980). *Maydica* **25,** 199–210.
Marks, M. D., and Larkins, B. A. (1982). *J. Biol. Chem.* **257,** 9976–83.
Martensson, P. (1980). in *Vicia faba Feeding Value, Processing and Viruses,* ed. D. A. Bond, pp. 159–72. ECSC, EEC, EAEC, Brussels.
Martynoff, G. de, Pays, E., and Vassart, G. (1980). *Biochem. Biophys. Res. Commun.* **93,** 645–53.
Matlashewski, G. J., Adeli, K., Altosaar, I., Shewry, P. R., and Miflin, B. J. (1982). *FEBS Lett.* **145,** 208–12.
Matlashewski, G. J., Fabijanski, S., Adeli, K., Robert, L. S., Garson, K., and Altosaar, I. (1983). in *Advances in Gene Technology: Molecular Genetics of Plants and Animals Proceedings Miami 15th Winter Symposium.*
Matta, N. K., and Gatehouse, J. A. (1982). *Heredity* **48,** 383–92.
Matta, N. K., Gatehouse, J. A., and Boulter, D. (1981). *J. Exp. Bot.* **32,** 1295–307.
Matthews, J. A., and Miflin, B. J. (1980). *Planta* **149,** 262–8.
Matthews, J. A., Brown, J. W. S., and Hall, T. C. (1981). *Nature* **294,** 175–6.
Maxam, A. M., and Gilbert, W. (1977). *Proc. Natl. Acad. Sci. U.S.A.* **74,** 560–4.
Mechan, D. K., Kasarda, D. D., and Qualset, C. O. (1978). *Biochem. Genet.* **16,** 831–53.
Meinke, D. W., Chen, J., and Beachy, R. N. (1981). *Planta* **153,** 130–9.
Mertz, E. T. (1975). Chapter 1 in *Protein Nutritional Quality of Foods and Feeds, Vol. 1, Part 2,* ed. M. Friedman, Marcel Dekker, New York.
Messing, J., Crea, R., and Seeburg, P. H. (1981). *Nucl. Acids Res.* **9,** 309–20.
Messing, J., Geraghty, D., Heidecker, G., Hu, N.-T., Kridl, J., and Rubenstein, I. (1983). in *Genetic Engineering of Plants,* ed. by T. Kosuge, C. P. Meredith, and A. Hollaender, pp. 211–27. Plenum Press, New York.
Metz, J. E., and Gurdon, J. B. (1977). *Proc. Natl. Acad. Sci. U.S.A.* **74,** 1502–6.
Miège, M. N. (1982). in *Encyclopedia of Plant Physiology, Vol. 14A, Nucleic Acids and Proteins in Plants. I. Structure, Biochemistry and Physiology of Proteins,* ed. D. Boulter and B. Parthier, pp. 291–345. Springer-Verlag, Berlin.
Miflin, B. J., and Burgess, S. R. (1982). *J. Exp. Bot.* **33,** 251–60.
Miflin, B. J., Field, J. M., and Shewry, P. R. (1983*a*). in *Seed Proteins,* ed. J.

Daussant, J. Mossé, and J. Vaughan, pp. 255–319. Academic Press, New York.

Miflin, B. J., Forde, B. G., Kreis, M., Rahman, S., Forde, J., and Shewry, P. R. (1984). *Phil. Trans. R. Soc. Lond.* [*Biol.*] **304,** 333–9.

Miflin, B. J., and Lea, P. J. (1984). *Nature* **308,** 498–9.

Miflin, B. J., Mathews, J. A., Burgess, S. R., Faulks, A. J., and Shewry, P. R. (1980). in *Genome Organisation and Expression in Plants,* ed. C. J. Leaver, pp. 233–43. Plenum Press, New York.

Miflin, B. J., Rahman, S., Kreis, M., Forde, B. G., Blanco, L., and Shewry, P. R. (1983*b*). in *Structure and Function of Plant Genomes,* ed. O. Ciferri and L. Dure, pp. 85–92. Plenum Press, New York.

Miflin, B. J., and Shewry, P. R. (1981). in *The Physiology and Biochemistry of Plant Productivity,* ed. J. D. Bewley. Martinus Nijhoff, The Hague.

Millerd, A. (1975). *Annu. Rev. Plant Physiol.* **26,** 53–72.

Millerd, A., and Spencer, D. (1974). *Aust. J. Plant Physiol.* **1,** 331–41.

Millerd, A., Spencer, D., Dudman, W. F., and Stiller, M. (1975). *Aust. J. Plant Physiol.* **2,** 51–9.

Millerd, A., Thomson, J. A., and Schroeder, H. E. (1978). *Aust. J. Plant Physiol.* **5,** 519–34.

Mirelman, D., Galun, E., Sharon, N., and Lotan, R. (1975). *Nature* **256,** 414–16.

Mishkind, M., Raikhel, N. V., Palevitz, B. A., and Keegstra, K. (1982). *J. Cell Biol.* **92,** 753–64.

Montanaro, L., Sperti, S., Mattioli, A., Testoni, G., and Stirpe, F. (1975). *Biochem. J.* **146,** 127–31.

Mooi. F. R., and Gaastra, W. (1983). in *Techniques in Molecular Biology,* ed. J. M. Walker and W. Gaastra, pp. 199–219. Croom Helm, London.

Moreira, M. A., Hermodson, M. A., Larkins, B. A., and Nielsen, N. C. (1979). *J. Biol. Chem.* **254,** 9921–6.

(1981*a*). *Arch. Biochem. Biophys.* **210,** 633–42.

Moreira, M. A., Mahoney, W. C., Larkins, B. A., and Nielsen, N. C. (1981*b*). *Arch. Biochem. Biophys.* **210,** 643–6.

Mori, T. (1984). *Kulturpflanz* **32.**

Mori, T., Nakamura, T., and Utsumi, S. (1982*a*). *J. Agric. Food Chem.* **30,** 828–31.

(1982*b*). *J. Fd. Sci.* **4,** 26–30.

Mori, T., Utsumi, S., and Inaba, H. (1979). *Agric. Biol. Chem.* **43,** 2317–22.

Morton, H., Evans, I. M., Gatehouse, J. A., and Boulter, D. (1983). *Phytochem.* **22,** 807–12.

Mossé, J. (1966). *Fed. Proc.* **25,** 1663–9.

Mossé, J., and Pernollet, J. C. (1983). in *Chemistry and Biochemistry of Legumes,* ed. S. K. Arora, pp. 111–93. Edward Arnold, London.

Müller, H. P. (1983). in *Seed Proteins: Biochemistry, Genetics, Nutritive Value,* ed. W. Gottschalk and H. P. Müller, pp. 308–51. Martinus Nijhoff, The Hague.

Murai, N., Sutton, D. W., Murray, M. G., Slightom, J. L., Merlo, D. J., Reichert, N. A., Sengupta-Gopalan, C., Stock, C. A., Barker, R. F., Kemp, J. D., and Hall, T. C. (1983). *Science* **222,** 476–82.

Murray, D. R. (1979). *Plant Cell Environ.* **2,** 221–6.
Murray, N. E. (1983). in *Lambda II,* pp. 395–432. Cold Spring Harbor Laboratory, New York.
Murray, N. E., Brammar, W. J., and Murray, K. (1977). *Mol. Gen. Genet.* **150,** 1469–70.
Murray, M. G., Hoffman, L. M., and Jarvis, N. P. (1983). *Plant Mol. Biol.* **2,** 75–84.
Musuho, Y., Kishida, K., and Hara, T. (1982). *Biochem. Biophys. Res. Commun.* **105,** 462–9.
Mutschler, M. A., and Bliss, F. A. (1977). *Plant Physiol.* **59,** 21–6.
Mutschler, M. A., Bliss, F. A., and Hall, T. C. (1980). *Plant Physiol.* **65,** 627–30.
Nakamura, T., Utsumi, S., Kitamura, K., Harada, K., and Mori, T. (1984). *J. Agric. Food Chem.* **32,** 647–50.
Nelson, O. E. (1980). in *Advances in Cereal Science and Technology, Vol. 3,* ed. Y. Pomeranz, pp. 41–71. American Association of Cereal Chemists, St. Paul, Minn.
Neucere, N. J., and Ory, L. (1970). *Plant Physiol.* **45,** 616–19.
Nielsen, N. C. (1984). *Phil. Trans. R. Soc. Lond.* [*Biol.*] **304,** 287–96.
O'Farrell, P. H. (1975). *J. Biol. Chem.* **250,** 4007–21.
Ohmiga, M., Hara, I., and Matsubara, H. (1980). *Plant Cell Physiol.* **21,** 157–67.
Ohtani, S., Odaka, T., Yoshizumi, H., and Kagamiyama, H. (1977). *J. Biochem.* (*Tokyo*) **82,** 763–7.
Okayama, H., and Berg, P. (1982). *Mol. Cell Biol.* **2,** 161–9.
Okita, T. W., and Greene, F. C. (1982). *Plant Physiol.* **68,** 778–83.
Olsnes, S., Haylett, T., and Refsnes, K. (1978). *J. Biol. Chem.* **253,** 5069–73.
Olsnes, S., and Pihl, A. (1977). in *Receptors and Recognition, Vol. 1,* ed. P. Cuatrecasas, pp. 129–73. Chapman and Hall, London.
Ornstein, L. (1964). *Ann. N.Y. Acad. Sci.* **121,** 532–9.
Osborne, T. B. (1924). *The Vegetable Proteins.* Longmans, Green, London.
Ozaki, Y., Wada, K., Hase, T., Matsubara, H., Nakanishi, T., and Yoshizumi, H. (1980). *J. Biochem.* **87,** 549–55.
Park, W. D., Lewis, E. D., and Rubenstein, I. (1980). *Plant Physiol.* **65,** 98–106.
Patey, A. L., Evans, D. T., Tiplady, R., Byfield, P. G. H., and Matthews, E. W. (1975). *Lancet* **II,** 718
Payne, P. I. (1983). in *Seed Proteins,* ed. J. Daussant, J. Mossé, and J. Vaughan, pp. 223–53. Academic Press, New York.
Payne, P. I., and Corfield, K. G. (1979). *Planta* **145,** 83–8.
Payne, P. I., Corfield, K. G., and Blackman, J. A. (1979). *Theor. Appl. Genet.* **55,** 153–9.
Payne, P. I., Corfield, K. G., Holt, L. M., and Blackman, J. A. (1981). *J. Sci. Fd. Agric.* **32,** 51–60.
Payne, P. I., Harris, P. A., Law, C. N., Holt, L. M., and Blackman, J. A. (1980*a*). *Ann. Technol. Agric.* **29,** 309–20.
Payne, P. I., Holt, L. M., Jackson, E. A., and Law, C. N. (1984). *Phil. Trans. R. Soc. Lond.* [*Biol.*] **304,** 359–71.

Payne, P. I., Holt, L. M., Lawrence, G. J., and Law, C. N. (1982). *Qualitas Plant. Plant Fd. Hum. Nutr.* **31,** 229–41.
Payne, P. I., Law, C. N., and Mudd, E. E. (1980*b*). *Theor. Appl. Genet.* **58,** 113–20.
Payne, P. I., and Lawrence, G. J. (1983). *Cereal Res. Commun.* **11,** 29–35.
Payne, P. I., and Rhodes, A. P. (1982). in *Encyclopedia of Plant Physiology, Vol. 14A, Nucleic Acids and Proteins in Plants. I. Structure, Biochemistry and Physiology of Proteins,* ed. D. Boulter and B. Parthier, pp. 346–65. Springer-Verlag, Berlin.
Pedersen, K., Bloom, K. S., Anderson, J. N., Glover, D. V., and Larkins, B. A. (1980). *Biochemistry* **19,** 1644–50.
Pedersen, K., Devereux, J., Wilson, D. R., Sheldon, E., and Larkins, B. A. (1982). *Cell* **29,** 1015–26.
Pernollet, J.-C., and Mossé, J. (1983). in *Seed Proteins,* ed. J. Daussant, J. Mossé, and J. Vaughan, pp. 155–91. Academic Press, New York.
Peumans, W. J., Stinisen, H. M., and Carlier, A. R. (1982). *Planta* **154,** 568–72.
Phillips, D. E., Eyre, M. D., Thompson, A., and Boulter, D. (1981). *J. Sci. Fd. Agric.* **32,** 423–32.
Phillips, R. L. (1983). in *Genetic Engineering of Plants: An Agricultural Perspective,* ed. T. Kosuge, C. P. Meredith, and A. Hollaender, pp. 453–6. Plenum Press, New York.
Pintor-Toro, J. A., Langridge, P., and Feix, G. (1982). *Nucl. Acids Res.* **10,** 3845–60.
Pleitz, P., Damaschun, G., Muller, J. J., and Schlesier, B. (1983*a*). *FEBS Lett.* **162,** 43–6.
Plietz, P., Damaschun, G., Muller, J. J., and Schwenke, K. D. (1983*b*). *Eur. J. Biochem.* **130,** 315–20.
Pleitz, P., Damaschun, G., Zirwer, D., Gast, K., and Schlesier, B. (1983*c*). *Int. J. Biol. Macromol.* **5,** 356–60.
Plietz, P., Zirwer, D., Schlesier, B., Gast, K., and Damaschur, G. (1984). *Biochim. Biophys. Acta* **784,** 140–6.
Polacco, J. C., Sparks, R. B., and Havir, E. A. (1979). in *Genetic Engineering, Vol. 1,* ed. J. K. Setlow and A. Hollaender, pp. 241–59. Plenum Press, New York.
Pomeranz, Y. (1980). in *Cereals for Food and Beverages: Recent Progress in Cereal Chemistry,* ed. G. E. Inglett and L. Munck, pp. 201–33. Academic Press, New York.
Porceddu, E., Lafiandra, D., and Scarascia-Mugnozza, G. T. (1983). in *Seed Proteins – Biochemistry, Genetics, Nutritive Value,* ed. W. Gottschalk and H. P. Müller, pp. 77–142. Martinus Nijhoff, The Hague.
Powers, J. R., and Culbertson, J. D. (1983). *Cereal Chem.* **60,** 427–9.
Przybylska, J., Hurich, J., and Blixt, S. (1981). *Pisum Newsl.* **13,** 44–5.
Pusztai, A. (1966). *Biochem. J.* **101,** 379–83.
Pusztai, A., Clarke, E. M. W., and King, T. P. (1979*a*). *Proc. Nutr. Soc.* **38,** 115–21.
Pusztai, A., Clarke, E. M. W., King, T. P., and Stewart, J. C. (1979*b*). *J. Sci. Fd. Agric.* **30,** 843–8.

Pusztai, A., Croy, R. R. D., Grant, G., and Stewart, J. C. (1983). in *Seed Proteins,* ed. J. Daussant, J. Mossé, and J. Vaughan, pp. 53–82. Academic Press, New York.

Pusztai, A., Croy, R. R. D., Grant, G., and Watt, W. B. (1977). *New Phytol,* **79,** 61–71.

Pusztai, A., and Duncan, I. (1971). *Biochim. Biophys. Acta* **229,** 785–94.

Pusztai, A., and Palmer, R. (1977). *J. Sci. Fd. Agric.* **28,** 620–3.

Pusztai, A., and Stewart, J. C. (1980). *Biochim. Biophys. Acta* **623,** 418–28.

Pusztai, A., and Watt, W. B. (1970). *Biochem. Biophys, Acta* **207,** 413–31.

(1974). *Biochim. Biophys. Acta* **365,** 57–71.

Rahman, S., Shewry, P. R., Forde, B. G., Kreis, M., and Miflin, B. J. (1983). *Planta* **159,** 366–72.

Ramshaw, J. A. M. (1982). in *Encyclopedia of Plant Physiology. Vol. 14A. Nucleic Acids and Proteins in Plants. I. Structure, Biochemistry and Physiology of Proteins,* ed. D. Boulter and B. Parthier, pp. 229–90. Springer-Verlag, Berlin.

Randall, P. J., Thomson, J. A., and Schroeder, H. E. (1979). *Aust. J. Plant Physiol.* **6,** 11–24.

Rasmussen, S. K., Hopp, H. E., and Brandt, A. (1983). *Carlsberg Res. Commun,* **48,** 187–99.

Rhighetti, P. G., and Drysdale, J. W. (1974). *J. Chromatog.* **98,** 271–321.

Rhighetti, P. G., Gianazza, E., Viotti, A., and Soave, C. (1977). *Planta* **136,** 115–23.

Rhodes, A. P., and Jenkins, G. (1978). in *Plant Protein,* ed. G. Norton, pp. 207–26. Butterworth, London.

Richardson, M. (1977). *Phytochem.* **16,** 159–69.

Rigby, P. W. J. (1982). in *Genetic Engineering, Vol. 3,* ed. R. Williamson, pp. 84–141. Academic Press, New York.

Roberts, T. M., Swanberg, S. L., Pateete, A., Ridell, G., and Backman, K. (1980). *Gene* **12,** 123–7.

Roberts, W. K., and Stewart, T. S. (1979). *Biochemistry,* **18,** 2615–20.

Romero, J., Sun, S. M., McLeester, R. C., and Hall, T. C. (1975). *Plant Physiol.* **56,** 776–9.

Rubenstein, I. (1982). in *Maize for Biological Research,* ed. W. F. Sheridan, pp. 189–95. Plant Molecular Biology Association.

Rubin, G. M., and Spradling, A. C. (1982). *Science* **218,** 348–53.

Saedler, H., Bonas, U., Daumling, B., Gupta, H., Hahlbrook, K., Harrison, B. J., Kreuzäler, F., Peterson, P. A., Reif, J., Schwarz-Sommer, Z., Shepard, N., Sommer, H., Ubben, D., and Wienand, U. (1983). in *Genetic Rearrangements,* ed. K. F. Chater, C. A. Cullis, D. A. Hopwood, A. Johnston, and H. W. Woolhouse, pp. 107–15. Croom Helm, London.

Saio, K., Kamiya, M., and Watanabe, T. (1969*a*). *Agric. Biol. Chem.* **35,** 890–8.

(1969*b*). *Agric. Biol. Chem.* **33,** 1301–8.

Salcedo, G., Sanchez-Monge, R., Argamenteira, A., and Argoncillo, C. (1980). *Plant Sci. Lett.* **19,** 109–19.

Sammour, R., Gatehouse, J. A., and Boulter, D. (1984). *Planta* **161,** 61–70.

Sanger, F., and Coulson, A. R. (1974). *J. Mol. Biol.* **94,** 441–8.
Sanger, F., Nicklen, S., and Coulson, A. R. (1977). *Proc. Natl. Acad. Sci. U.S.A.* **74,** 5463–9.
Sarkki, M.-L. (1980). in *Cereals for Food and Beverages–Recent Progress in Cereal Chemistry,* ed. G. E. Inglet and L. Munck, pp. 155–69. Academic Press, New York.
Schlesier, B., Manteuffel, R., Rudolph, A., and Behlke, J. (1978). *Biochem. Physiol. Pflanz.* **173,** 420–8.
Schmitt, J. M., and Svendsen, I. (1980). *Carlsberg Res. Commun.* **45,** 143–8.
Schroeder, H. E. (1982). *J. Sci. Fd. Agric.* **33,** 623–33.
(1984*a*). *J. Sci. Fd. Agric.* **35,** 191–8.
(1984*b*). *J. Exp. Bot.* **35,** 813–21.
Schuler, M. A., Doyle, J. J., and Beachy, R. N. (1983). *Plant Mol. Biol.* **2,** 119–27.
Schuler, M. A., Ladin, B. F., Polacco, J. C., Freyer, G., and Beachy, R. N. (1982*a*). *Nucl. Acids Res.* **10,** 8245–61.
Schuler, M. A., Schmitt, E. S., and Beachy, R. N. (1982*b*). *Nucl. Acids Res.* **10,** 8225–44.
Schwenke, K. D., Schultz, M., Linow, H.-J., Uhlig, J., and Franzke, C. (1974). *Nahrung* **18,** 709–13.
Schwinghamer, M. W., and Shepherd, R. J. (1980). *Anal. Biochem.* **103,** 426–34.
Seif, I., Khoury, G., and Dhar, R. (1980). *Nucl. Acids Res.* **8,** 2225–40.
Sengupta, C., Deluca, V., Bailey, D. S., and Verma, D. P. S. (1981). *Plant Mol. Biol.* **1,** 19–34.
Shapiro, S. Z., and Young, J. R. (1981). *J. Biol. Chem.* **256,** 1495–8.
Sharp, P. A., Berk, A. J., and Berget, S. M. (1980). *Methods Enzymol.* **65,** 750–68.
Shaw, C. H., Leemans, J., Shaw, C. H., Van Montagu, M., and Schell, J. (1983). *Gene* **23,** 315–30.
Shepherd, K. W. (1968). in *Proceedings 3rd International Wheat Genetics Symposium,* pp. 86–96. Australian Academy of Science, Canberra.
Shewry, P. R., Ellis, J. R. S., Pratt, H. M., and Miflin, B. J. (1978). *J. Sci. Fd. Agric.* **29,** 433–41.
Shewry, P. R., Faulks, A. J., and Miflin, B. J. (1980). *Biochem. Genet.* **18,** 133–51.
Shewry, P. R., Field, J. M., Lew, E. J.-L., and Kasarda, D. D. (1982). *J. Exp. Bot.* **33,** 261–8.
Shewry, P. R., Forde, J., Tatham, A., Field, J. M., Forde, B. G., Faulks, A. J., Fry, R., Parman, S., Miflin, B. J., Diether, M. D., Lew, E. J.-L., and Kasarda, D. D. (1983*a*). in *Proceedings 6th International Wheat Genetics Symposium,* Kyoto, 1983.
Shewry, P. R., Franklin, J., Parmar, S., Smith, S. J., and Miflin, B. J. (1983*b*). *J. Cereal Sci.* **1,** 21–31.
Shewry, P. R., and Miflin, B. J. (1983). in *Seed Proteins: Biochemistry, Genetics, Nutritive Value,* ed. W. Gottschalk and H. P. Müller, pp. 143–205. Martinus Nijhoff, The Hague.
Shewry, P. R., Miflin, B., and Kasarda, D. D. (1984). *Phil. Trans. R. Soc. Lond.* [*Biol.*] **304,** 297–308.

Shewry, P. R., Miflin, B. J., Lew, E. J.-L., and Kasarda, D. D. (1983*c*). *J. Exp. Bot.* **34,** 1403–10.
Shimamoto, K., Ackermann, M., and Dierks-Ventling, C. (1983). *Plant Physiol.* **73,** 915–20.
Shirsat, A. (1984). Ph.D. thesis, University of Durham.
Singer, M. F. (1979). in *Genetic Engineering–Principles and Methods, Vol. 1,* ed. J. K. Setlow and A. Hollaender, pp. 1–13. Plenum Press, New York.
Slightom, J. L., Sun, S. M., and Hall, T. C. (1983). *Proc. Natl. Acad. Sci. U.S.A.* **80,** 1897–901.
Smith, B. J., and Nicholas, R. H. (1983). in *Techniques in Molecular Biology,* ed. J. M. Walker and W. Gaastra, pp. 25–48. Croom Helm, London.
Smith, D. F., Searle, P. F., and Williams, J. G. (1979). *Nucl. Acids Res.* **6,** 487–506.
Smith, J. G. (1973). *Plant Physiol.* **51,** 454–8.
Soave, C., Reggiani, R., Di Fonzo, N., and Salamini, F. (1982). *Biochem. Genet.* **11,** 1027–37.
Soave, C., and Salamini, F. (1983). in *Seed Proteins,* ed. J. Daussant, J. Mossé, and J. Vaughan, pp. 205–18. Academic Press, New York.
(1984). *Phil. Trans. R. Soc. Lond* [*Biol.*] **304,** 341–7.
Soave, C., Tardani, L., Di Fonzo, N., and Salamini, F. (1981). *Cell* **27,** 403–10.
Soave, C., Viotti, A., Di Fonzo, N., and Salamini, F. (1980). in *Genome Organization and Expression in Plants,* ed. C. J. Leaver, pp. 219–26. Plenum Press, New York.
Sorenson, J. C. (1984). *Adv. Genet.* **22,** 109–44.
Sosinov, A. A., and Poperelya, F. A. (1980). *Ann. Technol. Agric.* **29,** 229–45.
Spradling, A. C., and Rubin, G. M. (1982). *Science* **219,** 341–7.
Spena, A., Viotti, A., and Pirrotta, V. (1982). *EMBO Journal.* **1,** 1589–94.
(1983). *J. Mol. Biol.* **169,** 799–811.
Spencer, D. (1984). *Phil. Trans. R. Soc. Lond* [*Biol.*] **304,** 275–85.
Spencer, D., and Higgins, T. J. V. (1980). *Biochem. Int.* **1,** 501–9.
Spencer, D., Chandler, P. M., Higgins, T. J. V., Inglis, A. S., and Rubira, M. (1984). *Plant Mol. Biol.* **2,** 259–68.
Sperti, S., Montanaro, L., Mattioli, A., Testoni, G., and Stirpe, F. (1976). *Biochem. J.* **156,** 7–13.
Staden, R. (1982). *Nucl. Acids Res.* **10,** 295–6.
Stafford, A., and Davies, D. R. (1979). *Ann. Bot.* **44,** 315–21.
St. Angelo, A. J., Yatsu, L. Y., and Altschul, A. M. (1968). *Arch. Biochem. Biophys.* **124,** 199–204.
Staswick, P. E., Hermodson, M. A., and Nielsen, N. C. (1981). *J. Biol. Chem.* **256,** 8752–5.
Staswick, P. E., and Nielsen, N. C. (1983). *Arch. Biochem. Biophys.* **223,** 1–8.
Steffens, R., Fox, F. R., and Kassell, B. (1978). *J. Agric. Fd. Chem.* **26,** 170.
Stinissen, H. M., Peumans, W. J., and Carlier, A. R. (1983). *Planta* **159,** 105–11.
Stirpe, F., Olsnes, S., and Pihl, A. (1980). *J. Biol. Chem.* **255,** 6947–53.
Stirpe, F., Pession-Brizzi, A., Lorenzoni, E., Strocchi, P., Montanaro, L., and Sperti, S. (1976). *Biochem. J.* **156,** 1–6.

Stirpe, F., Williams, D. G., Onyon, L. J., and Legg, R. F. (1981). *Biochem. J.* **195,** 399–405.

Strommer, J. N., Hake, S., Bennetzen, J., Taylor, W. C., and Freeling, M. (1982). *Nature* **200,** 542–4.

Struck, D. K., Lennarz, W. J., and Brew, K. (1978). *J. Biol. Chem.* **253,** 5786–94.

Sullivan, J. G. (1981). Ph.D. thesis, University of Wisconsin, Madison.

Sun, S. M., McLeester, R. C., Bliss, F. A., and Hall, T. C. (1974). *J. Biol. Chem.* **249,** 2118–20.

Sun, S. M., Mutschler, M. A., Bliss, F. A., and Hall, T. C. (1978). *Plant Physiol.* **61,** 918–23.

Sun, S. M., Slightom, J. L., and Hall, T. C. (1981). *Nature* **289,** 37–41.

Sung, M. T., and Slightom, J. L. (1981). in *Genetic Engineering in the Plant Sciences,* ed. N. J. Panopoulos, pp. 39–62. Praeger, New York.

Sussex, I. M., and Dale, R. M. K. (1979). in *The Plant Seed – Development, Preservation and Germination,* ed. I. Rubenstein, R. L. Phillips, C. E. Green, and B. G. Gengenbach, pp. 129–41. Academic Press, New York.

Svendsen, I., Martin, B., and Jonassen, I. (1980). *Carlsberg Res. Commun.* **45,** 79–85.

Talbot, C. F., and Etzler, M. E. (1978). *Biochemistry* **17,** 1474–9.

Thanh, V. H., and Sibasaki, K. (1977). *Biochim. Biophys. Acta* **490,** 370–84.
(1978). *J. Agric. Fd. Chem.* **26,** 692–5.
(1979). *J. Agric. Fd. Chem.* **27,** 805–9.

Thomas, P. S. (1980). *Proc. Natl. Acad. Sci. U.S.A.* **77,** 5202–5.

Thompson, R., and Casey, R. (editors) (1983). *Perspectives for Peas and Lupins as Protein Crops.* Martinus Nijhoff, The Hague.

Thompson, R. D., Bartels, D., Harberd, N. P., and Flavell, R. B. (1983). *Theor. Appl. Genet.* **67,** 87–96.

Thompson, W. F., and Murray, M. G. (1981). in *The Biochemistry of Plants – A Comprehensive Treatise, Vol. 6, Proteins and Nucleic Acids,* ed. P. K. Stumpf and E. E. Conn, pp. 1–8. Academic Press, New York.

Thomson, J. A., and Doll, H. (1979). in *Seed Improvement of Cereals and Grain Legumes, Vol. 1,* pp. 109–23. IAEF, Vienna.

Thomson, J. A., and Schroeder, H. E. (1978). *Aust. J. Plant Physiol.* **5,** 281–94.

Thomson, J. A., Schroeder, H. E., and Dudman, W. F. (1978). *Aust. J. Plant Physiol.* **5,** 263–79.

Thomson, J. A., Schroeder, H. E., and Tassie, A. M. (1980). *Aust. J. Plant Physiol.* **7,** 271–82.

Thorne, J. H. (1982). *Plant Physiol.* **70,** 953–8.

Thorpe, P., Brown, A. N. F., Ross, W. C. J., Cumber, A. J., Detre, S., Edwards, D. C., Davies, A. J. S., and Stirpe, J. (1981). *Eur. J. Biochem.* **116,** 447–545.

Toms, G. C., and Western, A. (1971). in *Chemotaxonomy of the Leguminosae,* ed. J. B. Harborne, D. Boulter, and B. L. Turner, pp. 367–462. Academic Press, New York.

Trewavas, A. (1982). *Plant, Cell Environment* **4,** 203–28.
Trowbridge, I. S. (1974). *J. Biol. Chem.* **249,** 6004–12.
Tsai, C. A. (1979). *Maydica* **24,** 129–40.
Tsai, C. Y., Larkins, B. A., and Glover, D. V. (1978). *Biochem. Genet.* **16,** 883–96.
Tumer, N. E., Thanh, V., and Nielsen, N. C. (1981). *J. Biol. Chem.* **256,** 8756–60.
Tumer, N. E., Thanh, V. H., and Nielsen, N. C. (1982). *J. Biol. Chem.* **257,** 4016–18.
Twigg, A. J., and Sherratt, D. (1980). *Nature* **283,** 216–18.
Uchida, T., Mekada, E., and Okada, Y. (1980). *J. Biol. Chem.* **255,** 6687–93.
Utsumi, S., Inaba, H., and Mori, T. (1980). *Agric. Biol. Chem.* **44,** 1891–4.
(1981). *Phytochem.* **20,** 585–9.
Utsumi, S., and Mori, T. (1980). *Biochim. Biophys. Acta* **621,** 179–89.
(1981). *Agric. Biol. Chem.* **45,** 2273–6.
(1983). *J. Biochem.* **94,** 2001–8.
Utsumi, S., Nakamura, T., and Mori, T. (1983). *J. Agric. Fd. Chem.* **31,** 503–6.
van Embden, J. (1983). in *Techniques in Molecular Biology,* ed. J. M. Walker and W. Gaastra, pp. 309–21. Croom Helm, London.
Vieira, J., and Messing, J. (1982). *Gene* **19,** 259–68.
Viotti, A., Albildsten, D., Pogna, N., Sala, E., and Pirrotta, V. (1982). *EMBO Journal.* **1,** 53–8.
Viotti, A., Sala, E., Alberi, P., and Soave, C. (1978). *Plant Sci. Lett.* **13,** 365–75.
Viotti, A., Sala, E., Marotta, R., Alberi, P., Balducci, C., and Soave, C. (1979). *Eur. J. Biochem.* **102,** 211–22.
Vodkin, L. O. (1980). *Plant Physiol.* **68,** 766–71.
Vodkin, L. O., Rhodes, P. R., and Goldberg, R. B. (1983). *Cell* **34,** 1023–31.
Wall, J. S., and Paulis, J. W. (1978). in *Advances in Cereal Science and Technology, Vol. 2,* ed. Y. Pomeranz, pp. 135–300. American Association of Cereal Chemists, St. Paul, Minn.
Wallace, R. W., and Dieckert, J. W. (1976). *Anal. Biochem.* **75,** 498–508.
Walton, D. C. (1980). *Annu. Rev. Plant Physiol.* **31,** 453–89.
Wang, J. L., Cunningham, B. A., Waxdal, M. J., and Edelman, G. (1975). *J. Biol. Chem.* **250,** 1490–502.
Weaver, R. F., and Weissmann, C. (1979). *Nucl. Acids Res.* **7,** 1175–93.
Weber, E., Ingversen, J., Manteuffel, R., and Puchel, M. (1981). *Carlsberg Res. Commun.* **46,** 383–93.
Weil, P. A., Luse, D. S., Segall, J., and Roeder, R. G. (1979). *Cell* **18,** 469–84.
Weinand, U., Bruschke, C., and Feix, G. (1979). *Nucl. Acids Res.* **6,** 2707–15.
Weinand, U., and Feix, G. (1980). *FEBS Lett.* **166,** 14–16.
Weinand, U., Langridge, P., and Feix, G. (1981). *Mol. Gen. Genet.* **182,** 440–4.
Weissman, C., Magata, S., Taniguchi, T., Weber, H., and Meyer, F. (1979). in *Genetic Engineering–Principles and Methods, Vol. 1,* ed. J. K. Setlow and A. Hollaender, pp. 133–50. Plenum Press, New York.

Wickens, M. P., Laskey, R. A. (1982). in *Genetic Engineering, Vol. 1,* ed. R. Williamson, pp. 163–7. Academic Press, New York.
Williams, B. G., and Blattner, F. R. (1979). *J. Virol.* **29,** 555–75.
(1980). in *Genetic Engineering, Vol. 2,* ed. J. K. Setlow and A. Hollaender, pp. 201–81. Plenum Press, New York.
Williams, J. G. (1981). in *Genetic Engineering, Vol. 1,* ed. R. Williamson, pp. 1–59. Academic Press, New York.
Williams, J. G., and Lloyd, M. M. (1979). *J. Mol. Biol.* **129,** 19–35.
Wilson, C. M., Shewry, P. R., Faulks, A. J., and Miflin, B. J. (1981). *J. Exp. Bot.* **32,** 1287–93.
Woo, S. L. C. (1979). *Methods Enzymol.* **68,** 389–95.
Worgan, J. T. (1978). in *Plant Proteins,* ed. G. Norton, pp. 191–204. Butterworth, London.
Wright, D. J., and Bumstead, M. R. (1984). *Phil. Trans. R. Soc. Lond.* [*Biol.*] **304,** 381–93.
Wrigley, C. W. (1980). *Ann. Technol. Agric.* **29,** 213–27.
Wrigley, C. W., Cros, D. L., Archer, M. J., Downie, P. G., and Roxburgh, C. M. (1980). *Aust. J. Plant Physiol* **7,** 755–66.
Wrigley, C. W. , and Shepherd, C. W. (1973). *Ann. N.Y. Acad. Sci.* **209,** 154–62.
Yamagata, H., Sugimoto, T., Tanaka, K., and Kasal, Z. (1982). *Plant Physiol.* **70,** 1094–100.
Yamauchi, F., and Yamagishi, T. (1979). *Agric. Biol. Chem.* **43,** 505–10.
Young, R. A., Hagenbuchle, O., and Schibler, U. (1981). *Cell* **23,** 451–8.
Zhao, W.-M., Gatehouse, J. A., and Boulter, D. (1983). *FEBS Lett.* **162,** 96–102.
Zimmerman, G., Weissman, G., and Yannai, S. (1967). *J. Fd. Sci.* **32,** 129–30.

9 Applications of genetic engineering to agriculture

M. G. K. JONES

Mankind uses a wide range of plants in agriculture and horticulture that have been selected over thousands of years, and agriculture is economically still the most important activity in the world. For each crop or species, specific breeding objectives and strategies for crop improvement can be defined. These depend on the nature of the crop and specific disease or cultural problems. Of the major world crops, three of the five major cereals and most of the important oil and protein crops (including cotton, soybean and pulses) that provide two-thirds of the world's food are inbreeders. Maize is the most important outbreeder, contributing 20% of world grain and a considerable proportion of forage production. Sunflower and many forage grasses and legumes are also outbreeders, as are vegetatively propagated crops such as potato, sugar cane, and orchard and soft fruit crops (Williams 1981). The possible applications of genetic engineering (excluding genetic manipulaton by protoplast and tissue culture techniques, as described in earlier chapters) to selected characters of this wide range of crop plants will be considered, following an outline of some breeding strategies, for it is in the context of current plant breeding practices that the contributions of genetic engineering will be judged.

Plant breeding

For the last 40 years, in developed countries the yield of agricultural crops has steadily increased, and it appears that the genetic component (i.e., the improvement of the crops by plant breeding) has contributed to about half of this increase. Improved agronomy and crop husbandry practices are responsible for the rest of the increase in yield. Application of genetic engineering techniques to agriculture clearly involves mainly the first (genetic) component, but engineered crops will undoubtedly require altered agronomic practices, as have "green revolution" cereals such as wheat and rice.

As supported by the evidence of increased crop yields, plant breeding has been highly successful. This success is based partly on an increased understanding of the parameters involved, but to a greater extent on improved and more efficient methods of plant selection. The usual procedures involve the production of genetic variability by crossing the best parental plants available to produce a population of plants that include superior genotypes, followed by selection of the superior genotypes, and eventually producing from them a new variety. Thus, apart from the thought that must be given to choosing the parents initially, most of the effort is put into examining large numbers of single plants and their progeny. Thus, a typical cereal breeding program may begin with more than 1,000 crosses each year, from which 1,500 plants are grown from each cross in the second (F_2) generation, such that about 1.6 million plants must be examined in the field for agronomic appearance and disease resistance (Bingham 1981). Selected progeny are grown as "ear rows" for three or four further generations, and, with further selecting and quality testing, promising lines may be released for statutory evaluations 9 years after crossing, and seed distributed commercially 12 years after crossing. The input of new genetic material therefore occurs during crossing, and when parental plants are chosen from within the breeding cycle, then concern has been expressed about the narrowness of the genetic base of the breeding program. In general, most genetic variability resides in land races, unadapted varieties and related wild species, and promising genes can be usefully introduced only over a longer time scale, using repeated back-crossing.

Some aims of plant breeding

The main objectives of breeding are improved yield, quality, disease resistance, and stress tolerance, and, of course, these are all interrelated.

Crop yield is related to total production of biomass by the partitioning of assimilates into harvestable organs ("harvest index"), and to disease resistance and stress tolerance, which affect potential losses. There is frequently a penalty to be paid in that increasing yield can lead to reduced quality. These breeding characteristics are frequently polygenic and genetically ill-defined. When considering possible application of genetic engineering to agriculture and crop improvement in the light of this general background, various criteria emerge. To be successful, new characters of value must be introduced, and more rapid methods of selection and testing must be developed. The characters should be inserted into advanced breeding material, so that other characters of the variety are not out of date by the time the variety is marketed. Molecular biologists should choose the correct genes (e.g., for disease or herbicide resistance) so that characters in varieties produced are durable and are not broken by new races of pathogen or weed biotype.

For outbreeding species, the production of F_1 hybrid varieties has been important, particularly for maize and also many *Brassica* species, onions, carrots, sorghum, and also many ornamentals (Williams 1981). Aspects of plant breeding systems and methodologies are discussed in more detail elsewhere (e.g., Bingham 1981; Simmonds 1979; Williams 1981).

Some areas in which genetic engineering techniques may be of value will be discussed in the following sections. The examples are necessarily selective, and more particularly relevant to European and North American agriculture.

Yield. Improvements in yield have frequently been associated with improved harvest index, that is, the ratio of harvested organ or product to total biomass. This is illustrated in Table 9.1.

Improvements of assimilate partitioning and harvest index to increase yield must clearly reach a limit for crops such as cereals. In

Table 9.1. *Varietal improvements in yield of crops grown in England and Wales relative to "baseline varieties" with yield value of 100%*

	Baseline variety	Year of introduction	Harvest index (1979)	Yield of best current variety (% of baseline yield)
Wheat	Bersee	1947	44–48	156
Barley	Plumage Archer	1947	48–52	132
Potatoes				
(early)	Home Guard	1943	75–85	120
(main)	King Edward	1902		114
Perennial	S24	1937	85–90	106
ryegrass	S23	1933		108

fact, for forage crops, for which a significant change in harvest index is not possible, there has been little increase in total biomass produced, and hence yield improvements have been small. Thus, there is a need to increase total biomass. Biomass production depends on photosynthesis, but selection based on photosynthetic rate by breeders has not been successful, and selection for yield has not produced varieties with high photosynthetic rates (Gifford and Evans 1981). The photosynthetic processes are therefore prime candidates for application of genetic engineering techniques and will be considered first.

Photosynthesis. Photosynthesis encompasses both the light reactions involved in converting energy from light to chemical energy and its utilization in a series of dark reactions that fix carbon dioxide.

Ribulose bisphosphate carboxylase (Rubisco) (Figure 9.1) is the key enzyme in fixation of CO_2. On a global basis this enzyme is responsible for fixing annually about 10^{11} tons of CO_2 and has therefore been the subject of considerable study. It is the major soluble leaf protein and exists as a complex of 16 subunits comprising equal numbers of two types: large (MW 52,680) and small (MW 14,480). The large subunit carries the catalytic site and is coded in the chloroplast genome; the small subunit is coded in the nucleus. In addition to catalyzing the fixation of CO_2, the large subunit also catalyzes an oxidative cleavage of ribulose bisphosphate (RuBP) that initiates the metabolic process of "photorespiration" (Figure 9.1) in C3 plants. This process reduces the efficiency of photosynthesis, because it dissipates energy and releases CO_2 while only recovering part of the carbon in glycolate formed as a result of the oxygenase activity, as 3-phosphoglycerate. The amount of photorespiration relative to CO_2 fixation is determined by O_2 and CO_2 partial pressures (Osmond 1981). In normal air the ration of carboxylation to oxygenation is 3:1 to 4:1 (Miziorko and Lorimer 1983), but decreasing photorespiration by growing plants in atmospheres enriched with CO_2 or depleted of oxygen has indicated that up to 50% increase in yields may be obtained (Keys 1983). The C4 photosynthetic plants have evolved a mechanism to decrease photorespiration that effectively raises the CO_2 concentration at the site of fixation. Because CO_2 and O_2 compete for the catalytic site, raising CO_2 levels favors the carboxylation reaction. The reactions and transport processes involved in C4 photosynthesis also require energy and appear to be of greatest advan-

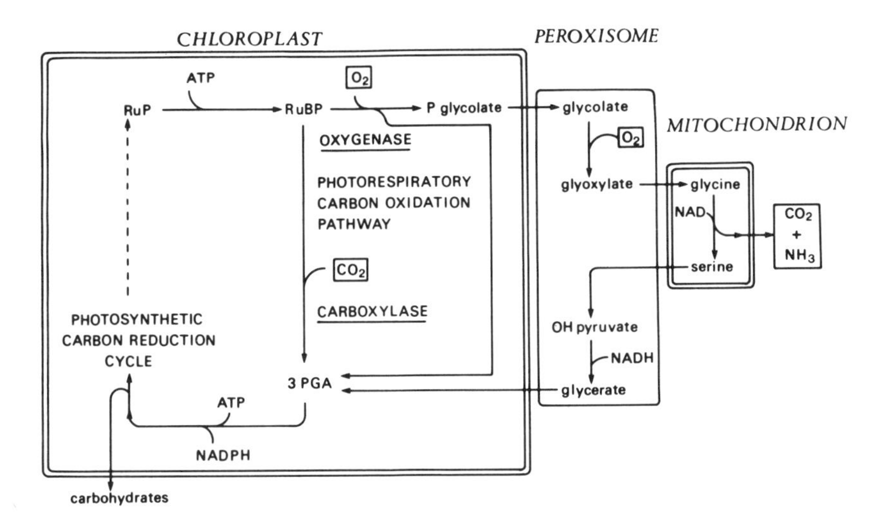

Figure 9.1. Ribulosebisphosphate carboxylase/oxygenase CO_2 assimilation and photorespiration pathways.

tage to crops such as sugar cane and maize in subtropical or tropical regions. The catalytic properties of Rubisco are similar in both C3 and C4 species. Modification of Rubisco to reduce or eliminate the oxygenase activity therefore appears a worthwhile objective that should lead to increased productivity of crops.

The large-subunit gene is present as a single copy per chloroplast DNA molecule. There are 10-100 DNA molecules per chloroplast, and 10-200 chloroplasts are present per cell. Thus, several thousand gene copies per cell may occur. The small subunit appears to be coded by a nuclear multigene family (about seven genes) (Miziorko and Lorimer 1983). Genomic sequences of both the large subunit (maize, barley, spinach, *Chlamydomonas, Anacystis, Synechococcus*) and small subunit (wheat, soybean) have been cloned and sequenced, and total or partial amino acid sequences have been obtained for many more species (Miziorko and Lorimer 1983; Shinozaki et al. 1983). One possible application of this knowledge is to increase the amount of Rubisco produced, by increased copy number or increased transcription. A further aim is to understand the mechanisms involved in carboxylase and oxygenase activities and to alter amino acids at or near the active site to favor carboxylation, either by increasing the affinity for CO_2 or by increasing the ration of activity of carboxylase to oxygenase. The same approach could be used to improve the turnover rate of the enzyme, which is rather low (Gutteridge and Lorimer 1983).

The genes for the active Rubisco polypeptide from a photosynthetic bacterium, when cloned in *Escherichia coli,* synthesize the enzyme in large amounts, allowing methods of site-specific or in vitro mutagenesis to be applied at the molecular level and the effects on the enzyme to be monitored (Somerville and Somerville 1984).

The function of the small subunit is unknown, but it may confer stability and induce a conformation change in the large subunit, and being nuclear-encoded, it will be easier to manipulate.

Two fundamental aspects remain that still must be resolved for successful application of these approaches: It has yet to be proved conclusively that the photorespiration pathway does not contribute to metabolism in unknown (but necessary) ways, and, at present, although it is possible to engineer the large-subunit gene, it is not possible to reinsert it back into the chloroplast genome. However, this problem may be bypassed in the future by inserting modified genes into the nuclear genome preceded by information for a chloroplast transit peptide, to direct nuclear-coded gene products into the chloroplast. An alternative but longer-term approach therefore might be to attempt to convert C3 plants to a C4 type of photosynthesis and hence to avoid the problems of modifying the chloroplast genome.

A further consideration is that, having learned how to increase photosynthesis or control photorespiration, it might be necessary to incorporate further genetic modification for plants to make full use of the additional capacity for assimilation (Keys 1983).

Nitrogen fixation. The roots of legumes normally possess nitrogen-fixing nodules induced by different strains of the bacterium *Rhizobium.* As long as they are infected with the correct strain, legumes can grow in the absence of added nitrogen. Arable crops require over 100 kg of nitrogen per hectare. If this is to be replaced completely by manipulation of N-fixing bacteria or their genes, then they must be capable of fixing 75–100 kg nitrogen per hectare without too great a drain on photosynthates to provide the necessary energy (Gutschick 1980).

Nitrogen fixation is catalyzed by the ATP-dependent six-electron reduction of dinitrogen (N_2) to ammonia by nitrogenase. In addition to *Rhizobium,* other free-living anaerobic (e.g., *Klebsiella*) and aerobic (e.g., *Azotobacter*) bacteria and cyanobacteria (e.g., *Anabaena*) also fix nitrogen. It is not known, at a biochemical level, what special features of leguminous plants or *Rhizobium* allow the symbiosis to occur. Nevertheless, a number of possible opportunities for manipulating the systems can be envisaged:

1. Transfer of nitrogen-fixing genes from bacteria to plants
2. Expansion of the host range of symbiosis to crops other than legumes
3. Increasing the efficiency of symbiotic bacteria to fix nitrogen

The possible transfer of N-fixing (nif) genes from free-living bacteria (e.g., *Klebsiella*) to higher plants that have been proposed frequently. There have been rapid advances in understanding the 17 nif genes in these bacteria, particularly their order, gene products, and regulation of expression (Roberts and Brill 1981) to produce active nitrogenase. But various complications to their transfer and expression in higher plants have emerged. Nitrogenases are oxygen-labile, a fact circumvented in *Azotobacter* by rapid respiration to reduce O_2 tension and the presence of leghemoglobin in *Rhizobium* nodules to facilitate O_2 diffusion to bacteroids without inactivating nitrogenase. Therefore, some method of protecting nitrogenase from oxygen inactivation must be included. In addition, extra host genes will be required to assimilate and transport the ammonia produced, as will be described later. The bacterial gene control sequences will also have to be modified for expression in higher plants. Thus, although this aim may be achieved at some time in the future, much work still has to be done.

Extension of the symbiotic host range of *Rhizobium* is perhaps more feasible. The processes of infection leading to nodulation in legumes follow a well-defined sequence: root-hair attachment, root-hair curling, development of the infection thread, stimulation of host cell divisions, development of bacteroids, leghemoglobin, nitrogen fixation. The symbiosis is remarkably specific (host and *Rhizobium* species), and genes for specificity (nodulation, nod) and nif genes have been located on plasmids in *Rhizobium* and mapped (Downie et al. 1984).

Cloned nodulation genes of *R. leguminosarum* that normally nodulate pea have been transferred to *R. phaseoli* (normally nodulates bean), so that the latter then nodulates pea. Despite the apparent complexity of the nodulation sequence, relatively few bacterial genes

(10 kb) are required (Downie et al. 1984). These can be analyzed in detail, and strains of *Rhizobium* with improved symbiotic performance can be produced.

There is less knowledge of host genes involved in the interaction, but more information is rapidly being gained. A series of nodule-specific proteins are produced by the host plant. Leghemoglobin is common to all hosts, as is glutamine synthetase, but other proteins may be concerned with subsequent export of the nitrogen (such as uricase and xanthine dehydrogenase for tropical legumes that transport ureides), and there is a series of about 30 nodule-specific proteins of unknown function ("nodulins"). Genes for leghemoglobin (present as a multigene family), nodule-specific glutamine synthetase (cDNA), and some nodulins have been cloned and sequenced (Cullimore and Miflin 1984; Verma and Nadler in press). An involvement of plant growth regulators is suggested from host cell responses during infection. If this is so, then it suggests that if the *Rhizobium* and host genes can be suitably manipulated to extend the host range, this will be confined at present to dicotyledonous crop plants.

One aspect of nitrogen fixation that may be amenable to manipulation is the reduced efficiency caused by ATP-dependent hydrogen evolution by nitrogenase. The gene for a H_2-uptake hydrogenase has been identified that can recapture hydrogen and regain some of the otherwise wasted energy (Evans et al. 1980). It has been transferred to other *Rhizobium* strains, with apparent increases in efficiency of N fixation and legume growth improvement.

A further useful modification would be to adjust the control mechanisms that prevent nodulation and fixation in the presence of added nitrogen, so that N fixation is not only confined to conditions of low nitrogen in the field. All engineered *Rhizobium* strains must also be able to compete effectively in the field with the natural populations of bacteria.

Nitrate and solute uptake. Although there has in fact been some improvement in the utilization of added fertilizer, particularly nitrogen, in more recently bred cereals (Bingham 1981), there is a strong case to be made for further improvement in uptake and utilization of added solutes. For most crops there is a linear increase in yield with increased nitrogen application. In the majority of crops, the most important pathway of entry of nitrogen into the plant is nitrate uptake, followed by reduction of nitrate (to nitrite to ammo nia) and assimilation of ammonia via glutamine synthetase. Higher plant nitrate and nitrite reductase have yet to be cloned, but are being studied intensively. Application of genetic engineering techniques could:

1. Improve NO_3^- carrier uptake efficiency
2. Increase copy numbers of NO_3^- reductase and NO_2^- reductase
3. Transfer NO_3^- reduction to leaves (when not already present) where NO_3^- reduction is energetically more favörable

Plant utilization of added phosphate and potassium might similarly be improved by manipulating uptake and transport parameters.

Quality. Although breeding for overall yield is still the main objective, breeders are now almost as concerned with improving quality, especially if the crop is in surplus, as in the European Economic Community (EEC) and North America. What this means in practice depends on the crop of interest. Thus, in the United Kingdom, for wheat, improvements in seed storage protein content and composition in relation to bread making are the aims. For barley, protein content and amino acid composition in relation to animal feeding and malting quality are important. The compositions of storage carbohydrates are also of interest. Similar considerations exist for other cereal grains. For oil crops such as oilseed rape (*Brassica napus*) or sunflower, oil composition (length of fatty acid chains, degree of unsaturation) and eradication of toxic or unpalatable compounds are important, whereas in root crops such as sugar beet the aim is to improve the sugar content and decrease impurity levels, and for potato the aims include increasing the dry matter and reducing amylase levels. In fruit and vegetable crops, consumer preferences, storage and harvesting properties, cooking properties, and taste are particularly important. This catalogue of diverse aims, for which our knowledge of the genetic basis is frequently lacking, shows that application of genetic engineering techniques will be limited to the better-defined areas.

Nutritional properties: seed storage proteins. Cereal and legume seeds contain proteins, carbohydrates, and lipids. Of the protein that constitutes about 10% (cereals) to 40% (soybean) of the grain, storage protein constitutes 50% (cereals) to 70% (legumes). The nutritional quality of the grain very much depends on the amino acid composition of the storage proteins, and being products directly coded by specific genes, unlike storage carbohydrates, these should be easier to manipulate. In terms of total storage protein production, that of the cereals (wheat, maize, rice, barley) is much more significant than that produced by legumes (except soybean). The properties of legume storage proteins are discussed in greater detail elsewhere in this volume, and so our discussion here will emphasize cereal storage proteins.

Of the 20 amino acids commonly found in protein, animals are unable to synthesize 10. When an amino acid is limiting in protein synthesis, other amino acids cannot be stored and are broken down. Lysine is the first limiting amino acid in wheat, barley, maize, sorghum, and triticale, and the second is theonine (barley, sorghum) or tryptophan (maize). Thus, in a pure cereal diet in which lysine is limiting, the quality will be poor, because the grain protein will not be metabolized by the animal efficiently (Bright and Shewry 1983). Similarly, legume storage proteins are deficient in sulfur amino acids. Seed storage proteins are a diverse group classified on the basis of solubility: albumins (water-soluble), globulins (salt-soluble), prolamins (soluble in alcohol/water mixtures), and glutelins. The major storage fractions of most cereals are prolamins, given trivial names such as gliadin (wheat), zein (maize), hordein (barley), and secalin (rye). In legumes the storage proteins are globulins. Storage proteins have been separated and analyzed for amino acids; this has shown that they are responsible for the poor quality of the whole grain (Bright and Shewry 1983).

The major use of barley, maize, and sorghum is for animal feed, and the aim is therefore to increase the relative amount of lysine (and threonine for barley, or tryptophan for maize). (For wheat, the emphasis is on bread making, as will be discussed later.) Two approaches have been suggested to achieve this (Shewry et al. 1981):

1. Insert extra codons for lysine, threonine, or tryptophan into cloned storage protein genomic DNA, followed by reintroduction of the gene into the plant
2. Modify the expression of existing genes so that proteins rich in limiting amino acids are preferentially synthesized

As described in another chapter, cloned cereal storage protein genes – gliadins, hordeins, zeins, and legume storage protein genes – have been sequenced, and knowledge is accumulating on the control of their expression and the number of genes involved. There should be no technical barrier to the addition or modification of codons for desired amino acids. Shewry and associates (1981) estimate that for barley an increase in total grain lysine from 3% to 5% would require about a 10-fold increase (0.5% to 5%) in the lysine content of B hordein (which contributes 40% of total grain proteins) by addition of 14 extra lysine residues to each 700-residue polypeptide. This type of modification must be made without affecting expression or other functions (e.g., deposition in protein bodies) of the proteins.

A specific problem is that prolamins are coded by multigene families (e.g., for zein, possibly up to 150 closely related genes); so relacement of a single modified copy would have little effect. Possible approaches to circumvent this problem (Bright and Shewry 1983) include the following:

1. To introduce a modified gene into a recipient that has a deletion lacking part of the normal gene family
2. To inactivate normal gene expression (without deletion), with expression of introduced modified genes
3. To insert a modified gene with a "strong" promoter such that it is transcribed more frequently than natural genes
4. To insert multiple copies of the modified gene, perhaps combined with approaches 1–3

At the present rate of progress, nuclear gene sequences either with modified promoter sequences to enhance expression, or modified to produce proteins with different amino acid composition, or both, will soon be constructed such that they will be transcribed in the plant after insertion. (The problem of gene insertion into cereals will be discussed later.) These goals are more likely to be achieved first in grain legumes, because there appear to be smaller gene families for storage proteins, and legumes can be transformed using *Agrobacterium* as a vector.

Soluble amino acids. It is also possible to improve the nutritional quality of cereals by increasing specific soluble amino acid levels. To this aim there has been some success in producing mutants with feedback-insensitive regulatory pathways, particularly that of lysine biosynthesis (Bright and Shewry 1983). The amino acids lysine, threonine, methionine, and isoleucine are derived from aspartic acid, and it is known for barley that there is a negative feedback to three isozymes of aspartate kinase, the first enzyme in the pathway, by end products lysine, threonine, and *S*-adenosyl methionine. In the presence of lysine plus threonine, aspartate kinase activity is reduced and plant growth inhibited because of methionine starvation. Mutants that can grow in the presence of added lysine plus threonine have aspartate kinase that is insensitive to feedback inhibition, and during normal growth they accumulate threonine. The cloning and modification of regulatory properties of key regulatory enzymes in this pathway (i.e., aspartate kinases and dihydrodipicolinic acid synthase, the first enzyme on the pathway branch to lysine) could, on reinsertion, lead to accumulation of free limiting amino acids and thus improve quality independent of modifying storage proteins (where quality is usually inversely related to yield).

Technological quality: bread making in wheat. Two of the oldest technologies, the baking of bread and the fermentation of alcoholic drinks, are based on the properties of cereal seeds.

The unique property of wheat, as compared with other cereals, that relates to bread making is the viscoelastic property of dough made from wheat flour. This property is determined mainly by the seed storage proteins (Miflin et al. 1983; Payne 1983), and the viscoelastic proteinaceous mass obtained after washing out starch, albumins, and globulins from dough is called gluten. Gluten is considered to consist of two major protein fractions, gliadin and glutenin, that are functionally equivalent to the storage proteins, and the major polypeptides of these fractions can be classified as prolamins after reduction of disulfide bonds. Because of the importance of bread making, a considerable body of information on the relevant proteins have been accumulated, but the situation is more complex for wheat than for other cereals because it is an allohexaploid with three genomes and storage protein genes located on more than one chromosome per genome. Terminologies have been somewhat confused, but now a clear picture of which proteins are important in bread-making quality is emerging. The major storage proteins of wheat (gliadin and glutenin) constitute 60–80% of the total grain protein. Gliadin consists of numerous different monomeric polypeptides, whereas glutenin has a large aggregate molecular weight (up to several millions) and is composed of a smaller number of different polypeptides cross-linked by disulfide bonds (Payne 1983; Thompson et al. 1983). Gliadin imparts the viscous property to dough, allowing the extensibility that lets dough rise during fermentation, whereas glutenin provides the elasticity that prevents dough being overextended and collapsing during fermentation and baking (Payne 1983). Bread-making quality has been correlated with the presence of high-molecular-weight (HMW) subunits of glutenin (95,000–150,000 apparent MW) that constitute the smaller proportion of the total glutenin aggregate. Payne and associates (1981) showed that differences in baking quality could be related to the presence or absence of specific HMW subunits, and this has already been used as a basis to select parental material for crosses in conventional wheat breeding programs (Payne 1983). Unlike the gene families that code for gliadins, it appears that the glutenin gene family is smaller (Thompson et al. 1983) and is therefore more amenable to manipulation. Although there are various theories on just how the HMW glutenins interact in a technologically useful way (e.g., the "linear glutenin" hypothesis) (Ewart 1977), the most important factors appear to be interactions leading to the presence of the large disulfide-linked aggregates (Miflin et al. 1983). The HMW subunits probably interact via disulfide bonds to form long flexible molecules, and genetically determined variations in properties (interactions with other glutenin or gliadin) may account for differences in dough strength between varieties. Although further work is needed to elucidate the amino acid sequences and sites of potential cross-linking (Shewry et al. in press) and processes that lead to synthesis of disulfide bonds during protein deposition, some applications of genetic engineering can be suggested:

1. Increase the copy numbers of various HMW glutenins
2. Insert genes coding for specific HMW glutenins known to be important in bread-making quality
3. Insert codons for cysteine at appropriate sites in cloned genes to modify cross-linking properties
4. Alter the expression of existing genes for HMW glutenin to enhance the required technological properties

The constraints and limitations on this type of approach are similar to those described in the previous section.

Malting and brewing: barley. The technology of producing beer involves the processes of malting and brewing. In malting, the barley grain is germinated under conditions that lead to enzymic hydrolysis or modification of starch and protein reserves and the production of flavor compounds, whereas brewing involves the fermentation of sugars to produce alcohol by yeast (Palmer 1980). Different varieties differ in regard to their suitability for malting, and in general barley suitable for malting should have a low protein content (Payne 1983). It is thought that storage proteins released from protein bodies during germination surround starch grains, and so reduce access to amylolitic enzymes and delay sugar release, and this leads to poor fermentation. Malting quality may be affected by the hordein fractions (Baxter and Wainwright 1979), particularly B hordeins and especially disulfide-linked components (Miflin et al. 1983). Hordeins rich in cysteine form disulfide-linked aggregates that may be less easily degraded when adhering to starch. However, malting quality is a complex character, and hordeins may affect other stages of brewing, such as filterability, foaming, and haze formation (Baxter, 1981), but with further knowledge it should be possible to identify specific proteins (e.g., disulfide-linked components) that could be manipulated or their expression reduced such that malting quality will be improved. Other aspects that could be manipulated include identification and removal of genes involved in polyphenol (proanthyocyanidins) production, to prevent the haze formed by interaction of polyphenols and protein (Erdal et al. 1980; Payne 1983), and increases in the copy numbers of α-amylase genes and those of other hydrolytic enzymes (e.g., β-glucanases) (Palmer 1980) to speed up breakdown of seed reserves during malting.

Fermentation. Another approach involving α-amylase that relates to fermentation has already been used. In fermentation of starch to ethanol, cereal starch is usually the source of sugars and oligosaccharides, and as outlined earlier, plant enzymes are usually used to degrade the starch. The wheat α-amylase gene, the major enzyme of starch degradation, has already been cloned (Rothstein et al. in press), and the approach of creating yeast strains capable of using starch substrates directly has been used by inserting the wheat α-amylase gene into a vector plasmid and transferring it into yeast. The plasmid construct included yeast regulatory DNA coding for a peptide that ensured the yeast would synthesize and export the plant α-amylase. The modified yeast cells, containing the plasmid, both synthesized plant α-amylase and recognized the export peptide, so that the α-amylase was actually excreted. When the engineered yeast was grown on a culture medium with starch, after staining with iodine, clear halos formed around the cells, showing that the starch

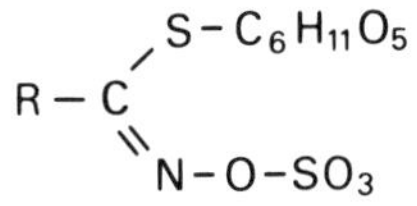

$$R-C(-S-C_6H_{11}O_5)(=N-O-SO_3)$$

Figure 9.2. Structure of glucosinolates found in *B. napus* seeds.

had indeed been broken down. This approach could easily be extended to include other genes (e.g., β-amylase, β-glucanase, etc.) to improve the efficiency of carbohydrate and also protein breakdown, and ethanol production via fermentation.

Beer protein. Up to 10% of the protein in beer is based on one protein (antigen 1) that is derived almost unchanged from one barley seed albumin, protein Z (Hejgaard and Kaergaard 1983). To a large extent this protein is responsible for beer haze (with polyphenols) and foam, and the type of beer produced. It might therefore be possible to modify the expression of protein Z, by the techniques outlined earlier, to tailor malting barley for brewing.

Oil seeds. Oilseed rape, *Brassica napus,* is the major temperate oil seed crop. Although grown mainly for its oil content, it is rich in protein (20–25% by weight) that is nutritionally of good quality. The use of rapeseed meal for animal feed is limited particularly by the presence of glucosinolates (Figure 9.2) that are degraded by myrosinase to release bitter, toxic isothiocyanates, goitrin, and cyanides (Payne 1983). Conventional plant breeding has reduced the levels of glucosinolates, but glucosinolate-free lines are desirable, and these could be obtained by deleting gene(s) involved in glucosinolate synthesis. This might be accomplished by in vivo site-directed mutagenesis, or site-directed integration of DNA, such as inactivated T-DNA or transposable elements, so that an enzyme in the synthetic pathway would be inactivated. Other undesirable components include tannins and saponins, particularly sinapine.

A similar approach can be considered for inactivating genes for enzymes responsible for the synthesis of erucic acid, which is undesirable in rape oil. Most of the oil (which constitutes about 40% by weight of the seed) is used for margarine and cooking fats. Erucic acid can produce heart defects in experimental animals. Its reduction by conventional breeding has been reasonably successful. The other breeding objectives, to increase oil content of the seed and to improve the chain length of the fatty acids, are amenable to genetic engineering techniques, and it can be envisaged that in time *B. napus* lines with defined oil properties will be produced.

Other oilseed crops, such as sunflower or oil palm, could be manipulated in the same way, and in the case of oil palm, clonal propagation of natural elite genotypes with specific oil and other properties is already a commercial reality.

Disease resistance. Yield losses caused by pests and diseases of agricultural and horticultural crops, and the costly control methods and energy expenditure required to reduce their effects, are well known. Breeding for disease resistance is an important part of all breeding programs, but there is a continual requirement for new sources of disease resistance. These may be present in nonagronomic lines or sexually compatible or incompatible wild species. Transformation of crop plants with disease resistance genes is therefore an attractive goal. Unfortunately, the molecular basis of resistance of plants to nearly all diseases is unknown. In some cases, information is accumulating on the cytological location of clustered resistance genes. Genes for resistance to mildew (*Erysiphe graminis hordei*) in barley are grouped on the short arms of chromosomes 4 and 5 (Day et al. 1983), and one resistance gene has been located between two structural loci (Hor-1 and Hor-2) coding for hordein seed storage proteins (Shewry and Miflin 1982). Similarly, the cytological locations of other resistance genes in barley, wheat, and maize are known (Day et al. 1983). The main approach that will be followed is to identify and clone such genes for resistance, so that they can subsequently be transferred between crop plants. The emphasis of future work in this area must therefore be, first, to devise methods that will identify the resistance genes of interest, for which, unlike seed reserve components, gene products are likely to be present only as a very small percentage of total mRNA. To complicate matters, it is possible in some cases that resistance is in fact due to the absence of a gene for susceptibility, or that the difference between susceptible and resistant-isogenic lines resides only in a one-base change leading to a single amino acid substitution. The ingenuity of molecular biologists will be tested in solving such problems.

Identification of resistance genes. Plant breeders and molecular biologists must first carefully identify the best resistance genes for study and obtain them in the most useful plant material that will allow the genes to be identified and cloned. Many different strategies to achieve this aim can be devised.

Although no products of resistance genes have been described, the same situation exists for transposable elements. However, these have now been identified, cloned, and sequenced by identifying mutant genes that have been mutated by insertion of an element (Döring et al. 1984), by comparing it with the sequence found in the wild type. The analogous technique of inserting transposable elements into the genome of a resistant plant at random, and then looking for loss of resistance, would (with probes available for the transposable element) allow clones containing the element within the gene to be isolated and the flanking sequences that hopefully code for the gene of interest to be obtained. Disarmed T-DNA could also be employed in this way. Bearing in mind that in some instances these elements will be inserted into controlling sequences or into some site involved in, but remote from, the resistance gene, this approach holds considerable promise.

Isogenic lines with and without a resistance gene may be used. In this case an approach would be to "subtract" the mRNA of the susceptible line from that of the resistant line, and hope that what is left is the product of the resistance gene. The cDNA from such message would be used to probe for the genomic sequence.

Other possible approaches have been described by Day and associates (1983), including the following: "genome walking," that is, making overlapping genomic clones that can be extended from a known DNA marker sequence into a region expected to contain a resistance gene; comparison of mRNA produced when a resistant host is challenged by a virulent race and an avirulent race of pathogen in which it is assumed that it is only in the incompatible reaction that the resistance gene mRNA is made, and this could be used to probe for the resistance genes.

Alternative ideas. Based on present biochemical understanding of some host-pathogen interactions, other approaches to plant defense can be pursued. For example, because most plants are not susceptible to most pathogens, there may be sequences, such as those coding for phytoalexin synthesis, that confer resistance and may be transferred to hosts normally susceptible to a pathogen. A "shotgun" transfer of cloned pieces of nonhost genome to the susceptible plant could be followed by selection of transformed plants for resistance to that pathogen (Day et al. 1983).

The phenomenon of "induced resistance," whereby a plant that has been inoculated with a pathogen once is more resistant to another infection some days later, correlates with the induction of various "pathogen-related" proteins (Van Loon 1980; White and Antoniw 1983). If these are causally involved in the "induced resistance," then their genes could be manipulated, perhaps by altering control sequences so that the proteins are expressed constitutively to improve pathogen resistance.

Resistance to insect or nematode pathogens might be induced by introducing genes whose products interfere either with digestion (e.g., protease inhibitors, bacteria-derived toxins) (Martin and Dean 1981) or with the nervous system (e.g., bacteria-derived inhibitors of acetylcholinesterase). If the latter course is adopted, the specificity of such inhibitors must be such that transformed plants are not lethal to humans or other organisms, and it might be safer to introduce such inhibitors into microorganisms of the soil rhizosphere where they could still inactivate nematode pathogens.

The possibility that cloned parts of viruses, particularly those with multipartite genomes, can be inserted into plants and confer resistance is being studied. The introduced virus genes might constitutively activate plant defense mechanisms or afford cross-protection or modify the symptoms of a viral infection.

Clearly, a better understanding of host-pathogen interactions is required before promising strategies can be developed. The only practical method of immediate application for transfer of disease resistance genes into and between plants is by fusion of protoplasts, where, as long as a suitable source of resistance has been identified, the molecular mechanisms of resistance do not need to be known.

Stress tolerance. One reason why average crop yields do not achieve their full potential is that their environmental growth conditions are not optimal; that is, they are stressed in some way. In the field, the major stresses are water, high salt, and high or low temperatures. Because it is possible both to select plants that are more stress-tolerant from populations and to subject individual plants to treatments that can increase stress resistance, this indicates that there are genetic aspects of stress that will be amenable to manipulation. Changes in protein synthesis in response to high temperature provide information on environmentally regulated gene expression. For example, when soybean seedlings are transferred to 45°C for 2 hr they die, but after pretreatment for 3 hr at 40°C before transfer to 45°C they survive (Schöffl and Key 1983). The 40°C treatment induces a new class of mRNA and proteins termed "heat-shock" proteins, and their presence correlates with heat-stressed plants in the field. Heat-shock genes have been cloned, and several multigenic families are involved (Schöffl and Key 1983) that are similar in a variety of crop plants.

In a similar manner, mRNA and proteins involved in defense mechanisms in plants can be rapidly induced by addition of cell wall fractions (elicitors) of a fungus or by UV light treatment of parsley cell suspensions. The proteins are involved in biosynthesis of related furano coumarins (antimicrobial, phytoalexin-like) and flavanoids that may protect plants from excess UV irradiation (Hahlbrock et al. 1976; Kruezaler et al. 1983). Techniques of genetic engineering may lead to an understanding of such stress-related responses that can be applied beneficially.

Herbicide resistance

Selective herbicides are routinely applied to control weeds among crop plants that would otherwise compete for available nutrients, space, and light, and thus reduce crop yield and quality. Because a single new gene may be all that is necessary to confer resistance to a herbicide, it can be predicted with confidence that plants resistant to specific herbicides will soon be produced. For example, the broad-spectrum weed killer glyphosate (phosphonomethylglycine) acts by inhibiting the enzyme 5-enolpyruvyl-3-phosphoshikimate synthetase that converts phosphoenol pyruvate and 3-phosphoshikimic acid to 5-enolpyruvyl-3-phosphoshikimic acid in the shikimic acid pathway in bacteria (Comai et al. 1983). Following mutagenesis of *Salmonella typhimurium,* an altered synthetase enzyme resistant to glyphosate has been identified, and the altered aroA gene that codes for it has been cloned and introduced into *E. coli,* where it confers resistance to glyphosate (Comai et al. 1983). The next step is to transfer this gene, with appropriate control signals, into crop plants. Similarly, propanil (3,4-dichloropropionanilide) is a selective photosynthetic herbicide used for postemergence weed control in rice fields. Resistance in rice is conferred by the mitochondrial enzyme arylacylamidase, which rapidly hydrolyzes acylamilide herbicides (Gaynor and Still 1983), and its gene could be moved into other crop plants.

Genes conferring resistance to herbicides may also be identified in weed species that have developed resistant biotypes under long periods in which one herbicide has routinely been applied, as has been found for S-triazines (atrazine and simazine), paraquat, and dichlorofop-methyl. Resistance genes may also be obtained from algae (e.g., *Chlamydomonas*) (Gressel in press).

Hormone engineering

The physiological properties of many crop plants are amenable to regulation by plant growth regulators. Until now, it has not been possible to alter endogenous levels of regulators such as auxins and cytokinins. With advances in our understanding and manipulation of the Ti plasmid of *Agrobacterium tumefaciens* and *A. rhizogenes* it is now possible to manipulate these regulators internally in dicotyledonous crops by altering expression of oncogenic loci of T-DNA. Potato is particularly amenable to this approach, because physiological properties such as tuber induction and photoperiodic response can be readily modified by hormones, and potato plants transformed by *Agrobacterium* have already been regenerated (Ooms et al. 1983) (Figure 9.3). In this case, the T-DNA of a shoot-inducing strain of *A. tumefaciens* apparently altered internal hormones such that the transformed potato plants readily produced aerial tubers. Similarly, carrot plants transformed by *A. rhizogenes* have been regenerated (David et al. 1984). Rapid advances and refinements to this approach, conferring more subtle changes by internal hormone manipulation and including a wide range of field, fruit, and horticultural plants, can undoubtedly be expected.

Screening: detection of virus infection

The application of genetic engineering to agriculture is not limited solely to engineering plants. Any technique that can speed up the assessment and selection procedures of plants in a breeding program will be useful, and practical examples already exist.

A major effort in any breeding program is directed toward finding genotypes resistant to viral pathogens, and the usual techniques

Figure 9.3. Potato plant (cv. Maris Bard) transformed by *Agrobacterium* T-DNA grafted onto normal rootstock. Hormone production encoded in T-DNA causes altered morphology. (Courtesy Dr. G. Ooms.)

employed are to grow the plants and look for symptoms of infection (which is not completely reliable) or to use enzyme-linked immunosorbent assays (ELISA) to detect viral proteins. The latter technique has, in general, proved effective to detect viruses that have coat proteins, but it is ineffective for viroids that lack them. Potato spindle tuber viroid (PSTV) is such a viroid, and an assay for PSTV based on hybridization of radioactive cDNA clones of PSTV to viroid RNA in potato sap bound to a nitrocellulose support has been developed (Owens and Diener 1981). The same approach has been used to make probes for detecting the presence of potato viruses X and Y and potato leaf roll virus (Flavell et al. 1983). These workers have compared the "sap-spot" technique with ELISA and found a good correlation. They suggest that there are a number of advantages over ELISA in the amount of sap required, the sensitivity, and the number of samples that can be screened in a given time. The method will be more widely used with the development of a non-radioactive biotin-labeled probe.

This approach to screening will undoubtedly have much wider application with the detection of other pathogens of crops and the refinement of probes, and it may be extended to detection of other pathogens. It could also be extended to probe for other known beneficial genes in plant breeding programs, and one example is given in the next section.

Cytoplasmic male sterility

In some crops, F_1 hybrid plants yield better than homozygous lines, and therefore F_1 hybrid seed sell at a premium (and must be purchased anew each year). In the production of hybrid seed, breeders can exploit male sterility, in which viable pollen is not produced in one parent, and this ensures that pollination occurs from another plant. The pollen sterility is often the result of changes in the mitochondrial genome. Thus, for maize, "Texas" cytoplasm confers male sterility and was almost universally used to produce F_1 hybrid maize seed in the United States, but it was susceptible to southern corn leaf blight, race T (*Drechslera maydis*). Work to isolate mitochondrial DNA sequences responsible for this sensitivity is in progress (Flavell et al. 1983), and suitable probes that can be used to screen populations for variants in the DNA resistant to *D. maydis* will be available.

Added value to agricultural products

Although not strictly a part of genetic engineering of plants, it is worth mentioning briefly that there is considerable interest in making useful products from agricultural biomass residues. The aims, for example, include characterization of pectinolytic and cellulolytic enzymes in plant pathogenic bacteria (e.g., *Erwinia* species) and cloning and engineering of these enzymes so that they can efficiently degrade waste products such as straw. Another aim is synthesis of higher-value sugar phosphates from starch or sucrose as glycosyl sources.

Some speculations

In the future, it seems probable that microorganisms will provide a rich source of useful genes with specific properties that can be introduced into plants. An example already cited is that of the resistance gene for the herbicide glyphosate (Comai et al. 1983). This is because the techniques of mutation and selection of mutants are so much easier than for higher plants. Thus, enzymes involved in production of secondary products, aphid alarm pheromones (Gibson and Pickett 1983), cyst-nematode hatching factors, and the like, may be obtained from microorganisms of diverse origin. Conversely, the ease of manipulation of soil microorganisms, such as *Rhizobium,* mycorrhizal fungi, and other organisms of the rhizosphere, may lead to their manipulation to benefit plant growth.

It is also possible to envisage transfer of genes coding for the production of pharmaceutical or pharmacological compounds into plants, such that the compounds can be extracted from transformed crops grown in the field, and thus bypass some of the expenses and problems of using fermentation technology.

Conclusions

The techniques of plant genetic engineering applied to agriculture and horticulture clearly hold great promise for the future. This confidence is reflected by the acquisitions of plant breeding and seed companies by many multinational corporations. However, plant genetic engineering is still in its infancy, for it was only in 1983 that successful genetic modification of plants by insertion of isolated, engineered genes was achieved (Flavell and Mathias 1983). The potential of inserting new genes into crops in a directed manner opens new horizons for plant breeders, but this potential is limited at present to the transfer of a single gene or a few genes, and many agricultural characters are polygenically controlled. In many instances, no known genotype displays the required character. Clearly, the limiting factor for proper exploitation of these techniques is lack of basic knowledge of plant molecular biology, biochemistry, and plant development, and a multidisciplinary approach is required, encompassing plant breeders, cell biologists, and molecular biologists. It is confidently predicted that the other major problem, that of transformation of cereals, will be solved in the near future, possibly by the use of transposable elements as gene vectors introduced into totipotent cereal cells by microinjection (Kleinhofs in press).

The time scale of application of new techniques of genetic manipulation is difficult to predict. It is clear that the nearer the manipulation is to intact plants, the more immediate is the application. Thus, techniques of tissue culture – micropropagation, storage of particular genotypes, anther culture to produce haploids, somaclonal variation of plant regenerated from culture – are everyday tools of plant breeders. Gene transfer by protoplast fusion is possible now, and the first fruits of plant genetic engineering (e.g., virus screening) are being realized. To achieve the potential of genetic engineering in agriculture will require identification of genuinely useful genes and understanding of their control sequences and their manipulation so that the genes will be expressed at the correct development stage, in the correct cellular compartment of the appropriate cells, tissues, and organs, so that there will be no adverse effects on other aspects of plant metabolism. It is clear that genetic engineering of plants will not replace or surplant current plant breeding practices, but over the next 10 to 20 years they will increasingly contribute to widening gene pools available to breeders, speed up the screening of plants, and eventually lead to the production of improved crops in the farmers' fields. The enthusiasm and originality of scientists working in this area and the rapid progress of research lead to the conclusion that many other new and useful applications will be found.

Acknowledgments

I think my colleagues in the Biochemistry Department, Rothamsted Experimental Station, for helpful discussions during the preparation of this manuscript.

References

Baxter, E. D. (1981). Hordein in barley and malt–a review. *Journal of the Institute of Brewing, London* **87,** 173–6.

Baxter, E. D., and Wainwright, T. (1979). *Journal of the American Society of Brewing Chemistry* **37,** 8–12.

Bingham, J. (1981). The achievements of conventional plant breeding. *Philo. Trans. R. Soc. Lond.* [*Biol.*] **293,** 441–55.

Bright, S. W. J., and Shewry, P. R. (1983). Improvement of protein quality in cereals, in *Critical Reviews in Plant Science, Vol. 1,* ed. B. V. Conger, pp. 49–93. CRC Press, Boca Raton, Fla.

Comai, L., Sen, L. C., and Stalker, D. M. (1983). An altered aroA gene product confers resistance to the herbicide glyphosate. *Science* **221,** 370–1.

Cullimore, J. V., and Miflin, B. J. (1984). Molecular cloning of plant glutamine synthetase from *Phaseolus* root nodules, in *Advances in Nitrogen Fixation Research,* ed. C. Veeger and W.E. Newton, p. 590. Martinus Nijhoff, The Hague.

David, C., Chilton, M. D., and Tempe, J. (1984). Conservation of T-DNA in plants regenerated from hairy root cultures. *Biotëchnology* 73–6.

Day, P. R., Barret, J. A., and Wolfe, M. S. (1983). The evolution of the host-parasite interaction, in *Genetic Engineering in Plants,* ed. T. Kosuge, T.P. Meredith, and E. Hollaender, pp. 419–30. Plenum, New York.

Döring, H. P., Tillmann, E., and Starlinger, P. (1984). DNA sequence of the maize transposable element Dissociation. *Nature* **307,** 127–30.

Downie, J. A., Ma, Q. S., Wells, B., Knight, C. D., Hombrecher, G., and Johnstone, A.W.B. (1984). The nodulation genes of *Rhizobium leguminosarum,* in *Advances in Nitrogen Fixation Research,* ed. C. Veeger and W.E. Newton, p. 678. Martinus Nijhoff, The Hague.

Erdal, K., Ahrenst-Larsen, B., and Jende-Strid, B. (1980). Use of proanthocyanidin-free barley in beer brewing, in *Cereals for Food and Beverage,* ed. G.E. Inglett and L. Munk, pp. 365–79. Academic Press, London.

Evans, H. J., Emerich, J. E., Lepo, J. E., Maier, R. J., Carter, K. R., Hanus, F. J., and Russell, S. A. (1980). The role of hydrogenase in nodule bacteroids and free-living *Rhizobia,* in *Nitrogen Fixation,* ed. W.D.P. Stewart and J.R. Gallon, pp. 55–81. Academic Press, London.

Ewart, J.A.D. (1977). Re-examination of the linear glutenin hypothesis. *J. Sci. Food Agric.* **28,** 191–9.

Flavell, R. B., Kemble, R. J., Gunn, R. E., Abbott, A., and Baulcombe, D. (1983). Applications of molecular biology in plant breeding; the detection of variation and viral pathogens, in *Better Crops for Food,* pp. 198–209. Pitman, London.

Flavell, R. B., and Mathias, R. (1983). Prospects for transforming monocot crop plants. *Nature* **307,** 108–9.

Gaynor, J. J., and Still, C. C. (1983). Subcellular localization of rice leaf aryl acylamidase activity. *Plant Physiol.* **72,** 80–5.

Gibson, R. W., and Pickett, J. A. (1983). Wild potato repels aphids by release of aphid alarm pheromone. *Nature* **302,** 608–9.

Gifford, R. M., and Evans, L. T. (1981). Photosynthesis, carbon partitioning and yield. *Annu. Rev. Plant Physiol.* **31,** 485–509.

Gressel, J. (in press). Herbicide tolerance and weed resistance: alteration of site of activity, in *Weed Physiology,* ed. S.O. Duke. CRC Press, Boca Raton, Fla.

Gutschick, V. P. (1980). Energy flows in the nitrogen cycle, especially in fixation, in *Nitrogen Fixation, Vol. 1,* ed. W.E. Newton and W. H. Orme-Johnson, pp. 17–27. University Park Press, Baltimore.

Gutteridge, S. G., and Lorimer, G. H. (1983). Genetic modification of the active site of ribulose bisphosphate carboxylase by site specific mutagenesis. *Poster Proc. Phytochem. Soc. Europe, Int. Symp.*, p. 10.

Hahlbrock, K., Knobloch, K.-H., Kreuzaler, F., Potts, J. R. M., and Wellmann, E. (1976). Coordinated induction and subsequent activity changes of two groups of metabolically interrelated enzymes. *Eur. J. Biochem.* **61,** 199–206.

Hejgaard, J., and Kaergaard, P. (1983). Purification and properties of the major antigenic beer protein of barley origin. *Journal of the Institute of Brewing* **89,** 402–10.

Kleinhofs, A. (in press). Cereal transformation: progress and prospects, in *Cereal Tissue and Cell Culture,* ed. S. W. J. Bright and M. G. K. Jones. Martinus Nijhoff, Amsterdam.

Keys, A. J. (1983). Prospects for increasing photosynthesis by control of photorespiration. *Pesticide Science* **19,** 313–16.

Kreuzaler, F., Ragg, H., Fautz, E., Kuhn, D. N., and Hahlbrock, K. (1983). UV-induction of chalcone synthase mRNA in cell suspension cultures of *Petroselinum hortense. Proc. Nat. Acad. Sci. U.S.A.* **80,** 2591–3.

Martin, P. A. W., and Dean, D. H. (1981). in *Microbial Control of Pests and Plant Diseases,* ed. H.P. Burgess. Academic Press, London.

Miflin, B. J., Field, J. M., and Shewry, P. R. (1983). Cereal storage proteins and their effect on technological properties, in *Seed Storage Proteins,* ed. J. Daussant, J. Mosse, and J. Vaughan, pp. 215–319. Academic Press, London.

Miziorko, H. M., and Lorimer, G. H. (1983). Ribulose-1,5-bisphosphate carboxylase-oxygenase. *Annu. Rev. Biochem.* **52,** 507–35.

Ooms, G., Karp, A., and Roberts, J. (1983). From tumour to tuber; tumour cell characteristics and chromosome numbers of crown-gall-derived tetraploid potato plants (*Solanum tuberosum* cv. Maris Bard). *Theoret. Appl. Genet.* **66,** 169–72.

Owens, R. A., and Diener, T. O. (1981). Sensitive and rapid diagnosis of potato spindle tuber viroid disease by nucleic acid hybridization. *Science* **213,** 670–2.

Osmond, C. B. (1981). Photorespiration and photoinhibition. Some implications for the energetics of photosynthesis. *Biochim. Biophys. Acta* **639,** 77–98.

Palmer, G. H. (1980). The morphology and physiology of malting barleys, in *Cereals for Food and Beverages,* ed. G. E. Inglett and L. Munk, pp. 301–538. Academic Press, London.

Payne, P. I. (1983). Breeding for protein quantity and protein quality in seed crops, in *Seed Proteins,* ed. J. Daussant, J. Mosse, and J. Vaughan, pp. 223–52. Academic Press, London.

Payne, P. I., Corfield, K. G., Holt, L. M., and Blackman, J. A. (1981). Correlation between the inheritance of certain high molecular weight subunits of glutenin and bread-making quality in progenies of six crosses of bread wheat. *J. Sci. Food Agric.* **32,** 51–6.

Roberts, P., and Brill, W. J. (1981). Genetics and regulation of nitrogen fixation. *Annu. Rev. Microbiol.* **35,** 207–35.

Rothstein, S. J., Lazarus, C. M., Smith, W. E., Baulcombe, D. C., and Gatenby, A. A. (in press). Secretion of wheat α-amylase expressed in yeast. *Nature.*

Schöffl, F., and Key, J. L. (1983). Identification of a multigene family for small heat shock proteins in soybean and physical characterization of one individual coding region. *Plant Mol. Biol.* **2,** 269–78.

Shewry, P. R., and Miflin, B. J. (1982). Genes for the storage proteins of barley. *Qual. Plant Foods Hum. Nutr.* **31,** 251–67.

Shewry, P. R., Forde, J., Tatham, A., Field, J. M., Forde, B. G., Faulks, A. J., Fry, R., Parmar, S., Miflin, B. J., Dietler, M. D., Lew, E. J.-L., and Kasarda, D. D. (in press). The biochemical and molecular genetics of high molecular weight gluten proteins of wheat, in *Proceedings 6th International Wheat Genetics Symposium, Kyoto, 1983.*

Shewry, P. R., Miflin, B. J., Forde, B. G., and Bright, S. W. J. (1981). Conventional and novel approaches to the improvement of the nutritional quality of cereal and legume seeds. *Science Progress, Oxford* **67,** 575–600.

Shinozaki, K., Yamada, C., Takahata, N., and Sugiura, M. (1983). Molecular cloning and sequence analysis of the cyanobacterial gene for the large subunit of ribulose-1,5-bisphosphate carboxylase/oxygenase. *Proc. Natl. Acad. Sci. U.S.A.* **80,** 4050–4.

Simmonds, N. W. (1979). *Principles of Crop Improvement.* Longman, London.

Somerville, C. R., and Somerville, S. C. (1984). Cloning and expression of *Rhodospirillum rubrum* ribulosebisphosphate carboxylase gene in *E. coli. Mol. Gen. Genet.* **193,** 214–19.

Thompson, R. D., Bartels, D., Harbard, N. P., and Flavell, R. B. (1983). Characterisation of the multigene family coding for HMW glutenin subunits in wheat using cDNA clones. *Theo. Appl. Genet.* **67,** 87–96.

Van Loon, L. C. (1980). Regulation of changes in proteins and enzymes associated with active defense against virus infection, in *Active Defense Mechanisms in Plants,* ed. R.K.S. Wood, pp. 247–73. Plenum Press, New York.

Verma, D. P. S., and Nadler, K. (in press). Legume-*Rhizobium*-symbiosis: host's point of view, in *Advances in Plant Gene Research.*

White, R. F., and Antoniw, J. F. (1983). Direct control of virus diseases. *Crop Protection* **2,** 259–71.

Williams, W. (1981). Methods of production of new varieties. *Philos. Trans. R. Soc. Lond.* [*Biol.*] **292,** 421–30.

Author index

Subject index